V.S. 20

CHICHESTER COLLEGE
WITHDRAWN
FROM LIBRARY STOCK

FOR REFERENCE ONLY

The Party Wall Casebook

Paul Chynoweth
BSc, LLB, Solicitor

Foreword by
The Earl of Lytton

Blackwell
Publishing

© Paul Chynoweth 2003

Blackwell Publishing Ltd
Editorial Offices:
9600 Garsington Road, Oxford OX4 2DQ, UK
  Tel: 01865 776868
108 Cowley Road, Oxford OX4 1JF, UK
  Tel: +44 (0)1865 791100
Blackwell Publishing USA, 350 Main Street, Malden, MA 02148-5018, USA
  Tel: +1 781 388 8250
Iowa State Press, a Blackwell Publishing Company, 2121 State Avenue, Ames, Iowa 50014-8300, USA
  Tel: +1 515 292 0140
Blackwell Munksgaard, 1 Rosenørns Allé, P.O. Box 227, DK-1502 Copenhagen V, Denmark
  Tel: +45 77 33 33 33
Blackwell Publishing Asia Pty Ltd, 550 Swanston Street, Carlton South, Victoria 3053, Australia
  Tel: +61 (0)3 9347 0300
Blackwell Verlag, Kurfürstendamm 57, 10707 Berlin, Germany
  Tel: +49 (0)30 32 79 060
Blackwell Publishing, 10 rue Casimir Delavigne, 75006 Paris, France
  Tel: +33 1 53 10 33 10

The right of the Author to be identified as the Author of this Work has been asserted in accordance with the Copyright, Designs and Patents Act 1988.

All rights reserved. No part of this publication may be reproduced, stored in a retrieval system, or transmitted, in any form or by any means, electronic, mechanical, photocopying, recording or otherwise, except as permitted by the UK Copyright, Designs and Patents Act 1988, without the prior permission of the publisher.

First published 2003

A catalogue record for this title is available from the British Library

ISBN 1-4051-0022-2

Library of Congress
Cataloging-in-Publication Data

Chynoweth, Paul.
  The party wall casebook / Paul Chynoweth.
    p. cm.
  Includes bibliographical references and index.
  ISBN 1-4051-0022-2 (Hardback : alk. paper)
  1. Party walls–Great Britain. 2. Construction contracts–Great Britain. I. Title.
KD921.C49 2003
343.41′078624–dc21
                                              2003005475

Set in 11/12½ Sabon
by DP Photosetting, Aylesbury, Bucks
Printed and bound in Great Britain by
MPG Books Ltd, Bodmin, Cornwall

For further information on
Blackwell Publishing, visit our website:
www.blackwellpublishing.com

Contents

Foreword by the Earl of Lytton — vii
Preface — viii
Abbreviations — x

An Introduction to Party Wall Case Law — 1

Digest of Cases — 21
Adams v Marylebone Borough Council (1907) — 23
Alcock v Wraith and Swinhoe (1991) — 34
Andreae v Selfridge & Company Ltd (1937) — 40
Apostal v Simons (1936) — 47
Barry v Minturn (1913) — 52
Bennett v Harrod's Stores Ltd (1907) — 58
Bond v Nottingham Corporation (1940) — 61
Bower v Peate (1876) — 64
Brace v South East Regional Housing Association Ltd (1984) — 69
Bradburn v Lindsay (1983) — 75
Burlington Property Company Limited v Odeon Theatres Limited (1938) — 80
Carlish v Salt (1906) — 85
Chartered Society of Physiotherapy v Simmonds Church Smiles (1995) — 88
Cowen v Phillips (1863) — 94
Crofts v Haldane (1867) — 97
Crosby v Alhambra Company Ltd (1907) — 101
Cubitt v Porter (1828) — 106
Dalton v Angus (1881) — 110
Dodd v Holme (1834) — 115
Drury v Army & Navy Auxiliary Co-operative Supply Ltd (1896) — 119
Emms v Polya (1973) — 125
Fillingham v Wood (1891) — 128
Frances Holland School v Wassef (2001) — 134
Frederick Betts Ltd v Pickfords Ltd (1906) — 143
Gyle-Thompson and Others v Wall Street (Properties) Ltd (1974) — 151
Hobbs, Hart & Co v Grover (1899) — 161

Holbeck Hall Hotel Ltd v Scarborough Borough Council (2000)	163
Hughes v Percival (1883)	169
J. Jarvis & Sons Ltd v Baker (1956)	175
Johnson (T/A Johnson Butchers) v BJW Property Developments Ltd (2002)	179
Johnston v Mayfair Property Company (1893)	186
Jolliffe v Woodhouse (1894)	190
Jones v Pritchard (1908)	194
Knight v Pursell (1879)	200
Leadbetter v Marylebone Corporation [No. 1] (1904)	204
Leadbetter v Marylebone Corporation [No. 2] (1905)	210
Leakey v National Trust for Places of Historic Interest or Natural Beauty (1980)	214
Lehmann v Herman (1993)	219
Lemaitre v Davis (1881)	225
Lewis & Solome v Charing Cross, Euston & Hampstead Railway Company (1906)	229
List v Tharp (1897)	235
London & Manchester Assurance Company Ltd v O & H Construction Ltd (1989)	239
London, Gloucester & North Hants Dairy Company v Morley & Lanceley (1911)	243
Loost v Kremer (1997)	249
Louis v Sadiq (1997)	257
Major v Park Lane Company (1866)	263
Marchant v Capital & Counties plc (1983)	266
Mason v Fulham Corporation (1910)	271
Matania v National Provincial Bank Ltd and The Elevenist Syndicate Ltd (1936)	276
Matts v Hawkins (1813)	284
Metropolitan Building Act *ex parte* McBride, Re (1876)	287
Midland Bank plc v Bardgrove Property Services Ltd and John Willmott (WB) Ltd (1992)	292
Moss v Smith (1977)	295
Observatory Hill Ltd v Camtel Investments SA (1997)	300
Orf v Payton (1905)	304
Phipps v Pears (1964)	307
Prudential Assurance Co Ltd v Waterloo Real Estate Inc (1999)	311
Ray v Fairway Motors (Barnstable) Ltd (1968)	317
Reading v Barnard (1827)	324
Rees v Skerrett (2001)	328
Riley Gowler Ltd v National Heart Hospital Board of Governors (1969)	336
Sack v Jones (1925)	340
Saunders v Williams (2002)	344

Selby v Whitbread & Co (1917)	347
Sims v The Estates Company (1866)	359
Solomons v R. Gertzenstein Ltd (1954)	364
Southwark & Vauxhall Water Company v Wandsworth District Board of Works (1898)	371
Spiers & Son Ltd v Troup (1915)	374
Standard Bank of British South America v Stokes (1878)	382
Stone and Hastie, Re (1903)	389
Thompson v Hill (1870)	394
Thornton v Hunter (1898)	398
Upjohn v Seymour Estates Ltd (1938)	402
Video London Sound Studios Ltd v Asticus (GMS) Ltd and Keltbray Demolition Ltd (2001)	406
Watson v Gray (1880)	414
Weston v Arnold (1873)	419
White v Peto Brothers (1888)	423
Whitefleet Properties Ltd v St Pancras Building Society (1956)	427
Williams v Bull (1890)	432
Wiltshire v Sidford (1827)	438
Woodhouse v Consolidated Property Corporation Ltd (1993)	442
Appendix: Party Wall etc. Act 1996	447
Glossary	466
Table of Cases	472
Table of Statutory Extracts	480
Subject Index	482

Foreword

The extent of property boundaries and the means of establishing workable relationships between neighbouring owners in the built environment are at the core of maintaining orderly arrangements for property title, value in use and certainty as to the parameters of ownership. The Party Wall etc. Act 1996 addresses these issues, which lie at the heart of one of the most efficient and cohesive property markets in the world, and builds on long and honourable past experience in its most dynamic arena, Central London.

The Act will soon be celebrating its seventh birthday. Like all infants, the learning process has been a matter of increasing wisdom through experience with some trial and error. The methodology has now grown up to the extent of needing some more serious education so I am delighted that this book has come from an author who is directly involved in the academic world.

The philosophy of dealing with party walls inevitably raises new questions as well as providing answers to existing problems. No single textbook can therefore hope to provide a one-hit solution and any attempt to do so would be out of date before it was printed. However, rounding up the available wisdom to be gleaned from past case law, together with guidance on its interpretation, is one of the great contributions that the academic world can give to the practitioner.

In this new book Paul Chynoweth brings together not just a digest of cases but the particular legislative background in point and a diagrammatic representation of some of the key issues. Case law thus becomes immediately more accessible and comprehensible. I therefore welcome *The Party Wall Casebook* as a worthy and valuable addition to the growing body of knowledge and understanding in this area.

The Earl of Lytton

Preface

Party wall procedures became a familiar part of the construction process with the enactment of the Party Wall etc. Act 1996. From that moment the statutory party wall code which had operated in London for over a century was extended to the whole of England and Wales.

Surveyors, architects, engineers and lawyers outside the capital soon acquired the necessary familiarity with the statutory code for practice in this new field. They also soon acquired an understanding of its complexities and an appreciation of some of the anomalies that lurk within it.

The words of the code themselves do not provide a complete account of the law in this area. When the Act passed into law it did so against a centuries-old backdrop of common law rules, many of which continue to apply. The meaning of the code itself has also been subjected to almost 150 years of judicial scrutiny due to its previous incarnation in the various London Building Acts. Practitioners seeking answers to the code's complexities and anomalies will therefore find many of these within the decided cases.

Reference to the original law reports or judgments of these cases will often be impractical for the busy practitioner. In common with other areas of practice – for example dilapidations or contract procedures – a ready source of reference to the key cases can often be all that is required to provide the necessary professional reassurance.

In view of the long history of party wall practice in London it is therefore surprising that no such publication has previously been published. Although there are some excellent books on the general law and practice of party walls there has never, to the author's knowledge, been a book dealing exclusively with party wall case law. This book aims to fill that gap in the practitioner's library. It provides an alphabetical digest of over 80 party wall cases. This format was adopted, rather than attempting to order the cases according to subject matter, as many of them provide lessons in a number of areas. It was also felt that this format would provide quicker access to individual cases, and that this was particularly important for a practitioner's reference book.

If this has produced any lack of continuity it is hoped that this is more than compensated for by the inclusion of a detailed introductory chapter on party wall case law. This provides the reader with a summary of the law of party walls and places the digested cases in their proper context within it. It is also hoped that this chapter will provide some food for thought, and a springboard for further study, for readers who are seeking a more detailed understanding of the law in this area.

The book is intended to be as self-contained as possible. For this reason a copy of the 1996 Act is included as an appendix and extracts from the earlier London Building Acts, and other statutes, appear within the text. An extensive glossary of legal and party wall terms has also been included.

Party walls are, of course, a technical as well as a legal subject. I am therefore indebted to Barry Walker who has produced the drawings which illustrate and enliven many of the cases throughout this book. I gratefully acknowledge his help in enabling the book to speak the language of the construction professional, as well as the lawyer. Thanks are also due to the many other people who have had an input into the production of the book. In particular my thanks go to Julia Burden, my publisher, for her patience and guidance as the book has slowly taken shape, and to Chris Waters for the opportunity to refine my understanding of the 1996 Act through working alongside his team at Hawley & Partners.

I am also grateful to all those surveyors, lawyers and others who have, at various times over the past six years, taken the time to debate the finer points of party wall practice with me. At the risk of offending those not mentioned I would particularly like to thank Robin Ainsworth, Stephen Bickford-Smith, David Bowden, Aidan Cosgrave, Elizabeth de Burgh Sidley, Lawrance Hurst, Donald Jessop, Sandra Laing, John Lytton, Graham North, Alistair Redler, Alex Schatunowski, Rosemary Silver, and Kaivin Wong in this regard.

Despite the contributions of others the views expressed in the book, and any errors, are entirely my own. I have endeavoured to state the law as at 1 March 2003.

Paul Chynoweth
University of Salford
March 2003

Abbreviations

The following abbreviations are used throughout this book to refer to the current party wall statute and the various London Building Acts on which its provisions are modelled:

1996 Act   Party Wall etc. Act 1996
1939 Act   London Building Acts (Amendment) Act 1939
1930 Act   London Building Act 1930
1894 Act   London Building Act 1894
1855 Act   Metropolitan Building Act 1855
1774 Act   "London Building Act" of 1774 (although the Act had no official short title)

AN INTRODUCTION TO PARTY WALL CASE LAW

An Introduction to Party Wall Case Law

What is a party wall?

The party wall is a popular, rather than a legal, concept. The term is generally used to describe a wall separating two terraced or semi-detached properties. It is also sometimes used more generally to describe any form of boundary wall which separates plots of land owned by different parties.

In either case, the term indicates that the wall is not exclusively owned and controlled by one party. Instead, the parties on either side of the wall each have an interest in its continued existence and in any future changes that may be made to it.

The law has responded to this popular concept by applying its own language of ownership, rights and obligations to the variety of factual situations that have come before the courts. The courts' approach to these situations is examined in detail throughout this book. Before we turn to the individual cases, this introductory chapter provides a brief summary of the law of party walls and places the various cases in their proper context within it.

Rights in a party wall

In *Watson v Gray* (1880) the court noted that the term "party wall" could be used in four different senses. However, it concluded that the term usually described a wall which was owned by the two parties as tenants in common. Indeed, in the absence of some contrary intention (*Matts v Hawkins* (1813)), this was presumed to be the case wherever there was common use of a wall (*Wiltshire v Sidford* (1827); *Cubitt v Porter* (1828)).

Ownership on this basis meant that the wall, and the land on which it stood, formed a separate title which belonged equally to the two parties. As common owners of this title they each had the necessary legal rights to safeguard their respective interests in the wall.

Neither owner was therefore entitled to dispossess (or "oust") the other from the wall, for example by permanently demolishing it. However, each was entitled to undertake construction work to any part of the wall, providing that this did not have the effect of dispossessing the other from it. The extent to which particular categories of work were permissible is explored in each of the four cases cited above and also in *Standard Bank of British South America v Stokes* (1878) and *Jolliffe v Woodhouse* (1894).

The Law of Property Act 1925

The law now describes the concept of the party wall in slightly different terms. Since the enactment of the Law of Property Act 1925 it has not been possible to own a legal estate as a tenancy in common[1]. On 1 January 1926 party walls were therefore deemed to be severed vertically with one half of the wall passing into the ownership of each party[2]. However, each party received rights of support and user over the other half of the wall so that the substance of their previous rights remained unchanged[3] (*Apostal v Simons* (1936)).

It is helpful to think of these rights as falling into two categories. Firstly, there are what we will call "fabric rights". These are an owner's rights in the fabric of the half of the wall which is owned by the other party. They ensure that each owner is protected from acts by the other which would otherwise deprive him of the use of the wall.

Secondly, there are what we will call "construction rights". These are rights to undertake construction works to the wall, either to facilitate some development activity on an owner's adjoining land, or for maintenance purposes related to the wall itself.

Fabric rights

The parties' fabric rights probably differ from ordinary easements. As the incidents of a party wall fabric rights have never been formally acquired as easements and it cannot be assumed that they are identical to easements in all respects. Nevertheless, they are usually regarded as part of the law of easements and are often described in these terms. They comprise rights of support, and also rights of user.

Rights under the Law of Easements

The rights of support are probably synonymous with easements of support. They will, for example, require the other owner to provide alternative support before demolishing his building (*Bond v Nottingham Corporation* (1940); *Bradburn v Lindsay* (1983); *Brace v South East*

---

[1] Law of Property Act 1925, section 1(6).
[2] Law of Property Act 1925, sections 38, 39 and schedule 1, Part V, para. 1
[3] *ibid*

Regional Housing Association Ltd (1984); *Rees v Skerrett* (2001)). However, as with any easement, he is probably under no obligation to keep his half of the wall in repair (*Jones v Pritchard* (1908)), even if this eventually results in a loss of support for the other half (*Sack v Jones* (1925); *Bond v Nottingham Corporation* (1940)).

Rights of user include all those other rights which are necessary to enable the wall to be used in the way which the parties originally intended. These can include the right to use a flue which runs through the other owner's half of the wall (*Jones v Pritchard* (1908)). However, based on the decision in *Phipps v Pears* (1964) – which related to easements – it has sometimes been suggested that they do not include a right to weather protection for one's own half of the party wall.

The courts have nevertheless been prepared to recognise a *de facto* right to weather protection in party wall situations by linking it to rights of support (*Bradburn v Lindsay* (1983); *Rees v Skerrett* (2001)). Other cases cast doubt on whether *Phipps v Pears* has ever applied to party wall fabric rights at all (*Upjohn v Seymour Estates Ltd* (1938); *Marchant v Capital & Counties plc* (1983)). In practice, the question has been placed beyond doubt by the Party Wall etc. Act 1996 (the 1996 Act). This now requires the provision of adequate weathering when exposing a party wall which was hitherto enclosed[4].

Apart from this limited intervention, the 1996 Act appears to have no effect on fabric rights[5]. The courts have previously ruled that the Metropolitan Building Act 1855 had no effect on easements (*Crofts v Haldane* (1867); *Re Metropolitan Building Act ex parte McBride* (1876)). Subsequent London Building Acts contained an express provision to this effect and this now also appears in the 1996 Act[6]. In view of the similarities between easements and party wall fabric rights there was presumably no intention to distinguish between them in this context.

The Law of Tort

Before moving on, the role of the law of tort should also be mentioned. Fabric rights, like easements, are rights in property. However, also like easements, they are intimately connected with the law of tort in their breach. An interference with a right of support will, for example, lead to liability in the tort of nuisance. An allegation that an easement has been infringed is also frequently accompanied by other allegations of trespass, nuisance or negligence.

---

[4] 1996 Act, section 2(2)(n).
[5] For the contrary view, see McCardie J's judgment in *Selby v Whitbread & Co* (1917)
[6] 1996 Act, section 9.

These issues are of particular importance due to the recent expansion of the principle established in *Leakey v National Trust for Places of Historic Interest or Natural Beauty* (1980). The extent to which they apply to the law of party walls is considered in more detail below, in the sections on damage to adjoining property and adjacent excavations.

Construction rights

The parties may enter into a party wall agreement which authorises one of them to undertake construction works to the wall. Should a dispute subsequently arise, the nature of the right to undertake the work will depend on the construction of this agreement (*Bennett v Harrod's Stores Ltd* (1907); *Brace v South East Regional Housing Association Ltd* (1984)).

In the absence of an agreement the right will depend on the parties' inherent rights in the wall, either at common law, or under statute. An owner's pre-1926 common law right to undertake work to the whole thickness of the wall has already been discussed. It has often been assumed that the 1925 legislation restricted these rights to the owner's half of the wall (*Moss v Smith* (1977)). This may not be correct in view of the express intention of the 1925 legislation to preserve all existing common law rights through the device of granting mutual rights of user over each half of the severed party wall[7] (*Alcock v Wraith and Swinhoe* (1991)).

Statutory construction rights

In practice the issue is now of less significance following the enactment of the 1996 Act. Whatever the parties' common law rights in a party wall, they will invariably seek to rely on the construction rights provided by the Act where these are available. Although these rights are extensive, the Act does not grant a general right to undertake construction work to a party wall. Instead it authorises particular types of work to particular named structure types.

The various types of authorised work are listed in the Act[8]. However, they generally do not include a right to permanently expropriate the property of an adjoining owner (*Weston v Arnold* (1873); *Burlington Property Company Limited v Odeon Theatres Limited* (1939); *Barry v Minturn* (1913); *Gyle-Thompson v Wall Street (Properties) Ltd* (1974); *Moss v Smith* (1977)).

---

[7] Law of Property Act 1925, sections 38, 39 and schedule 1, Part V, para. 1
[8] 1996 Act, section 2(2)

The relevant structure types are also listed[9]. Either party has a right to undertake work to these structures and the ownership of the structure is irrelevant to this entitlement (*Knight v Pursell* (1879); *Loost v Kremer* (1997)). A number of structure types are included and each has a particular meaning within the Act.

There can be no right to undertake work to a structure which does not fall within one of these meanings (*Johnston v Mayfair Property Company* (1893); *London, Gloucester & North Hants Dairy Company v Morley & Lanceley* (1911)). By way of example, the unbonded external walls of two adjoining buildings do not together constitute a "party wall" even where the two are in direct contact with each other (*White v Peto Brothers* (1888); *Thornton v Hunter* (1898)).

"Party wall" is only one of several listed structure types. The term was traditionally defined by reference to a wall performing a separating function between adjoining buildings (the type 'b' party wall). In this context it is well-established that a wall can be a party wall for only part of its height (*Weston v Arnold* (1873); *Drury v Army & Navy Auxiliary Co-operative Supply Ltd* (1896); *London, Gloucester & North Hants Dairy Company v Morley & Lanceley* (1911)) or indeed for part of its length (*Knight v Pursell* (1879); *Johnston v Mayfair Property Company* (1893)).

Since the 1894 Act (probably as a result of the decision in *Williams v Bull* (1890)) the term has been extended to include walls which straddle a legal boundary, even where they perform no separating function (the type 'a' party wall). It has recently been established that this type of structure can lose its party wall status where one party acquires title to the other party's half of the wall through adverse possession (*Prudential Assurance Co Ltd v Waterloo Real Estate Inc* (1999)).

Survival of common law construction rights?

If construction rights are not available under the Act an owner may wish to rely on a common law right to undertake the desired work. The extent to which this is possible remains unclear. In situations where the Act has codified an existing common law right the common law right presumably then ceases to exist (*Standard Bank of British South America v Stokes* (1878); *Lewis & Solome v Charing Cross, Euston, & Hampstead Railway Company* (1906); *Selby v Whitbread & Co* (1917)).

However, other common law party wall rights have not been repealed. The continued role of fabric rights has already been discussed. Common law construction rights might also continue to exist alongside the statutory rights. For example, the common law right to demolish a building and expose a party wall continued to exist despite the enactment of the

---

[9] *ibid*

London Building Acts (*Major v Park Lane Company* (1866); *Lewis & Solome v Charing Cross, Euston, & Hampstead Railway Company* (1906))[10].

An owner may also retain a common law right to undertake work to his own side of the party wall in circumstances where this does not interfere with his neighbour's fabric rights (*Upjohn v Seymour Estates Ltd* (1938)). It is even possible that some common law construction rights to undertake work to the whole thickness of the wall survive, despite the existence of parallel rights under the Act (*Jolliffe v Woodhouse* (1894)).

Service of notices

An owner cannot generally exercise any of the Act's construction rights without first serving a party structure notice on adjoining owners[11]. A failure to serve notice is probably not, in itself, an unlawful act (*Emms v Polya* (1973)). Nevertheless, works which are commenced without the appropriate authority will constitute a trespass and may be restrained by injunction (*Leadbetter v Marylebone Corporation [No. 1]* (1904); *Adams v Marylebone Borough Council* (1907); *London & Manchester Assurance Company Ltd v O & H Construction Ltd* (1989)). They may also lead to liability in nuisance for which injunctive relief will also be available (*Emms v Polya* (1973); *Louis v Sadiq* (1997)).

The notice must be served by all joint owners (*Lehmann v Herman* (1993)). It must contain sufficient information about the proposed works to enable the recipient to decide whether to consent to the works. Notices which are provisional in nature, or which otherwise fail to provide sufficient detail, will be invalid (*Hobbs, Hart & Co v Grover* (1899); *Spiers & Son Ltd v Troup* (1915)). If consent is forthcoming the works must commence within the statutory time limit or the notice will cease to have effect[12]. No time limit applies to a notice which has not been consented to (*Leadbetter v Marylebone Corporation [No. 2]* (1905)).

Definition of "owner"

Only persons falling within the Act's definition of "owner" need be served[13]. This only includes those owning significant property rights in

---

[10] It presumably no longer exists due to the effect of section 2(2)(n) of the 1996 Act.
[11] 1996 Act, section 3.
[12] 1996 Act, section 3(2)(b)(i)
[13] The current definition appears in section 20 of the 1996 Act. This differs slightly from those contained in the earlier London Building Acts. The reported cases should be understood in the context of these earlier definitions.

adjoining land. A tenant at will is therefore excluded (*Thompson v Hill* (1870)), even where some additional rights to use the land have been granted to him (*Orf v Payton* (1905)). For the same reason, a statutory tenant under the Rent Acts is excluded (*Frances Holland School v Wassef* (2001)), as is a receiver (*Solomons v R. Gertzenstein Ltd* (1954)), and a prospective purchaser who lacks any interest in the land (*Spiers & Son Ltd v Troup* (1915)).

However, once a purchaser enters into a contract to purchase the adjoining land, he thereby becomes an owner with an entitlement to be served with notice[14] (*Cowen v Phillips* (1863); *List v Tharp* (1897)). Ownership of an equitable interest has been held to be a sufficient interest (*Cowen v Phillips* (1863)), as has ownership of part-only of the adjoining land (*Fillingham v Wood* (1891)).

Where a number of different interests are owned in respect of the same adjoining property, notice must be served on all adjoining owners (*List v Tharp* (1897); *Crosby v Alhambra Company Ltd* (1907)). However, where a legal estate is owned by two or more co-owners, it is sufficient to serve the notice on only one of them (*Crosby v Alhambra Company Ltd* (1907)).

The Act specifically requires notices to be served on adjoining owners. Service on an adjoining owner's agent is acceptable (*Whitefleet Properties Ltd v St Pancras Building Society* (1956)). However, service will be ineffective if the agent has no authority to accept service of a particular notice (*Gyle-Thompson v Wall Street (Properties) Ltd* (1974)).

Reciprocal duties

An owner's construction rights in respect of a party wall are accompanied by reciprocal duties which he owes to adjoining owners and occupiers on a non-delegable basis (*Hughes v Percival* (1883); *Jolliffe v Woodhouse* (1894)). These exist, whether he exercises the rights at common law or under the Act (*Jolliffe v Woodhouse* (1894)). They impose an obligation on him to exercise his construction rights reasonably so as to avoid any unnecessary interference with the adjoining owners and occupiers (*Adams v Marylebone Borough Council* (1907)).

Types of reciprocal duties

There are two types of reciprocal duties, although there is inevitably some overlap between them. The first type is synonymous with a duty of care in

---

[14] 1996 Act, section 20.

the law of negligence. It imposes a duty on the owner to exercise reasonable care when undertaking the work so as to avoid injury or physical damage to the adjoining property and its adjoining owners and occupiers (*Hughes v Percival* (1883); *Alcock v Wraith & Swinhoe* (1991)).

The second type is synonymous with the law of nuisance. It requires the owner to avoid conduct which would unreasonably interfere with the use or enjoyment of the adjoining land. It is described, in the Act, as a duty to avoid unnecessary inconvenience[15] (*Adams v Marylebone Borough Council* (1907)). It requires the owner to undertake the work in the most appropriate manner so as to minimise its impact on the adjoining property (*Barry v Minturn* (1913)).

This will involve, for example, keeping noise and dust to a minimum (*Andreae v Selfridge & Company Ltd* (1937); *Matania v National Provincial Bank Ltd and The Elevenist Syndicate Ltd* (1936)), exercising proper supervision of the workforce (*Emms v Polya* (1973)) and undertaking the work as expeditiously as possible (*Jolliffe v Woodhouse* (1894); *Emms v Polya* (1973)).

The nature of lawful works

Work to a party wall will generally be lawful to the extent that it is undertaken in compliance with these duties. Where lawful work necessarily leads to some interference with the adjoining property the owner may, under the Act, be required to pay compensation or make good damage as a condition of undertaking the work (see below).

This will only be the case in situations where the Act makes specific provision for it (*Thompson v Hill* (1870); *Adams v Marylebone Borough Council* (1907); *Video London Sound Studios Ltd v Asticus (GMS) Ltd and Keltbray Demolition Ltd* (2001)).

Where no such provision is made the exercise of lawful works which are reasonably undertaken cannot lead to any liability (*Southwark & Vauxhall Water Company v Wandsworth District Board of Works* (1898)), either in negligence (*White v Peto Brothers* (1888)) or in nuisance (*Louis v Sadiq* (1997)).

Role of appointed surveyors

Reference has already been made to the requirement for service of a party structure notice before an owner can exercise construction rights under the Act. Where the adjoining owner provides written consent to the works

---

[15] 1996 Act, section 7(1).

within 14 days of service, the (building) owner is then free to undertake the works. He must commence these within 12 months of the date of service (*Leadbetter v Marylebone Corporation [No. 2]* (1905)).

However, where no such consent is forthcoming a dispute is deemed to arise between the parties[16]. This can then only be settled by the publication of an award by surveyors[17] who have been appointed by the parties for this purpose[18]. Until the dispute has been settled in this way the building owner has no right to commence the works (*Standard Bank of British South America v Stokes* (1878)).

The practical tribunal

The surveyors together constitute a tribunal with authority to resolve the dispute between the parties. The courts have therefore sometimes assumed this to be an arbitral tribunal (*Re Metropolitan Building Act ex parte McBride* (1876); *Re Stone and Hastie* (1903)). Most would now consider that the surveyors' role has more in common with an expert determination than with an arbitration (*Chartered Society of Physiotherapy v Simmonds Church Smiles* (1995)).

There is therefore no requirement that the surveyors should conduct a hearing before making an award (*Loost v Kremer* (1997)). It also seems unlikely that the individual party-appointed surveyors are required to act impartially of those who appoint them (*Chartered Society of Physiotherapy v Simmonds Church Smiles* (1995))[19]. In the absence of a conflict of interest there is therefore no impediment to a project architect accepting an appointment as surveyor by a building owner (*Loost v Kremer* (1997)).

The surveyors are charged with the role of facilitating the exercise of the building owner's construction rights whilst simultaneously safeguarding the interests of the adjoining owner (*Selby v Whitbread & Co* (1917); *Gyle-Thompson v Wall Street (Properties) Ltd* (1974)).

They do this in two ways. Firstly, they publish an award which describes how the building owner's work can lawfully be undertaken without causing unnecessary inconvenience to the adjoining owner. Secondly, through recording a schedule of condition and undertaking inspections, they determine the extent to which these lawful works cause damage to the adjoining owner's property. They have an ongoing

---

[16] 1996 Act, section 5.
[17] 1996 Act, section 10(10).
[18] 1996 Act, section 10(1).
[19] For a further discussion of this issue, see Chynoweth, P. (2001), Impartiality and the Party Wall Surveyor, *Construction Law Journal*, **17** (2), 127–137.

jurisdiction to resolve issues throughout the project by a series of addendum awards (*Selby v Whitbread & Co* (1917)).

Despite their function as members of a tribunal, the role is therefore a technical rather than a legal one, and for this reason they have been described as constituting a "practical tribunal" (*Adams v Marylebone Borough Council* (1907)).

Surveyors' authority

As members of the practical tribunal the surveyors perform a statutory role rather than acting as agents for their clients. Their awards will therefore only be valid to the extent that they comply with the terms of the Act. Specifically, the tribunal must have been properly constituted; it must only address matters within its competence; and it must only exercise powers which have been properly conferred upon it. Awards which contravene these principles are ultra vires and therefore invalid.

Requirement for proper constitution

The tribunal will only be properly constituted where the service of notices (discussed above) and the written appointment and selection of surveyors have been strictly in accordance with the Act (*Gyle-Thompson v Wall Street (Properties) Ltd* (1974)). The whole tribunal, rather than individual surveyor-members, must then join in making the award (*Frances Holland School v Wassef* (2001)).

Requirement for competence

The tribunal is only competent to make awards on matters which are in dispute between the parties and which are also connected with work to which the Act relates[20]. The requisite dispute will often be a deemed dispute and this will only arise where notices have been properly prepared and served (see above). In practice, awards will more often falter on the strict requirement that they confine themselves to matters which are connected with work to which the Act relates.

The two party-appointed surveyors will often be negotiating about a range of issues in addition to those connected with works under the Act. These may include issues relating to easements, crane oversailing, the general conduct of piling operations, or the impact of other general construction activities on adjoining owners. The surveyors act as their clients' agents in relation to these issues and have no authority to adjudicate upon

---

[20] 1996 Act, section 10(10).

them in their award (*Burlington Property Company Limited v Odeon Theatres Limited* (1939)).

Even where an award addresses matters related to work under the Act it will still be invalid if these have not been formally referred to the surveyors for them to adjudicate upon (*Re Stone and Hastie* (1903)). Hence, the surveyors have no jurisdiction to adjudicate upon works which are not described in the original notice (*Leadbetter v Marylebone Corporation [No. 1]* (1904)), or on matters which predate their appointment (*Woodhouse v Consolidated Property Corporation Ltd* (1993); *Louis v Sadiq* (1997)).

Surveyors' statutory powers

The surveyors' tribunal has power to determine "the right to execute" work referred to in the notice as well as the time and manner of its execution. It also has certain powers to determine a number of other matters which arise out of, or are in some way incidental to, the original dispute which has been referred to it[21].

In reality the right to execute work is determined by the Act. The surveyors' power in this respect is limited to determining the existence of some precondition – for example a defect or want of repair[22] – for the exercise of a right (*Gyle-Thompson v Wall Street (Properties) Ltd* (1974)). In practice the surveyors' award performs a declaratory role by setting out the works which the building owner is authorised to perform under the Act. Awards which purport to sanction works which are not authorised by the Act will be invalid (*Barry v Minturn* (1913); *Burlington Property Company Limited v Odeon Theatres Limited* (1939); *Gyle-Thompson v Wall Street (Properties) Ltd* (1974)).

The surveyors' power to determine the time and manner in which the work is executed is central to their role under the Act. By exercising this power they ensure that adjoining owners and occupiers are not subjected to unnecessary inconvenience, and that the works are therefore undertaken lawfully. As long as the tribunal is properly constituted and also acts within its competence, the courts are unlikely to interfere with decisions taken by the surveyors in this respect, even where these impose significant obligations on building owners (*Marchant v Capital & Counties plc* (1983)).

The power to determine other matters arising out of or incidental to the main dispute is strictly limited in its scope (*Woodhouse v Consolidated Property Corporation Ltd* (1993)). The surveyors have discretion to determine the liability for professional fees incurred in connection with the

---

[21] 1996 Act, section 10(12).
[22] 1996 Act, section 2(2)(b).

making of the award[23]. They also have power to adjudicate on liability for the cost of the work itself and may require a building owner to make good damage or to pay compensation to an adjoining owner. However, their power is limited to administering the Act's particular requirements in these areas (*Adams v Marylebone Borough Council* (1907); *Video London Sound Studios Ltd v Asticus (GMS) Ltd and Keltbray Demolition Ltd* (2001)).

The surveyors have no power to make decisions on matters of law. For example, although they may indeed have to make assumptions in this respect (*Loost v Kremer* (1997)), they have no power to determine whether a wall is a party wall (*Sims v The Estates Company* (1866)). They also have no power to make decisions authorising the interference with an easement (*Crofts v Haldane* (1867); *Re Metropolitan Building Act ex parte McBride* (1876)).

Challenges to surveyors' awards

The parties have a statutory right of appeal against a surveyors' award to the county court[24]. Appeals are not restricted to matters of law but involve a complete rehearing of all the issues determined by the surveyors. The court can therefore hear fresh evidence and, where appropriate, has the power to overturn the surveyors' original findings of fact (*Chartered Society of Physiotherapy v Simmonds Church Smiles* (1995)).

Appeals must be brought within 14 days of receipt of the award (*Riley Gowler Ltd v National Heart Hospital Board of Governors* (1969)). Challenges to the surveyors' technical decisions can only be made in this way and the right is therefore lost if no appeal is lodged within the 14 days (*Selby v Whitbread & Co* (1917)).

There is no such restriction where the validity of the award is itself challenged. Although the parties are entitled to raise such issues as part of an appeal to the county court, they may instead challenge the validity of the award in separate proceedings. They may do so, even after the expiry of the statutory 14-day appeal period (*Re Stone and Hastie* (1903); *Gyle-Thompson v Wall Street (Properties) Ltd* (1974)).

Effect of surveyors' awards on subsequent owners

Surveyors' awards create rights, liabilities and encumbrances. It is unclear whether these are personal to the parties or whether they pass to a

---

[23] 1996 Act, sections 10(12)(c) and 10(13).
[24] 1996 Act, section 10(17).

purchaser on the sale of the parties' properties[25]. A number of cases have produced apparently conflicting decisions on this point.

Where an adjoining owner is granted rights under an award in favour of his own property, this will be an encumbrance on the building owner's land. This type of right has been held to be binding on a purchaser of the building owner's land (*Selby v Whitbread & Co* (1917)). However, on another occasion it has been held not to be an interest in the building owner's land and therefore incapable of registration as a caution (*Observatory Hill Ltd v Camtel Investments SA* (1997)).

An owner may be subject to liabilities under an award – for example, an obligation to contribute to the cost of works. It is considered unconscionable for him to escape from these by transferring them to a man of straw. He therefore appears to remain responsible for them, even after transferring his land to a purchaser ((*Selby v Whitbread & Co* (1917)). However, apparently in contradiction to this, an owner's liability to contribute towards the cost of works under an award has also been held to be a material fact requiring disclosure to a purchaser in a contract of sale (*Carlish v Salt* (1906)).

A number of contingent rights can also arise under the Act. These include a building owner's right to receive a contribution towards the cost of raising a party wall if the adjoining owner subsequently makes use of the raised portion[26]. This right has been held to pass to a purchaser on a sale of the freehold (*Mason v Fulham Corporation* (1910)) but not on the grant of a 21-year lease (*Re Stone and Hastie* (1903)).

Liability for the cost of works

Where one party is responsible for damaging a party wall, he will be entirely responsible for the cost of repairing it under normal tortious principles (*Apostal v Simons* (1936); *Bradburn v Lindsay* (1983)). In the absence of fault by one party there appears to be no common law rule requiring the costs of repairs to be shared by both parties (*Orf v Payton* (1905); *Jones v Pritchard* (1908)), unless these became necessary due to the dangerous state of the wall (*Spiers & Son Ltd v Troup* (1915)).

---

[25] For a further discussion of the subject, see Bickford-Smith, S. & Sydenham, C. (1997) *Party Walls: The New Law*, Jordans, Chapter 14.
[26] 1996 Act, section 11(11).

The 1996 Act generally follows the principle that the cost of works should be discharged by the owner undertaking them[27]. However, where these have become necessary due to the other party's conduct, or where the other party will derive some benefit from them, he may be entitled to receive some contribution towards them[28].

In these circumstances both parties will generally be jointly and severably liable to the contractor for payment of his account (*J. Jarvis & Sons Ltd v Baker* (1956)). More usually, the party undertaking the work will settle the account himself and serve an account for the required contribution on the adjoining owner within two months of the work being completed[29]. This will be binding on the adjoining owner notwithstanding minor departures from the statutory requirements (*Reading v Barnard* (1827)). It is unclear whether time is of the essence for service of the account (*Spiers & Son Ltd v Troup* (1915); *J. Jarvis & Sons Ltd v Baker* (1956)).

Damage to adjoining property

Where an owner carries out work to a party wall there is a risk that damage will result to adjoining property. This may be due to some fault on the part of the contractor undertaking the work, or of the owner who employs him. Alternatively, it may simply be an inevitable consequence of the nature of the work and of the proximity of the two properties.

Lawful works

Where work is undertaken lawfully (for example, in compliance with a surveyors' award) and the damage is an inevitable consequence of the work, the owner will be protected from civil liability (*White v Peto Brothers* (1888); *Louis v Sadiq* (1997)). The exercise of a lawful right can never, in itself, be a wrongful act (*Southwark & Vauxhall Water Company v Wandsworth District Board of Works* (1898)).

An adjoining owner who suffers damage in these circumstances therefore has no remedy in the law of tort but is instead protected by the statutory machinery established by the 1996 Act. This generally makes the exercise of the statutory construction rights conditional upon making

---

[27] 1996 Act, section 11(1).
[28] 1996 Act, sections 11(3) to (11).
[29] 1996 Act, section 13(1).

good, or paying compensation for, certain consequential damage or other losses[30].

These remedies are administered by the appointed surveyors rather than the courts (*Adams v Marylebone Borough Council* (1907)). Their scope is limited by the words of the statute (*Thompson v Hill* (1870); *Video London Sound Studios Ltd v Asticus (GMS) Ltd and Keltbray Demolition Ltd* (2001)) and they will rarely be as extensive as an award for damages made by the courts (*Adams v Marylebone Borough Council* (1907)).

Unlawful works

In contrast, where the works are unlawful the adjoining owner will be entitled to recover damages for his losses, including consequential losses where these were reasonably foreseeable (*Louis v Sadiq* (1997); *Saunders v Williams* (2002)). These will generally be recoverable from the building owner personally, even where they result from some wrongful act by his contractor (*Bower v Peate* (1876); *Hughes v Percival* (1883); *Matania v National Provincial Bank Ltd and The Elevenist Syndicate Ltd* (1936); *Alcock v Wraith & Swinhoe* (1991)).

Unlawfulness may arise in a number of ways. Firstly, the works may have been undertaken without the appropriate authority. In this case they will lead to liability in trespass (*Leadbetter v Marylebone Corporation [No. 1]* (1904); *Frederick Betts Ltd v Pickfords Ltd* (1906); *Adams v Marylebone Borough Council* (1907); *London & Manchester Assurance Company Ltd v O & H Construction Ltd* (1989)).

Secondly, they may involve some interference with the adjoining owner's fabric rights (see above). For example, works which interfere with an adjoining owner's right to support from the building owner's half of the wall will lead to liability in nuisance (*Lemaitre v Davis* (1881); *Bond v Nottingham Corporation* (1940); *Brace v South East Regional Housing Association Ltd* (1984)). As discussed above, liability may also arise where the works deprive an adjoining owner of the weather protection previously afforded by the building owner's half of the wall (*Upjohn v Seymour Estates Ltd* (1938); *Marchant v Capital & Counties plc* (1983); *Bradburn v Lindsay* (1983); *Rees v Skerrett* (2001)).

Thirdly, the works might have been undertaken in breach of the reciprocal duties which are owed to an adjoining owner when exercising construction rights in respect of a party wall (see above). The common feature of such duties is an obligation to exercise the construction rights reasonably so as to minimise the interference with the adjoining property.

---

[30] 1996 Act, sections 2(3) and 7(2).

Depending on the circumstances, liability might arise in nuisance (*Louis v Sadiq* (1997); *Video London Sound Studios Ltd v Asticus (GMS) Ltd and Keltbray Demolition Ltd* (2001)), negligence (*Hughes v Percival* (1883); *White v Peto Brothers* (1888)), for breach of the statutory duty not to cause unnecessary inconvenience (*Jolliffe v Woodhouse* (1894)), or under some combination of these.

Finally, the works might involve the commission of some other tort, for example, under the rule in *Rylands v Fletcher* (1868) or under the common law rules relating to the spread of fire (*Johnson (T/A Johnson Butchers) v BJW Property Developments Ltd* (2002)). Liability between neighbouring owners is also now increasingly likely to arise in tort under the principle established in *Leakey v National Trust for Places of Historic Interest or Natural Beauty* (1980).

Under this principle owners have a duty to minimise risks to neighbouring property which arise from hazards on their own land. The principle has been applied in party wall cases (*Bradburn v Lindsay* (1983); *Rees v Skerrett* (2001)). However, it overlaps with the principles already discussed in respect of fabric rights. It is unclear whether it represents a significant change in the law in the context of party walls, as opposed to work affecting neighbouring property where no party wall or easement of support exists.

Line of junction works

Where a party wall already exists, each owner has a right to demolish it and to rebuild it in its original position, astride the line of junction. This right existed at common law (*Cubitt v Porter* (1828)) and also exists under the 1996 Act[31] (*Thompson v Hill* (1870)).

However, owners have no right to build an entirely new party wall without the other's consent, either at common law or under the Act. The act of placing a wall partly on the other owner's land would therefore constitute an act of trespass (*London & Manchester Assurance Company Ltd v O & H Construction Ltd* (1989)).

The construction of a new party wall is, of its very nature, a consensual act. The Act preserves this situation by providing a formal mechanism for obtaining an adjoining owner's consent. This commences with the service of a line of junction notice upon him[32] (*Orf v Payton* (1905)).

The Act also contains a right, when erecting a new wall entirely on an owner's own land, to place its projecting footings and foundations on an

---

[31] 1996 Act, sections 2(2)(b), (e), (l) and (m).
[32] 1996 Act, section 1(2).

adjoining owner's land[33]. The exercise of this right does not require the adjoining owner's consent but it is dependent on the prior service of a line of junction notice[34] and on compliance with the other requirements contained in the Act (*Thornton v Hunter* (1898)). Compensation is payable where damage is caused to an adjoining owner's property by line of junction works[35] (*Adams v Marylebone Borough Council* (1907)).

Adjacent excavations

Owners sometimes undertake excavations on their own land in close proximity to buildings on adjoining land. Where this occurs there may be a risk to the foundations, and therefore to the stability, of these adjoining buildings.

This sort of work may be entirely unconnected with work to a party wall. In practice, the two types of work are often undertaken contemporaneously and the law relating to adjacent excavations has therefore become associated with the law of party walls.

Adjoining owner's rights

The owner of the adjoining land has a number of rights in these circumstances. For instance, he has a natural right of support from the land being excavated (*Midland Bank plc v Bardgrove Property Services Ltd and John Willmott (WB) Ltd* (1992)). This exists in favour of his land in its natural state but does not extend to buildings which have been built upon it (*Dalton v Angus* (1881)). He can therefore only rely on his natural rights where he suffers damage to a building if he can demonstrate that the land would have subsided in any event, irrespective of the existence of the building (*Ray v Fairway Motors (Barnstable) Ltd* (1968)).

Where the damage would not have occurred but for the existence of the building, the adjoining owner will generally have to demonstrate that he is entitled to an easement of support which has been interfered with. This can be acquired by prescription (*Dalton v Angus* (1881)). Irrespective of the existence of a right of support, the excavating owner may be liable in negligence where he fails to exercise reasonable care in the conduct of the excavations (*White v Peto Brothers* (1888)). In the absence of some other fault there can presumably be no liability in negligence merely for

---

[33] 1996 Act, section 1(6).
[34] 1996 Act, section 1(2) or (5).
[35] 1996 Act, section 1(7).

removing support where no right of support exists (*Dodd v Holme* (1834); *Ray v Fairway Motors (Barnstable) Ltd* (1968)).

This assumption has become less certain following the development of the principle in *Leakey v National Trust for Places of Historic Interest or Natural Beauty* (1980) (referred to above in the section on damage to adjoining property). However, it may still represent good law as many of the recent decisions in this area relate to situations where a right to support already existed (*Bradburn v Lindsay* (1983); *Holbeck Hall Hotel Ltd v Scarborough Borough Council* (2000); *Rees v Skerrett* (2001)).

The 1996 Act

None of these common law rules have been affected by the 1996 Act[36]. The Act does however provide a statutory procedure for warning adjoining owners about impending excavations. It requires the excavating owner to serve a notice of adjacent excavation in appropriate circumstances[37]. The adjoining owner can then require him, at his own expense, to underpin or otherwise strengthen or safeguard the foundations of his adjacent building before commencing the excavations[38].

Equally importantly, the Act provides the excavating owner with the right to undertake these works to the adjoining owner's foundations, even in the absence of his consent[39]. This enables him to avoid the possibility of his excavations causing damage which could expose him to the risk of civil liability.

In common with the Act's other provisions, disputes between the parties are settled by appointed surveyors[40] and compensation will be payable by the excavating owner where loss or damage results from the exercise of his rights[41] (*Adams v Marylebone Borough Council* (1907)).

---

[36] 1996 Act, section 6(10).
[37] 1996 Act, sections 6(1) and (2).
[38] 1996 Act, section 6(3).
[39] *ibid*
[40] 1996 Act, sections 6(7) and 10(1).
[41] 1996 Act, section 7(2).

Digest of Cases

ADAMS v MARYLEBONE BOROUGH COUNCIL (1907)
Court of Appeal

In brief

- **Issues:** 1894 Act – nature of appointed surveyors' statutory role – adjoining owner's entitlement to compensation – surveyors' power to award compensation for lawful works – surveyors' power to award compensation for unnecessary inconvenience.

- **Facts:** The Council raised the party wall separating their building from Adams' restaurant. Adams asked the appointed surveyors to award compensation to her for loss of trade caused by the works. When they refused she appealed against the award.

- **Decision:** The Act provided for the payment of differing levels of compensation for the various types of work which it authorised. In each of these circumstances the appointed surveyors had power to determine the appropriate amount of compensation payable. However, an adjoining owner had no general right to compensation for work lawfully carried out under the Act in the absence of a specific provision to this effect. The only entitlement where a party wall was raised was to have consequential damage made good. Adams was not therefore entitled to compensation for loss of trade and the surveyors had no power to award this. The court also considered whether the surveyors also had a more general power to award compensation for works which resulted in unnecessary inconvenience and which were therefore unlawful. The power to determine what amounted to unnecessary inconvenience was central to the surveyors' role under the legislation. However, the court was unable to say whether they also had power to award compensation where this had occurred. In the event it proved unnecessary to determine this issue as it appeared that Adams' claim related only to compensation for lawful works and no allegation of unnecessary inconvenience had been made. Accordingly her claim for compensation was rejected.

- **Notes:** (1) Section 7(2) of the 1996 Act now contains an extensive right to compensation where loss or damage results from any work executed in pursuance of the Act. This right formerly applied only to works carried out under the legislation's adjacent excavation provisions. It would include an entitlement to compensation for loss of trade in the circumstances arising in this case. (2) See also: *White v Peto Brothers* (1888).

The facts

Mrs Adams ran a restaurant from 33 John Street, Edgware Road, London W2, and had a leasehold interest in the property. Marylebone Borough Council was the freehold owner of 34 John Street. The two properties were part of the same terrace and were separated from each other by a party wall.

In 1904, without serving the requisite party structure notice under the 1894 Act, the Council had started work to raise the party wall between the two properties. Mrs Adams successfully sued them for trespass and obtained an injunction to prevent any further such unauthorised work. She was also awarded damages to compensate her for the losses she had incurred as a result of the Council's actions, including damages for loss of trade.

As a result of the proceedings the Council stopped work and served a party structure notice on Mrs Adams. In due course surveyors were appointed and they published an award which authorised the Council to raise the party wall.

Before the award was published Mrs Adams had asked the surveyors to award compensation to her for her continuing loss of trade. Although she had been compensated by the courts for all losses incurred up to the date of service of the notice, the damage to her business caused by the original unlawful work was continuing.

She maintained that the Council had agreed to treat the notice as relating retrospectively to the unlawful work as well as to the work referred to in the notice. The surveyors were therefore competent to award compensation for her continuing losses which had been caused by the earlier work. She offered to produce evidence of her loss of trade to the surveyors but they refused to hear this, or to include a provision for compensation in their award.

Mrs Adams appealed against the award to the county court on the basis of the surveyors' refusal to consider her claim for compensation for loss of trade. The matter was ultimately decided by the Court of Appeal. Mrs Adams had never explicitly stated the legal basis on which she claimed to be entitled to compensation. Before deciding the appeal the Court of Appeal therefore reviewed the various provisions within the legislation which were capable of providing an entitlement to compensation, as well as the extent to which the surveyors had jurisdiction to award compensation in each case.

Line of junction compensation

In the context of the Act's line of junction provisions it referred to the building owner's right, under section 87(6)[1], to place projecting footings

---

[1] Broadly equivalent to section 1(6) of the 1996 Act.

on an adjoining owner's land following service of a notice. It noted that the right was conditional upon paying compensation for consequential damage[2].

The scope of this compensation was quite broad and, according to Fletcher Moulton LJ, extended to "damage or inconvenience" caused by the works[3]. The court noted that, under section 87(6), the surveyors were given jurisdiction to determine the amount of any compensation payable[4].

Adjacent excavation compensation

The court also examined section 93 which dealt with what would now be referred to as the Act's adjacent excavation provisions[5]. Where relevant excavations were undertaken, the building owner was given the right (and, if so required by the adjoining owner, the obligation) to underpin or otherwise strengthen the foundations of the adjoining owner's building.

The court noted that section 93(3) required the building owner to compensate adjoining owners and occupiers for losses caused by the exercise of this right. Compensation was payable for "inconvenience, loss or damage" so, once again, its scope was quite broad. However, the section was silent on the question of whether the surveyors were the proper tribunal to determine the amount of the compensation.

Laying open compensation

The Act's "laying open" provisions were then considered. As is the case today[6], a building owner had a draconian right under the 1894 Act to "lay open" an adjoining owner's building by demolishing and rebuilding a party wall to facilitate improvements to his own property[7]. The court examined section 95(2)(b), which dealt with the costs of such work.

Although the word "compensation" was not used in that section, it did provide that the building owner should pay "a fair allowance" to the adjoining owner for any "disturbance and inconvenience" caused to him by such works[8]. The precise scope of this payment was unclear as it was

---

[2] An equivalent provision now appears in section 1(7) of the 1996 Act.
[3] Under section 1 (7) of the 1996 Act, the scope of this compensation is now expressly limited to "damage" to "property".
[4] An equivalent provision now appears in section 1(8) of the 1996 Act.
[5] Contained in section 6 of the 1996 Act.
[6] Section 2(2)(e) of the 1996 Act.
[7] Section 88(7).
[8] This provision now appears as section 11(6) of the 1996 Act which refers to adjoining premises being "laid open" in exercise of the right contained in section 2(2)(e). The provision also now extends to adjoining occupiers as well as adjoining owners.

not defined in the legislation. Also, in common with the adjacent excavation provisions, the Act failed to state whether the surveyors had authority to determine the relevant amount payable.

Party wall compensation

The three provisions so far considered were helpful in understanding the Act's compensation regime. However, none of these could provide a basis for a compensation award by the surveyors in the present case as the Council's work involved the raising of a party wall rather than any of the work already described.

The court therefore considered whether section 88(6)[9], which authorised the raising of a party wall, could provide the necessary basis for such an award. It found that the compensation provisions within this section were more limited than elsewhere. Fletcher Moulton LJ distinguished the rights of adjoining owners in section 88(6) from those considered above in the following terms:

> "It is not that this part of the Act does not contemplate cases in which compensation is to be given for annoyance to the adjoining owner, but these cases are clearly defined in the Act, and are widely different in their nature from the case of a party wall, so that the express provision for compensation with regard to them, and the absence of any corresponding language with regard to the much more important and more frequently occurring case of a party wall greatly strengthen, in my opinion, the argument for the respondents."

Although the court was satisfied that the surveyors had the necessary authority to decide such issues, the scope of the compensation actually payable under section 88(6) was restricted to physical damage to adjoining premises. This was because the exercise of the right was expressed to be conditional, not on the payment of compensation, but only on "making good damage". The section was not therefore broad enough to accommodate the compensation for loss of trade which Mrs Adams was seeking.

Compensation for unnecessary inconvenience?

The surveyors could therefore only award compensation for lawful works where this was expressly provided for in the Act. However, the court also

---

[9] Broadly equivalent to sections 2(2)(a) and 2(3)(a) and (b) of the 1996 Act.

considered whether they had some inherent power to award compensation where losses resulted from unlawful works. In particular, could they award compensation for a breach of the building owner's obligation, under section 90(3)[10], not to cause unnecessary inconvenience to adjoining owners and occupiers?

In this context the court made reference to the surveyors' statutory powers in section 91(1)[11], which included the power to determine the "the right to do and the time and manner of doing any work and generally any other matter arising out of or incidental to such difference". It had no doubt that this wording was broad enough to enable the surveyors to determine what constituted an unnecessary inconvenience. Indeed Fletcher Moulton LJ was of the view that this went to the root of the surveyors' role under the Act:

> "In my view [sections 90 and 91] ... provide the practical machinery for working out the policy of the Legislature with regard to the rights given to [building] owners under the Act. Section 90 provides that, though these rights are given, they shall not be exercised without notice being given to the adjoining owner... It also provides for proper shoring for the protection of the adjoining premises, and for the work being carried out in such a manner and at such a time as not to cause unnecessary inconvenience to the adjoining owner or the adjoining occupier, or, in other words, that the building owner shall use his rights reasonably so as not to cause unnecessary damage to the adjoining owner or occupier. That provision clearly opens up the question as to what is unnecessary inconvenience in each particular case, and the Legislature has constituted a practical tribunal for deciding such questions."

The court was clearly of the view that these questions would normally be determined before work commenced and would be addressed by provisions within the surveyors' award dealing with such practical matters as the time and manner of carrying out the work. It was however less clear about the surveyors' powers where an alleged unnecessary inconvenience had already taken place and the building owner had exceeded his rights under the Act.

Fletcher Moulton LJ's comments suggest that where such an allegation is made the surveyors have no jurisdiction at all:

> "I confess that, as at present advised, I, personally, am disposed to think that, with regard to any act of the building owner outside the provisions of Part VIII, the special tribunal would have no jurisdiction; but at the same time I think

---

[10] An almost identical provision to section 7(1) of the 1996 Act.
[11] Broadly equivalent to sections 10(10) and 10(11) of the 1996 Act.

that it would have the widest powers of determining the manner and time of doing the work, so as not to cause unnecessary inconvenience to the adjoining owner or occupier."

Vaughan Williams LJ appears to have had no doubt that the surveyors have jurisdiction to adjudicate on allegations of unnecessary inconvenience:

"... I am disposed to think myself that, if a question arises as to whether or not the building owner is exercising the right given to him ... in such a manner as to cause unnecessary inconvenience to the adjoining owner, that is a question which it would be for the statutory tribunal to decide, if called upon to do so by the building owner or the adjoining owner ..."

However, he was less clear about the surveyors' powers once they have made such an adjudication:

"... but of course the fact that they had to decide that question would not on the face of it necessarily involve the further proposition that the tribunal having the right and the duty to decide that question should therefore have the power of saying whether any, and what, compensation should be paid, or what remedy should be given in respect of such unnecessary inconvenience, if any, to the adjoining owner or occupier. It may be that the remedy is that, if the special statutory tribunal decides that unnecessary inconvenience is being caused to the adjoining owner or occupier, the building owner thereupon loses the protection of the Act.

If that question had to be dealt with only in relation to future acts of the building owner, there might not be so much difficulty; but suppose the question comes before the statutory tribunal whether a building owner has in the past so exercised his rights as to cause considerable unnecessary damage to the adjoining owner, what then? Is the only remedy to be an action in the High Court, if such an action would lie? I think that is a question of considerable difficulty. For the purpose of dealing with it, one must consider what the provisions with regard to compensation are under the Act."

Vaughan Williams LJ then proceeds to examine the various sections that potentially provide expressly for the payment of compensation (see above). His conclusions following this examination are unfortunately ambiguous in the current context:

"Speaking for myself, I am disposed to think that in all the cases in which the Act provides for compensation the intention is that the amount of compensation is to be determined by the statutory tribunal, to which is also relegated the determination of the questions which form conditions precedent to the existence of the right to compensation.

> There are apparently no express words by which this is provided except in the case of section 87(6), but it would be extremely inconvenient if two inquiries were necessary, one by the special tribunal to ascertain whether the conditions precedent to the right to compensation existed and another before some other tribunal to determine the amount of the compensation."

It is unclear from these comments whether Vaughan Williams considers that section 90(3) is one of the situations in which the Act "provides for compensation" or whether he is simply referring to the four categories of compensation for lawful works, considered above. According to a strict reading of his words the latter would seem to be the case. However, his earlier comments on section 90(3) (particularly in the context of the separation of the condition precedent and quantum issues) do suggest that he might be sympathetic to the idea of the surveyors determining the amount of compensation in addition to the existence of an unnecessary inconvenience.

In fact the court side-stepped the issue of whether compensation could be awarded for unnecessary inconvenience. It concluded that Mrs Adams had never alleged unnecessary inconvenience as a cause of action. Her claim for compensation was simply a general statement about her losses which had been caused by works which had been properly authorised by the Act. As no claim had been made for compensation arising from unlawful works, there was no need for the court to decide whether the surveyors had authority to award compensation in these circumstances.

> Fletcher Moulton LJ: "... it appears to me to be unnecessary to decide these questions, because I hold that the adjoining owner has no right to compensation for any inconvenience caused by operations which the Act has given the building owner a right to carry out, so long as the latter acts within the provisions of the Act, for so long as he does so, his action is lawful."

> Vaughan Williams LJ: "For some time during the argument I thought that there was a claim made, which the evidence tendered might or might not substantiate, under section 90(3), in respect of what I may call excessive or unnecessary inconvenience caused by the operations of the respondents to the appellant. The claim made was, I think, simply a claim that by the exercise of their rights under section 88(6) of the Act, the building owners had caused damage and inconvenience to the adjoining owner. I have already said that, in my opinion, the mere exercise of such rights cannot afford any ground for compensation under the Act."

The decision

There was no right to receive compensation for losses caused by works which had been authorised by the Act unless the Act specifically provided for this. In the case of works to raise a party wall the right to compensation did not extend to business losses. Mrs Adams therefore had no right to compensation for the lawful carrying out of the work.

It was unclear whether the surveyors could award compensation for losses which resulted from works causing unnecessary inconvenience but, on the evidence, no such losses had been claimed by Mrs Adams.

Vaughan Williams LJ explained the court's decision in the following terms:

> "... we are in very considerable difficulty from not knowing exactly what the nature of the damage in respect of which the appellant claimed was, for it may very well be, so far as appears, that it was damage sustained by the appellant merely by reason of the exercise by the respondents of their rights under the London Building Act 1894, section 88. I can only say, if that was so, that, in my opinion, such damages not only cannot be claimed before the tribunal of arbitrators constituted under section 91 of the Act, but cannot be claimed before any tribunal whatsoever.
>
> ... in my opinion, the mere exercise of the rights which under the London Building Act 1894, are substituted for the rights which formerly existed at common law as between the tenants in common of a party wall cannot in itself give any right to compensation, because it is in its nature rightful.
>
> ... I have come to the conclusion that no claim was really made before the special tribunal that they should hear evidence of excessive or unnecessary inconvenience caused by the operation of the respondents to the appellant. The claim made was, I think, simply a claim that by the exercise of their rights under section 88(6) of the Act, the building owners had caused damage and inconvenience to the adjoining owner. I have already said that, in my opinion, the mere exercise of such rights cannot afford any ground for compensation under the Act."

For these reasons the court rejected Mrs Adams' claim for compensation and dismissed her appeal.

London Building Act 1894

Part VIII: RIGHTS OF BUILDING AND ADJOINING OWNERS

87. Rights of owners of adjoining lands respecting erection of walls on line of junction

Where lands of different owners adjoin and are unbuilt on at the line of junction and either owner is about to build on any part of the line of junction the following provisions shall have effect:

(6) Where ... the building owner proceeds to build an external wall on his own land he shall have a right at his own expense at any time after the expiration of one month from service of the notice to place on the land of the adjoining owner below the level of the lowest floor the projecting footings of the external wall with concrete or other solid substructure thereunder making compensation to the adjoining owner or occupier for any damage occasioned thereby the amount of such compensation if any difference arise to be determined in the manner in which differences between building owners and adjoining owners are hereinafter directed to be determined.

88. Rights of building owner

The building owner shall have the following rights in relation to party structures (that is to say):

(6) A right to raise and underpin any party structure permitted by this Act to be raised or underpinned or any external wall built against such party structure upon condition of making good all damage occasioned thereby to the adjoining premises or to the internal finishings and decorations thereof and of carrying up to the requisite height all flues and chimney stacks belonging to the adjoining owner on or against such party structure or external wall.

(7) A right to pull down any party structure which is of insufficient strength for any building intended to be built and to rebuild the same of sufficient strength for the above purpose upon condition of making good all damage occasioned thereby to the adjoining premises or to the internal finishings and decorations thereof.

90. Rules as to exercise of rights by building and adjoining owners

(3) A building owner shall not exercise any right by this Act given to him in such manner or at such time as to cause unnecessary inconvenience to the adjoining owner or to the adjoining occupier.

91. Settlement of differences between building and adjoining owners

(1) In all cases (not specifically provided for by this Act) where a difference arises between a building owner and an adjoining owner in respect of any matter arising with reference to any work to which any notice given under this part of this Act relates unless both parties concur in the appointment of one surveyor they shall each appoint a surveyor and the two surveyors so appointed shall select a third surveyor and such one surveyor or three surveyors or any two of them shall settle any matter from time to time during the continuance of any work to which the notice relates in dispute between such building and adjoining owner with power by his or their award to determine the right to do and the time and manner of doing any work and generally any other matter arising out of or incidental to such difference but any time so appointed for doing any work shall not unless otherwise agreed commence until after the expiration of the period by this part of the Act prescribed for the notice in the particular case.

93. Building owner to underpin adjoining owner's building

Where a building owner intends to erect within ten feet of a building belonging to an adjoining owner a building or structure any part of which such ten feet extends to a lower level than the foundations of the building belonging to the adjoining owner he may and if required by the adjoining owner shall (subject as hereinafter provided) underpin or otherwise strengthen the foundations of the said building so far as may be necessary and the following provisions shall have effect:

(3) The building owner shall be liable to compensate the adjoining owner and occupier for any inconvenience loss or damage which may result to them by reason of the exercise of the powers conferred by this section.

(4) Nothing in this section shall relieve the building owner from any liability to which he would otherwise be subject in case of injury caused by his building operations to the adjoining owner.

95. Rules as to expenses in respect of party structures

(1) [Expenses to be borne jointly by the building owner and adjoining owner]

(2) As to expenses to be borne by the building owner:

(b) if any party structure which is of proper materials and sound or not so far defective or out of repair as to make it necessary or desirable to pull it down be pulled down and rebuilt by the building owner the expense of pulling down and rebuilding the same and of making good any damage by this part of the Act required to be made good and a fair allowance in respect of the disturbance and inconvenience caused to the adjoining owner shall be borne by the building owner.

ALCOCK v WRAITH AND SWINHOE (1991)
Court of Appeal

In brief

- **Issues:** Work to roof covering above a party wall – entitlement to undertake work at common law – trespass – nuisance – negligence – non-delegable duties – extra-hazardous acts – withdrawal of support – work to party walls – reciprocal duties to adjoining owners at common law.

- **Facts:** Wraith was a building contractor who replaced the roof covering of Mr and Mrs Swinhoe's terraced house. The joint between the new covering and that of the neighbouring property, owned by Alcock, was inadequately constructed. As a result damage was caused to Alcock's property.

- **Decision:** Mr and Mrs Swinhoe were liable for the damage which had been caused by their independent contractor. Work at the joint between roof coverings was no different from work to a party wall. In both situations a building owner had a common law right to interfere with the fabric of his neighbour's property whilst undertaking repairs to the shared structure. He owed a reciprocal duty of care to his neighbour whilst undertaking this work. As it involved a risk of damage to the neighbour's property the duty was owed on a non-delegable basis. Mr and Mrs Swinhoe were in breach of this duty and were therefore responsible for Alcock's losses.

- **Notes:** See also (on a building owner's entitlement to undertake work to his neighbour's side of a party wall at common law): *Standard Bank of British South America v Stokes* (1878) – (on non-delegable duties): *Bower v Peate* (1876); *Dalton v Angus* (1881); *Hughes v Percival* (1883); *Matania v National Provincial Bank Ltd and The Elevenist Syndicate Ltd* (1936).

The facts

Mr Alcock was the owner-occupier of a terraced property at 47 East View, Wideopen, Newcastle upon Tyne. Mr and Mrs Swinhoe were owner-occupiers of the adjoining terraced property at 50 East View (Figure 1).

The two properties were separated by a party wall and each had a slate roof which extended as a single uninterrupted structure across the top of the party wall. Mr and Mrs Swinhoe obtained an improvement grant from the local authority to reroof their property and instructed Mr Wraith, a builder, to undertake the work.

Mr Wraith carried out the reroofing work in January/February 1984. Because the grant was not sufficient to allow for the use of slates, the new roof covering consisted of Marley concrete tiles which had to be joined to the existing slate covering on Mr Alcock's side. The joint between the two coverings encroached 200 mm on to Mr Alcock's premises. Furthermore the joint was inadequately constructed. Mr Wraith had effected it by "stuffing newspapers in the gap [between the two coverings] and covering the joint with a cement fillet with no metal flashing".

The joint was not weather-tight and, towards the end of 1984, damp patches appeared on the ceiling of one of Mr Alcock's first floor bedrooms. Mr Alcock eventually, at his own expense, arranged for the necessary remedial work to be carried out. This included trimming the slates and tiles at the intersection and installing a secret gutter under the roof covering to direct the roof water straight to the eaves gutter. The cost of this work was £663.95.

Mr Alcock brought proceedings against Mr Wraith (the first defendant) and Mr and Mrs Swinhoe (the second defendants) for damages for the cost of the repairs and also for stress, inconvenience and anxiety. The judge in the Newcastle upon Tyne County Court held that both defendants were liable for negligence, trespass and nuisance and awarded special damages of £663.95 for the repairs and general damages of £200 for the stress,

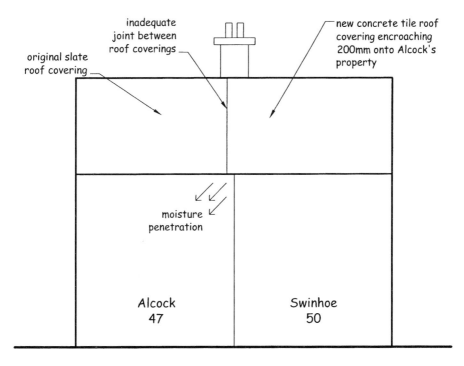

Figure 1: Alcock v Wraith and Swinhoe

inconvenience and anxiety. Although the judge also ordered that Mr and Mrs Swinhoe should be wholly indemnified by Mr Wraith, this was effectively worthless as Mr Wraith had already been declared bankrupt.

Mr and Mrs Swinhoe appealed to the Court of Appeal on the basis that they could not be held liable for Mr Wraith's actions as he had been employed by them as an independent contractor.

Liability for the torts of an independent contractor

In his judgment Neill LJ noted the general principle that where someone employs an independent contractor to do work on his behalf he will not be liable for torts committed by the contractor in the course of the execution of the work. He then went on to identify the seven main exceptions to this general principle:

(1) Cases where the employer is under some statutory duty which he cannot delegate.
(2) Cases involving the withdrawal of support from neighbouring land.
(3) Cases involving the escape of fire.
(4) Cases involving the escape of substances, such as explosives, which have been brought on to the land and which are likely to do damage if they escape; liability will attach under the rule in *Rylands v Fletcher* (1868).
(5) Cases involving operations on the highway.
(6) Cases involving non-delegable duties of an employer for the safety of his employees.
(7) Cases involving extra-hazardous acts.

He was satisfied that both the general principle and the seven exceptions must apply equally to the torts of nuisance, trespass and negligence. Furthermore, the theoretical distinction between the three torts was of little relevance to the main issue in the present case. This was simply whether either the extra-hazardous acts exception or the withdrawal of support exception applied to the damage to the roof so as to make Mr and Mrs Swinhoe liable for Mr Wraith's torts.

Cases involving extra-hazardous acts

Neill LJ considered some of the authorities for the "extra-hazardous acts" exception to the general rule. He noted Slesser LJ's definition of extra-hazardous acts in *Honeywill & Stein v Larkin Brothers Ltd* (1934) as "acts which, in their very nature, involve in the eyes of the law special danger to others; of such acts the causing of fire and explosion are obvious and established instances".

He also made reference to Talbot J's statement of the relevant principle in *Brooke v Bool* (1928):

> "The principle is that if a man does work on or near another's property which involves danger to that property unless proper care is taken, he is liable to the owners of the property for damage resulting to it from the failure to take proper care, and is equally liable if, instead of doing the work himself, he procures another, whether agent, servant or otherwise, to do it for him."

The distinction between work which is inherently dangerous and which will automatically involve danger unless some positive step is taken to prevent it and work which may simply become dangerous if negligently undertaken was made in the case of *Salisbury v Woodland* (1970), which Neill LJ also referred to. In that case a contractor damaged some telephone wires and caused an obstruction on a roadway whilst felling a tree. The Court of Appeal was not satisfied that felling a tree was an extra-hazardous act within the terms of the exception. Neill LJ quoted Harman LJ's observations in that case:

> "The act of felling the tree did involve danger to others because it was negligently done. But it was a perfectly simple job to remove this tree without causing any danger to anybody and it was not work which was inherently dangerous so as to come within that exception."

Neill LJ set out his own understanding about the extent of the exception:

> "It is not possible in my judgment to provide a list of activities which will be regarded as 'extra-hazardous' so as to fall within this exception, but it is clear that the activity must involve some special risk of damage, or – to use the test approved by Atkin LJ in *Belvedere Fish Guano Co Ltd v Rainham Chemical Works Ltd* (1920) at page 504 and by Harman LJ in *Salsbury's* case at page 345 – the work must be work 'which from its very nature is likely to cause danger'. I would only add that in many cases it will be more appropriate to substitute 'damage' for 'danger' in Atkin LJ's test. What one looks for is to see whether there is some special risk or whether the work from its very nature is likely to cause danger or damage."

On this basis, he held that the work to the roof in the present case did not constitute an extra-hazardous act within the meaning of the exception:

> "It seems to me that though a waterproof joint of this kind may have presented difficulties it should have been well within the competence of an ordinary building contractor. It was not work which in the ordinary way involved some special risk. The general rule that an employer is not liable for the torts of an

independent contractor is of long-standing, and, while this rule remains in its present form, it is important not to extend the exception of special risk too far."

Cases involving withdrawal of support

Neill LJ then made reference to the various authorities which are often grouped together as constituting the "withdrawal of support" exception. This exception was established in *Bower v Peate* (1876) in the context of excavation work which interfered with an easement of support for neighbouring land. It was approved in *Dalton v Angus* (1881) and extended to work affecting (pre-1925) party walls in *Hughes v Percival* (1883).

In *Hughes v Percival* Lord Blackburn had explained the imposition of the non-delegable duty on each owner of a party wall in terms of reciprocality for the rights to undertake works to the whole thickness of the wall, which each owner possessed:

> "... The defendant had a right to utilize the party-wall, for it was his property as well as the plaintiff's; a stranger would not have had such a right. But I think that the law cast upon the defendant, when exercising this right, a duty towards the plaintiff. I do not think that duty went so far as to require him absolutely to provide that no damage should come to the plaintiff's wall from the use he thus made of it, but I think that the duty went as far as to require him to see that reasonable skill and care were exercised in those operations which involved a use of the party-wall, exposing it to this risk. If such a duty was cast upon the defendant he could not get rid of responsibility by delegating the performance of it to a third person."

Neill LJ placed great emphasis on the extension of the *Bower v Peate* principle to work affecting party walls in *Hughes v Percival*, to the extent of thereafter referring to the withdrawal of support exception as the "exception relating to party walls". He also described the decision in *Matania v National Provincial Bank* (1936) in terms of being a further extension of this same principle to work affecting floors between flats.

He was therefore, by analogy, able to extend this principle still further, to the facts of the present case:

> "The question therefore arises: if work on a party-wall between two houses or on the floor beneath an adjoining flat or maisonette imposes a duty of care which cannot be delegated to an independent contractor, is there any valid reason why a similar non-delegable duty should not be imposed on the owner of a terrace house who, in order to repair his roof, has to interfere with the integrity of his neighbour's roof by replacing the joint between the two roofs?"

He explained the basis of the "exception relating to party walls", not in terms of the withdrawal of support, but in terms of reciprocal rights:

> "But, as it seems to me, the true basis for the exception in the party-wall cases is that where the law confers a right to carry out work on a wall or other division between two properties, and that work involves a risk of damage to the adjoining property, the law also imposes a duty on the party carrying out the work to ensure that it is carried out carefully.
>
> ... Mr and Mrs Swinton had the right to interfere with the joint between the two roofs and, because of its structure, to intrude slightly on to the slate roof of Mr Alcock's house. But if they exercised this right they were under a duty, as I see it, to see that reasonable skill and care were used in the operation. Moreover, this duty could not be delegated to an independent contractor. In this context I can see no satisfactory distinction between interfering with a party-wall or with a floor on the one hand and interfering with the edge of a contiguous roof on the other hand.
>
> ... In the end I have come to the conclusion that the judgment should be upheld on the ground that because the work necessarily involved interference with the joint between the two roofs and the removal of some of Mr Alcock's slates Mr and Mrs Swinhoe were under a non-delegable duty to ensure that the roofing work was done with proper care. The work was not so done."

The decision

Mr and Mrs Swinhoe were liable for Mr Wraith's torts, notwithstanding the fact that they had employed him as an independent contractor. The work to the roof involved an interference with Mr Alcock's property which carried the likelihood that damage would be caused to it. In these circumstances they owned a non-delegable duty of care to him which had been breached. Accordingly, the appeal was dismissed and the original judgment was upheld.

ANDREAE v SELFRIDGE & COMPANY LTD (1937)
Court of Appeal

In brief

- **Issues:** Nuisance – unlawful interference with the enjoyment of land – unnecessary inconvenience – noisy building operations – generation of dust and grit from building operations – damages for loss of trade.

- **Facts:** Andreae's hotel business suffered disruption from Selfridges' neighbouring building operations. This resulted in a loss of trade. The disruption included noisy night-time operations and the generation of dust and grit in connection with demolition activities.

- **Decision:** Building and demolition activities were "a common and ordinary use of land". Neighbouring owners therefore had to tolerate a level of inconvenience from them providing they were undertaken with reasonable skill and care. Andreae's losses were only recoverable to the extent that they had been caused by Selfridges having failed to exercise reasonable skill and care. This was the case in respect of the night-time operations (which should have been avoided entirely) and also in respect of the dust and grit. The impact of the latter should have been minimised by the use of water, the boarding up of windows and the staggering of the demolition process. Andreae was entitled to damages for the losses caused by these activities. However, the majority of her losses resulted from Selfridges' lawful building activities which were lawfully undertaken. No damages could be awarded for these losses.

- **Notes:** (1) See also: *Video London Sound Studios Ltd v Asticus (GMS) Ltd and Keltbray Demolition Ltd* (2001). (2) The concept of unlawful interference with the use or enjoyment of land in the tort of nuisance is synonymous with that of unnecessary inconvenience under section 7(1) of the 1996 Act.

The facts

The case concerned a large 'island site' in the West End of London bounded by Wigmore Street to the north, Oxford Street to the south, Orchard Street to the west and Duke Street to the east (Figure 2). The freehold in the whole site was owned by the Portman Estate.

Selfridges were developing the site for occupation by their business on a rolling basis. As each of the buildings on the site became available they acquired its lease and developed that part of the site. Under an agreement with the Portman Estate a new lease of the whole site would be granted to Selfridges upon completion of the development.

Mrs Andreae ran a hotel business from numbers 119 and 121 Wigmore Street which formed a part of the site. In 1931 she had agreed to assign her lease to Selfridges, with completion to take place on her retirement from the business in 1936. In the meantime, she continued to trade from the premises whilst the remainder of the site was developed. During this period neighbouring building operations proved to be a serious deterrent to hotel guests and, as a result, she suffered a serious loss of trade.

The allegation of nuisance

Mrs Andreae issued proceedings against Selfridges, alleging that the neighbouring building operations constituted a nuisance. She provided evidence of significantly reduced profits over the five years since commencement of the works and sought damages for the £4500 loss of profits suffered during this period. Her complaint related to two separate building operations undertaken during the five-year period.

The first operation related to demolition and reconstruction works carried out at the south-east corner of the site between November 1931 and February 1932. Mrs Andreae alleged that the work had been excessively noisy. The work had included basement excavations to depth of 60 ft, the erection of a steel frame for the new building and the use of a noisy crane. For a month the work had continued throughout the night without interruption. Following complaints the work paused between 10 PM and 7 AM each night but Mrs Andreae still found this unacceptable.

The second operation related to the demolition of buildings at the south-west corner of the site between July and September 1935. Mrs Andreae's compliant was two-fold. Firstly, she complained about excessive noise from the use of pneumatic hammers. These had been used for a period of six days, to break up the reinforced concrete roof of one of the buildings. Secondly, she complained about the ongoing generation of dust and grit throughout the second operation.

The court at first instance found in Mrs Andrea's favour and awarded damages to her for all her losses. Bennett J considered the very nature of the works to have been intrusive and made his decision on this basis, as he explained in his judgment:

> "I cannot regard what the defendants did on the site of the first operation as having been commonly done in the ordinary use and occupation of land and houses. It is neither usual nor common, in this country, for people to excavate a site to a depth of 60 ft and then to erect upon that site a steel framework and fasten the steel frame together with rivets.
>
> Nor is it, I think, a common or ordinary use of land, in this country, to act as the defendants did when they were dealing with the site of their second

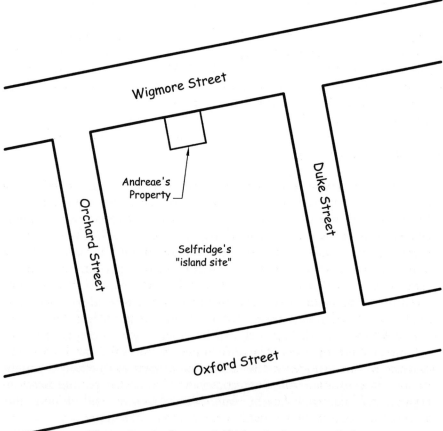

Figure 2: Andreae v Selfridge & Company Ltd

operation, namely, to demolish all the houses that they had to demolish, five or six of them I think, if not more, and to use for the purpose of demolishing them pneumatic hammers."

Selfridges appealed against Bennett J's judgment to the Court of Appeal.

The common and ordinary use of land

The Court of Appeal was critical of the trial judge's approach to the case. It described a distinction in the law of nuisance between common and ordinary uses of land and other categories of use. Where a common and ordinary use of land nevertheless causes interference to a plaintiff's reasonable and comfortable enjoyment of his land, a defendant will only be liable to the extent that he has failed to exercise reasonable care and skill. Where such an interference results from some other category of use a defendant is automatically liable, without proof of negligence.

As demonstrated by the extracts from his judgment quoted above, Bennett J had found that Selfridges' building operations did not constitute a common and ordinary use of their land. The very fact of undertaking such operations therefore amounted to an actionable nuisance upon proof that Mrs Andreae had suffered loss as a result.

Sir Wilfred Green MR found that the judge had failed to apply the rule correctly in the context of temporary operations such as demolition and building:

> "... when one is dealing with temporary operations, such as demolition and building, everybody has to put up with a certain amount of discomfort, because operations of that kind cannot be carried on at all without a certain amount of noise and a certain amount of dust. Therefore, the rule with regard to interference must be read subject to this qualification, and there can be no dispute about it, that, in respect of operations of this character, such as demolition and building, if they are reasonably carried on, and all proper and reasonable steps are taken to ensure that no undue inconvenience is caused to neighbours, whether from noise, dust, or other reasons, the neighbours must put up with it.
>
> It seems to me that it is not possible to say, nor do I think that there is any evidence in this case which would warrant it being said, that the type of demolition, excavation and construction in which the defendant company was engaged in the course of these operations was of such an abnormal and unusual nature as to prevent the qualification to which I have referred coming into operation. It seems to me that, when the rule, as indeed it is a rule, speaks of the common or ordinary use of land, it does not mean that the methods of using land and building on it are in some way to be stabilised for ever. As time goes on, new inventions and new methods enable land to be more profitably used...
>
> ... but it is part of the normal use of land to make use, upon your land, in the matter of construction, of what particular type and what particular depth of foundations and particular height of building, may be reasonable, in the circumstances, and in the development of the day. I am unable to take the view that any of these operations was of such an abnormal character as to justify treating the disturbance created by it, and the whole of the disturbance created by it, as constituting a nuisance."

As Selfridges' building operations were a common and ordinary use of their land the correct test, which should have been applied by the trial judge, was whether they had undertaken them with reasonable care and skill.

Reasonable care and skill

The court considered the nature of the requirement to exercise reasonable care and skill and this was described by Sir Wilfred Green MR in the following terms:

> "The use of reasonable care and skill in connection with matters of this kind may take various forms. It may take the form of restricting the hours during which work is to be done; it may take the form of limiting the amount of a particular type of work which is being done simultaneously within a particular area; it may take the form of using proper scientific means of avoiding inconvenience. Whatever form it takes it has to be done, and those who do not do it must not be surprised if they have to pay the penalty for disregarding their neighbour's rights."

In the present case the court found that Selfridges had indeed failed to exercise reasonable care and skill in a number of respects. With regard to the first operation the court found that they should not have carried out noisy operations throughout the night.

> Sir Wilfred Greene MR: "I am satisfied that, so far as the previous night operations of the cranes were concerned, there was sufficient interference with the reasonable comfort of the plaintiff in carrying on her establishment to constitute an actionable nuisance.
> ... that the complaints were substantial complaints I, for one, am satisfied, and I certainly protest against the idea that, if persons, for their own profit and convenience, choose to destroy even one night's rest of their neighbours, they are doing something which is excusable. To say that the loss of one or two nights' rest is one of those trivial matters in respect of which the law will take no notice appears to me to be quite a misconception, and, if it be a misconception existing in the minds of those who conduct these operations, the sooner it is removed the better."

They were also at fault because of the manner in which they had conducted the second operation. Although their use of the pneumatic hammers had been reasonable, they had failed to take adequate precautions to minimise the spread of dust and grit by, for example, the use of water, the boarding up of windows or the staggering of demolition operations.

> Sir Wilfred Green MR: "I am satisfied, from the evidence, that the dust produced by this operation was something which is not to be endured unless some proper explanation is given of why it came. The burden of establishing that is upon the defendant company ... I am not satisfied that it has discharged the burden of proof upon it.

It appears that ... there was a flat roof ... and that flat roof had to be broken up, consisting, as it did, of reinforced concrete. For that purpose, the defendant company's contractors used two pneumatic hammers... The actual use of those hammers extended over six days only; there is no suggestion that they were used by night, and [Mrs Andreae's counsel] did not feel himself able to say that the use of the hammers, in those circumstances, constituted an actionable nuisance. The complaint, therefore with regard to the second operation consists of a complaint in respect of dust and grit."

Assessment of damages

The trial judge had taken the view that Selfridges' building operations, by their very nature, were unjustified. He therefore awarded damages based on the entirety of the losses which Mrs Andreae had suffered as a consequence of those operations.

Although the Court of Appeal also found that there had been an actionable nuisance this was not on the basis that the entirety of the works was unjustifiable. It was because, in the particular circumstances of this case, reasonable care and skill had not been exercised in their execution.

The Court of Appeal therefore took a different approach in determining Mrs Andreae's recoverable losses. The correct assessment of damages should not be based on the whole of the losses suffered as a result of the building operations but simply the proportion of the losses which had been caused by the objectionable operations:

Sir Wilfred Green MR: "I think that, on the evidence, there is proof of a substantial loss of actual customers, with, in the background, the inevitable repercussion that has on the reputation of a hotel. On the other hand, in this case ... one must be careful not to penalise the defendant company by throwing into the scales against it the effect of the loss of clients caused by operations which it was legitimately entitled to carry out. It can be made liable only in respect of matters on which it has crossed the permissible line."

Regarding the actual calculations, Sir Wilfred Green MR concluded thus:

"I think [Mrs Andreae] is entitled, not to a nominal sum, but to a substantial sum, based on these principles. What that sum is to be is a matter as to which the individual mind can only satisfy itself as to what is fair; but, in arriving at the sum which I consider to be the proper one ... I have discounted any loss of custom, or endeavoured to discount, so far as I can, any loss of custom, which might be due to the general loss of amenities owing to what was going on at the back, and I have tried to give what, in my opinion, would be a fair measure of the loss which, on the evidence, I infer that the plaintiff has suffered from that part of the defendant company's operations of which she has legitimate cause for complaint over the period in question.

I would say that, in a case of this kind, where the defendant company has, at any rate at certain stages, and in certain respects, shown, in my judgment, a reprehensible lack of regard for the duty which it owes to its neighbours, I should not be disposed, in drawing inferences, to draw inferences with regard to loss of custom in the defendant company's favour. It is a very difficult thing to attribute loss of custom to a particular cause in a case of this kind.

This is eminently a case where a jury, or a judge sitting alone, should use common sense and their knowledge of affairs in relation to the evidence which is given. I think the evidence in this case does establish a substantial injury, and the sum at which I think, on that basis, the damages should be fixed is £1000."

The decision

The trial judge's judgment was varied and the award of damages was reduced from £4500 to £1000.

APOSTAL v SIMONS (1936)
Court of Appeal

In brief

- **Issues:** Nature of party wall at common law – effect of Law of Property Act 1925 – damage to party wall – liability for damage – effect of agreement to withdraw claim.

- **Facts:** The tenants of business premises caused damage to a party wall by storing timber against it over a prolonged period of time. The owner of the adjoining property agreed to withdraw his claim against them for the damage. Subsequently further damage became apparent and the adjoining owner sought to recover the entire cost of repairs from them.

- **Decision:** The adjoining owner only had a potential claim for the cost of repairing the half of the wall within his ownership. This potential claim was limited to damage which had occurred after the date of the agreement withdrawing the earlier claim. The evidence suggested that the subsequently discovered damage had actually occurred prior to the date of the agreement. Consequently the tenants were not liable for any of the damage to the wall.

- **Notes:** See also: *Matts v Hawkins* (1813); *Watson v Gray* (1880); *Moss v Smith* (1977).

The facts

Apostal (the plaintiff) owned the freehold of 117 Hackney Road, London E2. The freehold of the adjoining property at 1A Gorsuch Place was owned by Simons (the first defendant). This property was let to Cave and Co (the second defendants), who were a firm of timber merchants. The two properties were separated by a party wall.

Cave caused damage to the party wall through their longstanding practice of storing timber against it. Apostal had complained to them about this in 1933 and had threatened to sue them for the cost of repairing the wall. However, on 12 November 1934 the dispute between the parties was settled and the terms of the agreement between them was set out in a letter of that date.

In the letter Apostal confirmed that he would withdraw his claim against Cave in consideration of them repairing the top of the wall, paying him the sum of £2 10s (which they promptly paid on the signing of the agreement) and undertaking not to store timber against the wall in future.

Service of dangerous structure notice

Three days after the agreement was signed (on 15 November) the District Surveyor visited the premises. From his inspection he concluded that the wall did not present a danger which would justify the service of a dangerous structure notice.

About a month later (on the night of 12 and 13 December) Cave undertook certain operations on their side of the wall, including the removal of the timber which had been stacked against it. Following these operations there was concern about the stability of the wall and the District Surveyor was again called in. He reinspected the wall on 17 December and now concluded that the wall was unsafe. He considered that it was badly built and noted that some of the bricks were extremely loose. On the basis of his recommendation the London County Council then served a dangerous structure notice on Apostal in respect of the wall.

On receipt of the notice Apostal issued proceedings against Cave and also against Simons. He alleged that they had wilfully, wrongfully and/or negligently damaged the party wall by placing heavy loads of timber against it.

The judge, at first instance, held that Simons, as freeholder, was not liable for the damage and dismissed the action against him. However, he found Cave liable. The damage had been caused by the stacking of timber against the wall, both before and after the date of the agreement, and they were responsible for this. He therefore ordered Cave to pay damages to Apostal representing the cost of repairing the wall together with the costs incurred in respect of the dangerous structure notice. Cave appealed to the Court of Appeal.

Entitlement to damages for repairs to party wall

The appeal raised two issues. Firstly, ownership of the party wall had been severed by the Law of Property Act 1925[1]. The court therefore had to consider whether Apostal, even if successful, was entitled to damages for repairs to the whole thickness of the wall, or simply for those to the portion within his ownership.

It drew a distinction between claims against the other party to the party wall arrangement and claims against a third party. It concluded that where the claim for damage was against a third party, a plaintiff could only recover for damage to the portion of the wall within his ownership. Apostal's claim against Cave, as tenant, fell into this category. It was

---

[1] Section 39 & Schedule 1, Part V.

therefore restricted to one half of the cost of the damage, rather than to the full amount awarded by the trial judge.

The situation would have been different had Simons also been liable. As the freeholder he was under a duty to provide rights of support to Apostal over the portion of the wall within his (Simons') ownership. A successful claim against Simons would therefore have provided Apostal with an entitlement to damages for the cost of repairing this portion of the wall, in addition to that in respect of his own half.

However, the present proceedings were confined to the issues between Apostal and Cave. Slesser LJ explained the legal situation as follows:

> "The subject of party walls has been dealt with in recent legislation, and by the Law of Property Act 1925, Schedule 1, Part V, it is provided that where, immediately before the commencement of that Act, a party wall is held in undivided shares the ownership shall be deemed to be severed vertically and the owner of each part shall be given such rights of support and user as shall be requisite for conferring rights corresponding to those subsisting before the commencement of that Act.
>
> The effect of the Schedule, which is brought into operation by section 39, is that after the coming into operation of that Act, only one half of this wall remains the property of the plaintiff. In so far as that part which was the property of the plaintiff was injured ... he may recover from the person who committed the wrongful act.
>
> As regards the other half, which is the property of Mr Simons, while Mr Apostal has rights of support, any remedy for the interference with such rights of support must lie against Mr Simons, who is put under an obligation by that section and, in my view, no action can be supported against a third party such as the second defendants in this action. It follows, therefore, of the amount which the learned judge may think fit to give, only one half can be attributed to Mr Apostal's section of the party wall."

Effect of agreement to withdraw claim for cost of repairs

Having decided that Apostal's claim could only relate to his own half of the party wall, the court then considered the second issue. This related to whether Cave could be liable for the damage at all, in view of Apostal's prior agreement to withdraw the claim against them. The trail judge had dismissed this agreement as Cave had failed to provide any consideration.

The Court of Appeal took a different view. Consideration was not itself relevant, as Apostal's claim was not founded in breach of contract. In any event, any failure of consideration had not been total as Cave had paid the agreed sum and had also cleared the timber away from the wall. Consequently Apostal was barred from claiming for damage which pre-dated the agreement. He could therefore only claim for damage which could be proved to have occurred after 12 November. At the very least this required a substantial reduction in the damages awarded by the trial judge.

Although the trial judge had referred to damage being caused "both before and after November 12" it was not clear how much, if any, damage had occurred after this date. The Court of Appeal was not convinced that any further damage had been caused. On the contrary, the evidence suggested that the removal of the timber had simply exposed damage which had occurred prior to 12 November. In this context Scott LJ noted that "... the probability is that when the support to the wall given by the timber was withdrawn, the wall came to pieces by its own rottenness".

The decision

Cave was not liable for the damage to the wall. Liability could only possibly have extended to one half of the thickness of the wall following severance of party walls by the Law of Property Act 1925. As Apostal had agreed to withdraw his existing claim on 12 November this liability must be further restricted to damage which could be proved to have occurred after this date. There was no evidence that any further damage had occurred after this date. Cave's appeal was therefore upheld and Apostal was ordered to repay all damages which had been ordered by the trial judge.

Law of Property Act 1925

1. Legal estates and equitable interests

(6) A legal estate is not capable of subsisting or of being created in an undivided share in land or of being held by an infant.

38. Party structures

(1) Where under a disposition or other arrangement which, if a holding in undivided shares had been permissible, would have created a tenancy in common, a wall or other structure is or is expressed to be made a party wall or structure, that structure shall be and remain severed vertically as between the respective owners, and the owner of each part shall have such rights to

support and user over the rest of the structure as may be requisite for conferring rights corresponding to those which would have subsisted if a valid tenancy in common had been created.

(2) Any person interested may, in case of dispute, apply to the court for an order declaring the rights and interests under this section of the persons interested in any such party structure, and the court may make such order as it thinks fit.

39. Transitional provisions

For the purpose of effecting the transition from the law existing prior to the commencement of the Law of Property Act 1922 to the law enacted by that Act (as amended), the provisions set out in the First Schedule to this Act shall have effect.

SCHEDULE 1

PART V: PROVISIONS AS TO PARTY STRUCTURES AND OPEN SPACES

1. Where, immediately before the commencement of this Act, a party wall or other party structure is held in undivided shares, the ownership thereof shall be deemed to be severed vertically as between the respective owners, and the owner of each part shall have such rights to support and of user over the rest of the structure as may be requisite for conferring rights corresponding to those subsisting at the commencement of this Act.

3. Any person interested may apply to the court for an order declaring the rights and interests under this Part of this Schedule, of the persons interested in any such party structure or open space, or generally may apply in relation to the provisions of this Part of this Schedule, and the court may make such order as it thinks fit.

BARRY v MINTURN (1913)
House of Lords

In brief

- **Issues:** 1894 Act – appropriation of adjoining owner's property – meaning of "defect" – meaning of "repair" – relevant considerations when making award – unnecessary inconvenience.

- **Facts:** Damp penetrated from the exposed face of a type 'a' party wall into Minturn's building. She wished to eradicate the damp from the wall and proposed to erect a damp-proof barrier on Barry's side of the wall. She served notice of her intention to repair the wall under the 1894 Act.

- **Decision:** The wall had formerly been a garden wall, and continued to fulfil this function on Barry's side. However, the fact that it now allowed damp to penetrate into a building meant that it was defective within the meaning of the Act. Minturn therefore had a right, under the Act, to repair it. The surveyors were required to select the most appropriate method of repair so as to avoid unnecessary inconvenience being caused to Barry. Specifically, the past history of the wall as a garden wall was not a relevant consideration in this context. The Act did not authorise the permanent appropriation of property so Minturn had no right to erect the damp-proof barrier on Barry's side. For the same reason the surveyors had no authority to require her to erect it on her side although she was of course free to do so. She was however entitled to insert the barrier within the thickness of the wall under the Act.

- **Notes:** See also (on the meaning of unnecessary inconvenience): *Thompson v Hill* (1870); *Jolliffe v Woodhouse* (1894) – (on the permanent appropriation of property): *Reading v Barnard* (1827); *Burlington Property Co Ltd v Odeon Theatres Ltd* (1938); *Gyle-Thompson v Wall Street (Properties) Ltd* (1974).

The facts

Sir John Woolfe Barry was the owner of 15 Chelsea Embankment, London SW3. Eliza Theodora Minturn was the owner of the next door property at 14 Chelsea Embankment. The properties were separated by a party wall. Both properties also had back gardens which were separated by a party fence wall (Figure 3).

In 1883 a previous owner of Minturn's property had extended it to the rear, to form a servants' hall in the basement, a billiards room on the ground floor and two first floor bedrooms. The work had involved excavating to a depth of 8 ft and, after thickening, making use of the

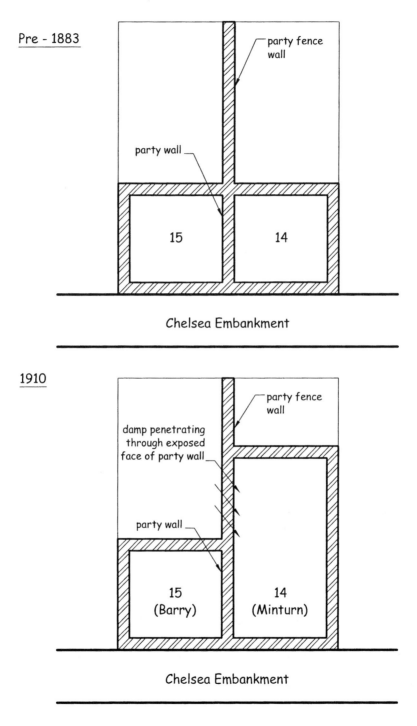

Figure 3: Barry v Minturn

existing party fence wall between the two gardens to form the lower part of the new extension wall. This wall therefore became a party wall.

By 1910, due to the damp condition of the exposed wall, moisture percolated into Minturn's basement. She therefore served a party structure notice on Barry to undertake repairs to his side of the wall under section 88(1) of the 1894 Act.[1]

The right to execute work under the Act

Barry objected to the proposed work as it would involve a considerable intrusion on to his property. In due course surveyors were appointed and they published an award under the Act. This decided that Minturn had no right to undertake the work as the wall was not "defective or out of repair" within the meaning of section 88(1). The wall was damp but, as it was essentially a garden wall, this did not amount to a defect.

Minturn appealed against the award to the county court. After protracted litigation the court decided that the wall was, in fact, "defective" within the meaning of the Act because it permitted damp to percolate from Barry's garden into Minturn's basement. However, having so decided it then had to consider the most appropriate remedy for the defect from the three possibilities suggested at the hearing:

- Minturn wished to install a 2-in thick vertical damp-proof barrier on Barry's side of the wall.
- Barry suggested that Minturn should instead construct a lining wall, with appropriate damp-proofing qualities, in the basement on her side of the wall.
- As an alternative, Barry also suggested that Minturn could insert a vertical asphalt damp-proof course, $\frac{3}{4}$ in thick, in the centre of the wall, with access for the work being obtained from her basement.

The court dismissed Minturn's preferred remedy on three grounds. Firstly, the intrusion on to Barry's property would cause great inconvenience to him which could not be justified in the circumstances. Secondly, the Act could not authorise the permanent appropriation of the 2 in of Barry's land which would be required to accommodate the vertical damp-proof barrier. Thirdly, having regard to the past history of the wall it would not be appropriate for works to be carried out on Barry's side. In particular, the wall would have been a perfectly effective wall if Minturn's predecessor in title had not utilised it for a purpose for which it was not originally designed.

---

[1] Broadly equivalent to section 2(2)(b) of the 1996 Act.

The court did not feel able to make an order on the basis of Barry's first suggested proposal (the lining wall in Minturn's basement). Although this might indeed provide an appropriate solution to the problem, the court had no power under the Act to order an appropriation of Minturn's property, any more than it had with respect to Barry's property.

The court therefore ordered that Minturn should be at liberty to insert a vertical asphalt damp-proof course into the party wall from her side of the wall in accordance with Barry's second alternative proposal.

Minturn appealed against this decision and the matter was ultimately decided by the House of Lords. She cited two grounds of appeal.

Definition of "defect" and "repair"

Under her first ground of appeal she argued that the works ordered by the county court could not adequately remedy the defect to the party wall. The installation of a damp-proof course in the middle of the wall would only affect half its thickness. The other half would be left in a damp state. It did not therefore amount to making good or repairing a defect within the meaning of the 1894 Act.

The House of Lords rejected this argument. It considered that it was not the dampness itself which made the wall defective. It is only where the dampness renders the wall less effective for the purposes for which it is used or intended to be used that it becomes defective. The installation of the damp-proof course in the middle of the wall was perfectly adequate to remedy this particular defect in Minturn's basement, notwithstanding the fact that the dampness would remain in the other half of the wall.

> Lord Parker: "... it was contended before your Lordships' House that the real defect in the wall is its dampness, and that the vertical damp course directed to be inserted in the centre of the wall will at most prevent this dampness on [Minturn's] side, and therefore only partially remedy the defect. In my opinion, dampness in a wall is not a defect within the meaning of the Act unless its existence renders the wall less effective for the purposes for which it is used or intended to be used. If a wall between two gardens is damp, it is not on that account defective, for its dampness is immaterial.
>
> ... There is no reason to suppose that the stability of a wall is imperilled by damp. Were this so the usual method of constructing walls of a house with either a hollow or a damp-proof course in the centre would be wrong.
>
> The county court judge has found that the only defect in the wall is that part of it allows damp to percolate into [Minturn's] premises, and that this defect will be effectively remedied by the work he directed. This finding as a finding of fact is conclusive between the parties, and I do not think that it involves any error of law. [This] ground of appeal fails."

> # London Building Act 1894
>
> Part VIII: RIGHTS OF BUILDING AND ADJOINING OWNERS
>
> **88. Rights of building owner**
>
> The building owner shall have the following rights in relation to party structures (that is to say):
>
> (1) A right to make good underpin or repair any party structure which is defective or out of repair
>
> **90. Rules as to exercise of rights by building and adjoining owners**
>
> (3) A building owner shall not exercise any right by this Act given to him in such manner or at such time as to cause unnecessary inconvenience to the adjoining owner or to the adjoining occupier.

Unnecessary inconvenience and past history

Under her second ground of appeal Minturn maintained that the previous history and user of a party wall were not relevant considerations in determining the appropriateness of works under the 1894 Act. The county court had therefore misdirected itself in holding that, as a matter of law, it could have regard to these matters.

The House of Lords first considered the nature of matters which surveyors should consider when adjudicating on works under the Act. Its conclusions were described by Lord Parker in the following terms:

> "For this purpose the tribunal must obviously determine whether there be a defect such as alleged, and what is the proper way to make good the party structure qua such defect. Inasmuch as a building owner is not entitled to exercise any right given by the Act in such manner as to cause unnecessary inconvenience to the adjoining owner, the tribunal must, in determining the proper way of making good the defect, if there be one, have due regard to the convenience of the adjoining owner.
>
> ... These being the functions of the tribunal constituted by the Act, it is, in my opinion, extremely difficult to see how the past history of the party structure is in any way relevant to any question which the tribunal has to decide. Certainly such past history cannot be invoked for the purpose of

showing that one or other of the parties is to blame for any defect which, in fact, exists."

So, whilst the surveyors' tribunal must have regard to the convenience of the adjoining owner, the House of Lords was quite clear that it should not take the previous history of the wall into account in this context.

However, despite this finding, the House of Lords did not consider that it would have made any difference in the present case. The past history of the wall was only one of the matters considered by the county court. It had also, quite properly, addressed other matters in reaching its decision. The ultimate decision was correct and its consideration of the wall's past history had not affected this. There were not therefore sufficient grounds to order a retrial.

> Lord Parker: "After carefully perusing the evidence before the county court so far as it was admissible, it seems to me that the ultimate decision was right, however it may have been arrived at.
>
> I am not satisfied that it was, on the true construction of the Act, open to the county court [judge], at any rate under the particular circumstances of this case, to make any award which would involve the permanent appropriation of any land either of the appellant or the respondent.
>
> I am satisfied that it was not open to him to direct any works which would entail great inconvenience to the appellant, if it were possible to direct other works which, while equally effective and not involving any considerable extra cost so far as the respondent was concerned, would be accompanied by no such inconvenience.
>
> The county court judge adopted, and I think rightly adopted, the only other alternative open to him on the evidence by directing the works mentioned in his award, which works he finds, as a fact, will be perfectly effective, and can be carried out without entry on the appellant's land."

The decision

The House of Lords therefore dismissed Minturn's appeal and upheld the decision of the county court. She had no right to undertake work on Barry's side of the wall but was at liberty to insert a vertical damp-proof course into the centre of the wall.

BENNETT v HARROD'S STORES LTD (1907)
King's Bench Division

In brief

- **Issues:** 1894 Act – raising a party fence wall – party wall agreement – interpretation of agreement – requirement for service of party structure notice.

- **Facts:** The parties entered into a written agreement in respect of a party fence wall between their two properties. This permitted Harrods to raise the height of the wall. Harrods proceeded to raise the whole thickness of the wall rather than simply the half within their ownership.

- **Decision:** On a true construction of the agreement Bennett had given Harrods permission to raise the whole thickness of the wall, as was their right under the 1894 Act. The existence of the agreement obviated the need for them to comply with the Act's procedural requirements before exercising the right.

- **Notes:** See also (on the right to raise the whole thickness of a party wall): *Matts v Hawkins* (1813); *Watson v Gray* (1880); *Moss v Smith* (1977); *Alcock v Wraith and Swinhoe* (1991) – (on the nature of party wall agreements): *Mason v Fulham Corporation* (1910); *Brace v South East Regional Housing Association Ltd* (1984).

The facts

Bennett owned a leasehold interest in premises at 40 and 42 Hans Crescent in Knightsbridge (Figure 4). The area within his lease also included the whole width of a passageway at the rear of the premises. The far side of the passageway (to the north-east) was bounded by a freestanding wall. Harrods owned a leasehold interest in a large plot of land on the far side of this wall which they intended to develop.

Prior to Harrods' development taking place the parties entered into an agreement which addressed the effect of the proposed works on the freestanding wall. This provided that it would be lawful for Harrods "to use and make use of half of the wall as and for a party wall, and to thicken, raise, and underpin the same as required at their own expense, the height not to exceed 30 ft from the level of the passageway". Bennett also produced plans of the site to Harrods which he indicated had been approved by the common freeholder of their two properties.

Based on the approved plans Harrods then developed their plot. This included the raising of the whole thickness of the free-standing wall to a

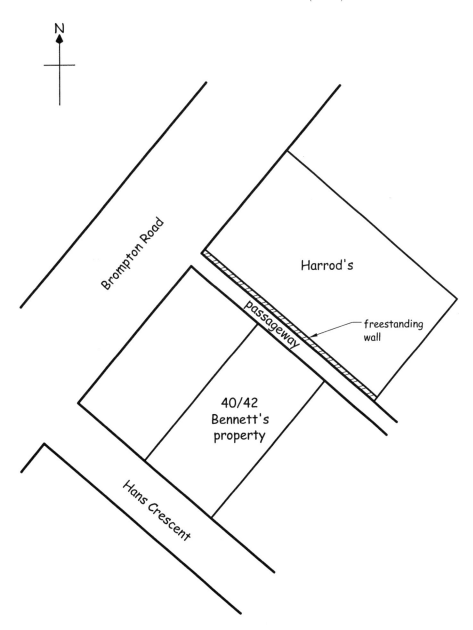

Figure 4: Bennett v Harrod's Stores Ltd

height of 27 ft and incorporating it into the new building which they erected on their side of the wall.

Interpretation of party wall agreement

Bennett then commenced proceedings against the Harrods for trespass. He argued that the whole wall was owned by him although the agreement had given them a right to raise half its thickness. In raising the whole thickness Harrods had committed a trespass for which he now claimed damages and an injunction.

Harrods argued that the wall straddled the boundary line. It was therefore originally a party fence wall within the meaning of the 1894 Act and they had the right to raise the whole thickness of the wall under that Act. Although the statutory notice procedures had not been followed, Harrods argued that they were nevertheless entitled to exercise their statutory rights by virtue of the agreement entered into with Bennett. They maintained that the works they had undertaken to the wall were fully described in the agreement, that they were in accordance with the plans produced by Bennett, and that Bennett had "stood by and without any plaint and objection had allowed them to do the work".

The decision

The court considered the terms of the agreement and accepted Harrods' argument. It concluded that the agreement authorised Harrods to use one half of the wall for the permanent use of girders but that it also gave them the right to raise and underpin the whole thickness of the wall. Harrods were therefore entitled to exercise their rights under the 1894 Act by virtue of this agreement without the requirement that they also comply with its procedural requirements. Judgment was therefore entered for Harrods.

BOND v NOTTINGHAM CORPORATION (1940)
Court of Appeal

In brief

- **Issues:** Easement of support – demolition of servient building – servient owner's duties – effect of statutory obligation to demolish – requirement to provide shoring.

- **Facts:** The local authority wished to demolish a building under a statutory power to do so. The building was subject to an easement of support in respect of an adjoining building.

- **Decision:** The existence of an easement of support did not impose any repairing obligations on the servient owner. This was the case even where the deterioration of the servient building deprived the dominant building of support. However, if the servient owner demolishes his building he is under an obligation to provide equivalent support for the dominant building. The local authority was in exactly the same position when demolishing under a statutory power.

- **Notes:** (1) This case must now be read in the context of the following cases: *Bradburn v Lindsay* (1983); *Holbeck Hall Hotel Ltd v Scarborough Borough Council* (2000); *Rees v Skerrett* (2001). (2) See also: *Lemaitre v Davis* (1881); *Southwark & Vauxhall Water Company v Wandsworth District Board of Works* (1898); *Selby v Whitbread* (1917); *Brace v South East Regional Housing Association Ltd* (1984).

The facts

Bond owned a house in Aspley Place, Alfreton Road, Nottingham. His property had the benefit of an easement of support from some cottages which were owned by Norman.

The cottages were included in a clearance scheme by Nottingham Corporation who had recently served a demolition notice on Norman under the Housing Act 1936. He had refused to comply with the notice so the Corporation now planned to demolish the cottages itself, in pursuance of its statutory powers under the same Act.

Before it could do so Bond obtained a court order restraining it from demolishing the cottages without providing equivalent support for his property, as was his right under the easement of support. The Corporation appealed against the order. The Court of Appeal therefore had to decide whether demolishing the cottages under the statute had the effect of extinguishing Bond's common law easement of support.

Easements of support: the servient owner's duties

The court first examined the nature of an easement of support, and the nature of the duties owed by Norman, as the servient owner, in respect of this. The following well-known extract from the judgment of Sir Wilfred Greene MR summarises the law in this regard:

> "The nature of the right of support is not open to dispute. The owner of the servient tenement is under no obligation to repair that part of his building which provides support for his neighbour. He can let it fall into decay. If he does so, and support is removed, the owner of the dominant tenement has no cause for complaint.
>
> On the other hand, the owner of the dominant tenement is not bound to sit by and watch the gradual deterioration of the support constituted by his neighbour's building. He is entitled to enter and take the necessary steps to ensure that the support continues by effecting repairs and so forth to the part of the building which gives the support.
>
> What the owner of the servient tenement is not entitled to do, however, is by an act of his own to remove the support without providing an equivalent. There is the qualification upon his ownership of his own building that he is bound to deal with it subject to the rights in it which are vested in his neighbour, and can deal with it subject only to those rights."

Effect of statutory obligation to demolish servient tenement

The court was of the view that Norman's statutory obligation to demolish should be viewed in the same way as his right to do so at common law. Just as his common law rights could only be exercised subject to Bond's easement of support, so too could his statutory obligations. In reaching this conclusion the court had particular regard to the presumption that, unless compensation was provided for this, statutes should be interpreted so as not to interfere with private property rights[1].

The court therefore concluded that Norman was under an obligation to provide shoring or other support as a condition of the exercise of his statutory obligation to demolish the cottages.

> Sir Wilfred Greene MR: "In those circumstances, it appears to me that the proper construction to put upon the word 'demolish' involves that the manner of demolition must be such as not to interfere with the neighbour's rights.
>
> That does not mean that the demolition is necessarily to stop short at the point where the existing building begins to provide support, for the reason that

---

[1] Sir William Brett, MR, J in *Attorney General v Horner* (1884).

it is not the right of the owner of the dominant tenement to insist upon the maintenance of that individual and particular building. His right is given effect to, if that building is taken away, by providing that some substitute is put there for it. He is entitled to say to his neighbour: 'Either you leave your building, or the necessary part of it, as it is, or, if you choose to pull it down, you must provide a substitute.'

In this case, Norman being confronted with that alternative, it seems to me that the method of demolition which he is ordered to carry out must be one which gives effect to the rights of his neighbour..."

The decision

Having considered the nature of Norman's obligations the court then addressed the nature of the obligations owed by Nottingham Corporation. It concluded that these were no different. The Corporation's role was to perform a duty primarily imposed on Norman but not carried out by him.

As such, the court could find no justification, in the statute, for imposing different requirements on the Corporation. It must also respect Bond's rights and could only therefore demolish the cottages upon condition that it provided alternative means of support for his property. The Corporation's appeal was therefore dismissed.

BOWER v PEATE (1876)

Queen's Bench Division

In brief

- **Issues:** Adjacent excavations – easement of support – nuisance – building owner's responsibility for damage caused to adjoining owner's building – non-delegable duties.

- **Facts:** Peate's independent contractor carried out excavations adjacent to Bower's building. Bower was entitled to an easement of support from the soil being excavated. Peate had required the contractor to provide adequate support but he had failed to do so. As a result, Bower's building suffered damage.

- **Decision:** Peate was responsible for the damage to Bower's building. Injurious consequences would inevitably have followed from the excavations unless adequate steps were taken to provide alternative support. In these circumstances Peate could not escape liability by delegating responsibility for taking these steps to his contractor. Judgment was therefore entered for Bower.

- **Notes:** The court explained the basis of Peate's liability in two ways. Firstly, he was taken to have authorised the commission of the tort and was therefore himself liable for it. Secondly, he was in breach of a non-delegable duty to Bower because of the nature of the work being undertaken. The concept of non-delegable duties was further developed in later cases. See: *Dalton v Angus* (1881); *Hughes v Percival* (1883); *Matania v National Provincial Bank Ltd and The Elevenist Syndicate Ltd* (1936); *Alcock v Wraith and Swinhoe* (1991).

The facts

The parties were the owners of two adjoining houses in Liverpool. Although in physical contact with each other, each house had its own external wall rather than sharing a party wall. Bower's house was of more recent construction than Peate's and had deeper foundations.

Peate now wished to demolish his existing house and to excavate the site in order to rebuild the property with deeper foundations. Bower's house was entitled to an easement of support from the soil which Peate was proposing to excavate.

Peate entered into a contract with Rae, to undertake the demolition works. The contract contained the following clause in which Rae became responsible for safeguarding Bower's existing support:

"... and the said Thomas Rae further agrees to take upon himself the risk and responsibility of shoring and supporting, as far as may be necessary, the adjoining buildings affected by this alteration during the progress of the works, and to make good any damage which may be sustained by the said buildings during progress or in consequence of the said works hereby contracted for, and to satisfy any claims for compensation arising therefrom which may be substantiated."

Rae then carried out the demolition work. Unfortunately he failed to provide adequate support for Bower's house which consequently suffered damage. Bower therefore sought to recover this from Peate in the present action. He based his claim in nuisance on the basis of the wrongful removal of support to which he was entitled by virtue of the easement.

Rae's culpability was not in doubt but he was not a party to the action. The issue before the court was whether Peate could be held responsible for Rae's tortious acts, notwithstanding the fact that Rae was an independent contractor.

Legal status of adjacent excavations

The general rule is that a person will not be liable for the torts of their independent contractor although there are a number of exceptions to this. Bower maintained that Peate fell within one of these exceptions in that he had authorised the commission of the nuisance by instructing Rae to excavate the land providing support to his building.

Peate argued that he had only instructed Rae to undertake a lawful act and that, under the general rule, he could not therefore be held responsible for the nuisance which Rae had committed.

He cited the doctrine in *Bonomi v Backhouse* (1858) in support of this. That case had decided that the excavation of soil which provides support to which an adjoining owner is legally entitled is not, in itself, a wrongful act (a tort). It only becomes so if it results in damage to the adjoining owner's property. Therefore, if damage can be avoided, for example by the provision of alternative support, no tort is committed.

On this basis, Peate argued that he had not authorised Rae to commit an unlawful act. He had entered into a contract in which he required him to excavate the soil providing support and in the same contract had also required him to provide alternative support to maintain the lawfulness of the activity. If Rae had undertaken the work in conformity with the contract no damage would have been suffered. The damage which Bower had suffered was therefore caused entirely by Rae's omissions which Peate had certainly not authorised.

The court was unimpressed by this argument. It considered that, where damage actually occurred, an act of excavation which interfered with an entitlement to support was indeed an unlawful act and this had been authorised by Peate.

Cockburn CJ explained the court's reasoning in the following terms:

> "It is true, that according to the doctrine in *Bonomi v Backhouse*, the removal of the soil, to the support of which an adjacent building or land may be entitled, is not in itself wrongful, and becomes so only when damage to the adjoining property results; whence it follows that if by artificial means of support the damage can be prevented, no cause of action arises. But it is equally clear that if effectual means of prevention fail to be applied, and damage once results, the act of removal becomes wrongful, and an action can at once be maintained."

The court placed particular emphasis on the fact that Peate had not stipulated the nature of the preventive works but had simply sought to pass this responsibility on to Rae and obtain an indemnity for himself.

> Cockburn CJ: "In other words he directs an act to be done from which injurious consequences will result unless means are taken to prevent them in the shape of additional work, but omits to direct the latter to be done as part of the work to be executed, contenting himself with securing to himself a pecuniary indemnity in the event of any claim arising from damage to the adjoining property."

This combination of authorising work which would constitute a tort unless adequate preventive measures were taken, and the failure to ensure that such measures were taken, in effect, amounted to authorising the commission of a tort.

> Cockburn CJ: "In the present instance preventive measures adequate to the occasion having failed to be provided, the removal of the soil was followed by actual damage to Bower's house, and the act of removal was therefore wrongful as causing a wrong done to Bower. But the act of removal was an act done by the order and authority of the defendant; and no man can get rid of liability for injury occasioned to another by a wrongful act by seeking to throw the responsibility on an agent whom he has employed to do the act. The agent may no doubt be responsible, but the responsibility of the principal is none the less."

The court therefore found that Peate had authorised Rae to commit a tort. Because this was an exception to the general rule, Peate was liable, notwithstanding the fact that Rae was his independent contractor.

Adjacent excavations and non-delegable duties

The court had rejected Peate's argument on specific issues relating to the interference with rights of support by adjacent excavation. However, this case was one of the first to introduce the more general notion that a person might continue to be responsible for the performance of certain duties, even where these had been properly delegated to an independent contractor. The court suggested that Peate's arguments might equally have been answered by reference to a more general principle governing what later became known as non-delegable duties.

> Cockburn CJ: "The answer to the defendant's contention may, however, as it appears to us, be placed on a broader ground, namely, that a man who orders a work to be executed, from which, in the natural course of things, injurious consequences to his neighbour must be expected to arise, unless means are adopted by which such consequences may be prevented, is bound to see to the doing of that which is necessary to prevent the mischief, and cannot relieve himself of responsibility by employing some one else – whether it be the contractor employed to do the work from which the danger arises or some independent person – to do what is necessary to prevent the act he has ordered to be done from becoming wrongful."

This dictum unfortunately lacks precision and, in one sense, appears to suggest that non-delegable duties are owed wherever dangerous acts are undertaken[1]. Whilst similar non-delegable duties are indeed now imposed in the case of "extra-hazardous" acts[2] this rule had not been clearly established at the time that this case was heard. Cockburn CJ's use of the words "injurious" (from *injuria* – a legal wrong) and "wrongful" suggest that his notion of non-delegable duties may have been based, not so much on dangerous (or even, extra-hazardous) acts, as on those which were unlawful.

The non-delegable duty he refers to therefore appears to arise because the performance of the duty (to provide adequate alternative support) is necessary to prevent the commission of the tort of nuisance, rather than to avoid an act which is simply inherently risky and therefore potentially more likely to cause harm to neighbouring owners. Whilst this broad principle could, no doubt, be applied to other circumstances and other

---

[1] This is presumably the reason why Lord Blackburn expressed doubts in *Hughes v Percival* [1883] 8 App Cas 443, at p. 447 as to whether Cockburn CJ had stated the principle too broadly.
[2] See *Honeywill & Stein Ltd v Larkin Bros Ltd* (1934).

torts[3], in practice it is usually confined to the duty to avoid nuisance by the withdrawal of support which to which a neighbour is entitled[4].

The decision

The court concluded that judgment should be entered for Bower. This was justified, both by the specific considerations relevant to an interference with a right of support, and by the more general principle of non-delegable duties which had also been identified in some other recent authorities[5].

---

[3] See the general statement of principle in Lord Blackburn's judgment in *Dalton v Angus* [1881] 6 App Cas 740, at p. 829: "... a person causing something to be done, the doing of which casts on him a duty, cannot escape from the responsibility attaching on him of seeing that duty performed by delegating it to a contractor. He may bargain with the contractor that he shall perform the duty and stipulate for an indemnity from him if it is not performed, but he cannot thereby relieve himself from liability to those injured by the failure to perform it."

[4] "Cases involving the withdrawal of support from neighbouring land" are often now described as one of the exemptions from the general rule that people are not responsible for the torts of an independent contractor employed by them. See *Alcock v Wraith and Swinhoe* (1991).

[5] In its deliberations, the court had made reference to two other cases where duties had been held to be non-delegable. In *Gray v Pullen* (1864) the defendant's statutory duty to reinstate the highway after laying a drain was held to be non-delegable. In *Tarry v Ashton* (1876) an owner of premises was held to be subject to a non-delegable duty to maintain a lamp which projected over the highway to prevent it becoming a nuisance to highway users.

BRACE v SOUTH EAST REGIONAL HOUSING ASSOCIATION LTD (1984)

Court of Appeal

In brief

- **Issues:** Easement of support – demolition of servient building – evaporation of moisture from vacant plot – subsidence damage caused to adjoining owner's building – right to removal of underground water – interpretation of party wall agreement – effect of agreement on common law rights of support.

- **Facts:** Brace and the Housing Association owned adjoining terraced houses which were entitled to mutual easements of support. The Housing Association demolished its house in accordance with a party wall agreement previously entered into with Brace. Moisture evaporated from the vacant plot, causing the desiccation of the clay subsoil. This resulted in subsidence damage occurring to Brace's house.

- **Decision:** The Housing Association was liable for the damage. The damage had occurred due to a failure to provide support and the court was unconcerned about the precise causal mechanism by which this had occurred. The Housing Association was unable to escape liability on the basis that it had a natural right to remove underground water from the plot. This right related to percolating water to be used for particular purposes and was quite different to the drying out of the subsoil which had occurred in the present case. The party wall agreement had no effect on the Housing Association's liability for interfering with Brace's easement of support. It contained no express term surrendering Brace's common law rights and, on the facts, the court was unwilling to imply one.

- **Notes:** See also: *Lemaitre v Davis* (1881); *Southwark & Vauxhall Water Company v Wandsworth District Board of Works* (1898); *Selby v Whitbread* (1917); *Bond v Nottingham Corporation* (1940); *Bradburn v Lindsay* (1983); *Rees v Skerrett* (2001).

The facts

Mrs Brace was the owner of 19 Stroud Gate, Harrow, Middlesex. In 1969 the South East Regional Housing Association purchased the adjoining (end of terrace) property at number 20 Stroud Gate with a view to demolishing it and redeveloping a site at the rear. The two properties were separated by a party wall and enjoyed mutual easements of support.

Although the properties were outside the area regulated by the London Building Acts the parties each appointed a surveyor in connection with the proposed demolition of number 20. Once the surveyors' negotiations had been concluded they prepared a document setting out the terms of the agreement between the parties which each party then signed. This document was described as a "party structure agreement" and broadly followed the format of an award under the London Building Acts.

The agreement provided (inter alia) that:

"(a) The building owner shall be at liberty at his own cost to pull down no. 20 Stroud Gate and to strengthen, repair or underpin the party wall and generally to carry out the various works described in the attached schedule marked 1A'.

(b) During the progress of the works mentioned in the Schedule marked 'A' the building owner shall adequately shore, uphold, maintain and protect the adjoining owner's premises."

The demolition took place during 1970 and, in due course, Mrs Brace's surveyor confirmed that the works had been completed in accordance with the terms of the agreement.

However, during the spring of 1976, Mrs Brace noticed some cracks in the walls of her property. These were the result of subsidence which had been caused by the desiccation of the clay subsoil. This had, in turn, taken place because of evaporation of moisture from the soil of the now vacant site of the adjoining property at number 20.

When her insurance company disclaimed liability under the terms of her policy, she issued proceedings against the Housing Association to recover her losses from it. Her claim was in nuisance[1], on the basis of the wrongful removal of the support to which she was entitled by her easement.

Her claim was successful at first instance but the Housing Association then appealed to the Court of Appeal. It raised three grounds of appeal and each of these is discussed below.

Cause of damage to adjoining owner's property

Under its first ground of appeal the Housing Association argued that the damage to Mrs Brace's property had not been caused by a loss of support from the adjacent building at number 20. It had been caused because of evaporation of moisture from the site of the Housing Association's demolished building. Mrs Brace's claim was therefore, in reality, a claim

---

[1] The claim also raised negligence and breach of contract as alternative causes of action. These parts of the claim were dismissed by the court at first instance for reasons which are not explored in the published law report.

for an infringement of a supposed right to have the subsoil under her property protected from the atmosphere. As the law did not recognise any such right her claim must fail.

The court rejected this argument. It did not accept that Mrs Brace's claim related to anything other than an interference with her easement of support. It was satisfied that she had an easement of support. It was satisfied that the Housing Association's actions had resulted in a loss of that support and was unwilling to explore the precise causal mechanism by which this had occurred. It was satisfied that the loss of support had led to the damage which Mrs Brace had actually suffered. On this basis it had no difficulty in finding that the Housing Association was liable in nuisance.

> Eveleigh LJ: "It is important to bear in mind that the plaintiff has not acquired any right to the continued existence of moisture in the clay, to the continued presence of the clay or indeed to the continued presence of the wall: the right that had been acquired was a right of support, and the defendants could remove the particular means to support that existed provided that they replaced it with an equally effective support. It does not matter from a plaintiff's point of view how that support is provided or upon what mechanism it depends, but the plaintiff is entitled to have her right against the defendants' land protected against interference.
>
> On the facts of this case the defendants had changed their land in a way that in fact weakened the support. The chain of causation passed through a stage where the moisture in the clay was dried out, but that does not alter the fact, as I see it, that the defendants acted in such a way as to interfere with the support that had been provided, and they interfered with it not, as I see it, by draining off water but by altering the conditions which operated to afford the support."

The court dismissed the Housing Association's argument that the right being claimed by Mrs Brace was the same as that rejected by the court in *Phipps v Pears* (1965). In that case the court had held that the law would not recognise an easement of protection from the weather in favour of a building erected against another building. Eveleigh LJ distinguished the two cases on the basis that the plaintiff in *Phipps* had not acquired an easement of support and was attempting to prove the existence of an independent right of protection from the weather. In the present case an existing right of support was being relied on and no independent right of protection was being claimed.

Law relating to removal of underground water

Under its second ground of appeal the Housing Association argued, in the alternative, that Mrs Brace's claim related to the removal of underground water from the subsoil beneath her property. There was established authority that an owner is free to remove underground water from his land even if this results in some loss of support to neighbouring property. The Housing Authority referred to the following passage in *Clerk and Lindsell on Torts* (15th edn, p. 1177) which summarised the legal position in this respect:

> "There is no [natural] right to have land supported by water and such a right cannot be acquired by prescription. Therefore one who by draining his own land withdraws from an adjoining owner the support of water theretofore lying beneath the land of that owner and thereby causes the surface of that land to subside is not liable for the damage inflicted."

However, the court distinguished the present case from the authorities cited by the Housing Association. It noted that all the authorities which sanctioned the removal of underground water, even where this had some detrimental effect on adjoining land, related to the use of percolating water for some purpose.

In particular, the cases all sought to protect the proprietary rights of a land owner in the percolating water under his property. The immunity which was granted to him in respect of adjoining land was simply a necessary consequence of protecting this primary right. This was quite different from the process of drying out an area of land possessing a finite amount of moisture, as was the situation in the present case.

> Eveleigh LJ: "But it does seem to me, considering those cases, that the court was dealing with a situation where the defendant is draining off water and dealing with the water in manageable quantities and, furthermore, dealing with it for purposes where water in quantity can be used: drinking, husbandry, preserving the condition of land for agricultural purposes, and so on. I just cannot equate in my mind the drying out of clay through atmospheric conditions, in particular by heat, as bearing any resemblance to the kind of activity that is granted immunity in the cases to which we have referred.
>
> It seems to me that there is a world of difference in the action of a defendant who takes water in some quantity from land where it is found to be percolating and the action of someone who dries out the moisture content of a mineral, namely clay in this case. If I dry out a piece of wood or I dry out a twig I cannot regard myself as taking water from it in any ordinary sense of that term, and it seems to me that it is straining the language of the cases to equate the action referred to in them with that of evaporating moisture from clay or from anything else. Therefore I could not accept that the defendants were entitled to do what they did on the authority of the cases dealing with the removal of water."

Effect of party structure agreement on common law rights

The Housing Association's third ground of appeal related to the effect of the party structure agreement on Mrs Brace's existing common law rights. There is authority that rights defined in a statutory party wall award replace all previously existing rights at common law[2]. The Housing Association therefore argued that the rights of the parties must be defined entirely by reference to the terms of the party structure agreement which they had entered into. In particular, by entering into the agreement, Mrs Brace had waived her previous entitlement to an easement of support.

The court noted that the London Building Acts were of no relevance to the present case in view of the location of the properties. The parties' rights were not regulated by an award, but by a contractual party structure agreement. The question of whether Mrs Brace had waived her common law right to support therefore depended entirely on the construction of that agreement.

The agreement contained no express term surrendering the right of support. The court therefore considered whether one should be implied. According to ordinary contract law principles this would only be the case if such a term was necessary to give business efficacy to the agreement, and the "officious bystander test" described in *Shirlaw v Southern Foundries* (1939)[3] could be a useful method of determining whether this is the case.

After considering the terms of the agreement the court held that there was no such implied term.

> Parker LJ: "If one applies the officious bystander test and suppose that in the course of negotiations someone had said: 'What is to happen if the wall bulges and falls down the day after the completion of the works provided by the agreement?' I cannot contemplate that both sides would have answered: 'Well of course there can be no remedy'. That is too plain for argument and we need not bother to say it. Certainly Mrs Brace and her advisers would, in my view, have given no such answer and I very much doubt whether [the Housing Association's surveyor] would either. Therefore, I would reject the argument based on the agreement."

> Eveleigh LJ: "... it does not seem to me that the agreement can be read as an undertaking by the plaintiff to abandon her rights, and there are no grounds

---

[2] Jessel MR in *Standard Bank of British South America v Stokes* (1878) 9 Ch 68 at p. 73; McCardie J in *Selby v Whitbread & Co* [1917] 1 KB 736 at p. 752.

[3] According to MacKinnon J, a term which was "so obvious that it goes without saying" would be implied into a contract "so that, if while the parties were making their bargain an officious bystander were to suggest some express provision for it in their agreement, they would testily suppress him with a common, 'Oh of course!'.": *Shirlaw v Southern Foundries (1926) Ltd* [1939] 2 All ER 113, at p. 124.

whatsoever for implying such a term. This is not a case where terms are necessary to give business efficacy to it, and if a party wishes to rely upon a term in a contract it is for that party to establish it, and that has not been done in this case. So, I would reject the argument that Mrs Brace's rights are restricted to such as can be found in the agreement."

The decision

The court rejected each of the Housing Association's three grounds of appeal. Mrs Brace had an easement of support which had been infringed by the demolition of the Housing Association's adjoining building. There was nothing in the party structure agreement which indicated an intention to waive this easement. Furthermore, the precise manner in which the support had failed and the various authorities on the removal of underground water were irrelevant to the present case. The Housing Association was liable in nuisance for the damage which Mrs Brace had suffered. Its appeal was therefore dismissed.

BRADBURN v LINDSAY (1983)
Chancery Division

In brief

- **Issues:** Easement of support – neglect of servient building – damage to dominant building – demolition of servient building by local authority – weather protection – negligence – nuisance.

- **Facts:** Bradburn and Lindsay owned adjoining terraced houses which were entitled to mutual easements of support. Lindsay allowed her house to become derelict. As a result dry rot spread to Bradburn's house and ultimately Lindsay's house had to be demolished by the local authority.

- **Decision:** Lindsay was liable for the damage to Bradburn's house and for providing alternative support following the demolition of her own house. The remedial measures could properly include a rendered and weatherproofed finish, notwithstanding the absence of a separate right to protection from the weather. She was unable to escape liability simply because of the lack of a repairing obligation within the law of easements. Liability arose in tort where a building owner was aware of a danger to adjoining property but failed to take any action to prevent it. It was also irrelevant that the demolition had actually been undertaken by the local authority rather than Lindsay. Furthermore, Bradburn's right to abate the nuisance was in addition to an entitlement to damages and not instead of it.

- **Notes:** See also: *Jones v Pritchard* (1908); *Bond v Nottingham Corporation* (1940); *Phipps v Pears* (1964); *Leakey v National Trust for Places of Historic Interest or Natural Beauty* (1980); *Rees v Skerrett* (2001).

The facts

Numbers 53 and 55 Kennerley Road, Stockport, were two semi-detached houses. They had been built towards the end of the nineteenth century and were separated by a 9-in brick wall. The two properties passed into separate ownership in 1919. The conveyance at that time contained an agreement and declaration that the dividing wall was a party wall which belonged to and was repairable by the two owners in equal shares. As a result of the Law of Property Act 1925[1] the wall subsequently became a longitudinally divided party wall.

---

[1] Section 39 and Schedule 1, Part V.

Number 53 was owned by Mrs Lindsay, an "eccentric lady" who had allowed the property to fall into disrepair. Mr and Mrs Bradburn bought number 55 in 1970 and soon afterwards discovered the extent of the dereliction at number 53. There were holes in the roof and the property was uninhabited apart from the presence of "drop-outs" who used it for shelter. Pigeons were nesting in the roof space and there was evidence of dry rot in the property.

Mr and Mrs Bradburn were concerned that the dry rot, in particular, represented a threat to their property and regularly requested Mrs Lindsay to undertake the necessary repairs to number 53. She failed to respond to any of these requests and the condition of her property continued to deteriorate, allowing dry rot to spread through the party wall and to attack the timbers at number 55.

In 1977 the council served a demolition order on Mrs Lindsay under the Housing Act 1957. She failed to comply with it but agreed that the council could demolish the property themselves. The council therefore carried out the demolition works, leaving the whole thickness of the party wall standing. However, no support or weather protection was provided to Mrs Lindsay's side of the party wall which continued to be infested with dry rot.

Mr and Mrs Bradburn therefore sought to protect their own property by issuing proceedings against Mrs Lindsay. They claimed for damages for the existing injury to number 55 as well as for the provision of support and weather protection to protect it from further injury.

They based their claim in negligence and nuisance and on Mrs Lindsay's failure to maintain her property in accordance with an obligation to do so.

Adjoining owner's obligation to repair

Mrs Lindsay denied liability on three grounds. Firstly, she denied that she was subject to any legal obligation to maintain her property.

The court accepted that the agreement and declaration in the 1919 conveyance imposed no such obligation in the present circumstances. This was intended to define the responsibility for repair in non-contentious situations where neither party was clearly at fault. As the repairs in the present case had only become necessary because of Mrs Lindsay's default, the agreement and declaration were of less significance than other considerations.

Mrs Lindsay also argued that she owed no obligation to repair within the law of easements, even though she acknowledged that she was subject to an easement of support in favour of Mr and Mrs Bradburn's property. Her counsel cited the decisions in *Jones v Pritchard* (1908) and *Bond v Nottingham Corporation* (1940) in support of this position.

In *Jones*, Parker J had stated that "Apart from negligence or want of reasonable care and precaution, neither party is, in my judgment, subject to any liability to the other in respect of nuisance or inconvenience caused by an exercise of the rights or easements impliedly granted or reserved". In *Bond*, Greene MR had explained: "The nature of the right of support is not open to dispute. The owner of the servient tenement is under no obligation to repair that part of his building which provides support for his neighbour. He can let it fall into decay. If it does so, and support is removed, the owner of the dominant tenement has no cause for compliant".

However, the court rejected Mrs Lindsay's reliance on the law of easements. She had obligations in the law of negligence and nuisance, in addition to any responsibilities she may also have in the law of easements. The court relied on the principle established in *Leakey v National Trust for Places of Historic Interest or Natural Beauty* (1980) and found that she had failed to take reasonable steps to prevent the spread of dry rot to number 55.

> Blackett-Ord VC: "In the present case I have already said that I find the defendant knew of the dry rot at an early stage. In the *Bond* case Greene MR was dealing with the nature of the right of support, but the plaintiffs put their case on the wider ground of negligence or nuisance, and rely on the case of *Leakey v National Trust for Places of Historic Interest or National Beauty*...
>
> ... I find that the defendant should reasonably have appreciated the danger to number 55 from the dry rot and from the lack of repair of number 53, and that there were steps which she could reasonably have taken to prevent the damage occurring. In my judgment she owed a duty to the plaintiffs to take reasonable steps. She failed to do so and, therefore, she is liable for the damage caused."

The right to abate a nuisance

Mrs Lindsay's second defence relied on Mrs and Mrs Bradburn's entitlement, under the law of nuisance, to enter her property to abate the nuisance which was emanating from it. She argued that, because of this entitlement, they were primarily responsible for undertaking the repairs.

This argument relied on the words of Greene MR in *Bond* where he stated that "... the owner of the dominant tenement is not bound to sit by and watch the gradual deterioration of the support constituted by his neighbour's building. He is entitled to enter and take the necessary steps to ensure that the support continues by effecting repairs and so forth to the part of the building which gives the support".

The court, in the present case, noted Greene MR's description of the dominant owner's *entitlement* to undertake repairs. However, it rejected the notion that this absolved Mrs Lindsay from her *obligation* to do so.

On the contrary, the existence of a right of abatement, by its nature, highlighted the fact that the servient owner must be subject to some obligation. This must, by definition, be capable of enforcement by other means as well as by abatement. Blackett-Ord VC referred to the following statement by Megaw LJ in *Leakey* in support of this:

> "If it is said that I have in such circumstances a remedy of going on my neighbour's land to abate the nuisance, that would, or might, be an unsatisfactory remedy. But in any event, if there were such a right of abatement, it would, as counsel for the plaintiffs rightly contended, be because my neighbour owed me a duty. There is, I think ample authority that, if I have a right of abatement, I have also a remedy in damages if the nuisance remains unabated and causes me damage or personal injury. That is what Scrutton LJ said in the *Job Edwards*[2] case with particular reference to *Attorney-General v Tod Heatley*[3]. It is dealt with also in the speech of Viscount Maughan in *the Sedleigh-Denfield* case[4], and in the speech of Lord Atkin[5]."

Mrs Lindsay was not therefore freed from liability because of Mr and Mrs Bradburn's entitlement to abate the nuisance.

Liability following demolition by local authority

Mrs Lindsay's final defence relied on the fact that it was not she, but the local authority, who had demolished her property and exposed Mrs and Mrs Bradburn's property to danger. As she had not undertaken the act of demolition, she could not be held liable for its consequences.

The court quickly dismissed this argument. As the local authority had undertaken the demolition with her consent and because of her failure to do it, she was as liable as if she herself had undertaken the demolition.

Obligation to provide support and weather protection

As the court had rejected each of Mrs Lindsay's three defences it found her liable for the damage which had been caused to Mrs and Mrs Bradburn's property. However, it also had to consider Mrs Lindsay's ongoing liability (if any) to provide support and weather protection to the party wall following the demolition.

---

[2] *Job Edwards Ltd v Birmingham Navigations* [1924] 1 KB 341 at 359.
[3] *Attorney-General v Tod Heatley* (1897).
[4] *Sedleigh-Denfield v O'Callaghan* [1940] AC 880 at 893–894.
[5] *ibid*, at 899–900.

Although there was no dispute that Mr and Mrs Bradburn enjoyed an easement of support, there was disagreement as to the extent of the liability that arose from this. The dispute related largely to a difference of opinion between the parties' experts as to the most appropriate mechanisms for providing support, but there was also a legal question about the wall's finish.

Mrs Lindsay's counsel cited *Phipps v Pears* (1964) in support of the argument that Mr and Mrs Bradburn had no entitlement to have the wall protected from the weather. The court distinguished *Phipps* from the present case and decided that a reasonable level of finish must be applied to the wall.

> Blackett-Ord VC: "That was a case where a house had been built right against an old house so that the wall of the new house could not be properly pointed on the outside, and then, the old house having been pulled down, damp got into the wall of the new house. There was no question there of support and no question of party walls. If I am right that the plaintiffs are entitled to require the defendant to provide support for the gable wall which I have mentioned they must, in my judgment, be entitled to have the work done to a reasonable specification so as to make good, within reason, the damage done by the defendant's neglect. This includes reasonable steps to prevent a recurrence of the dry rot from number 53..."

The decision

The court awarded damages to Mr and Mrs Bradburn for the cost of dry rot treatment which they had already undertaken. After hearing the expert evidence from both parties the court also held that they were entitled to have the party wall treated with dry rot preservation fluid from the outside. They were further entitled to have the party wall supported by piers at either end and by three buttresses along its length. Finally they were entitled to have the whole face of the wall, including the new buttresses and piers, rendered so as to provide a weatherproof surface.

BURLINGTON PROPERTY COMPANY LIMITED v ODEON THEATRES LIMITED (1938)

Court of Appeal

In brief

- **Issues:** 1930 Act – building owner's right to demolish and rebuild a party wall – permanent appropriation of adjoining owner's property – surveyor's jurisdiction – validity of surveyors' award.

- **Facts:** A type 'a' party wall separated Odeon's building from a public courtyard owned by Burlington. Odeon planned to demolish the wall and to rebuild it with arches to provide access to their building from the courtyard. A surveyors' award purported to authorise the works.

- **Decision:** Odeon had no right to make permanent changes to the design of the party wall. The Act did not authorise the permanent appropriation of an adjoining owner's property and the wall had to be rebuilt in substantially its original form. The surveyors had acted outside their jurisdiction and their award was invalid.

- **Notes:** See also: *Leadbetter v Marylebone Corporation [No.1]* (1904); *Barry v Minturn* (1913); *Gyle-Thompson v Wall Street (Properties) Ltd* (1974).

The facts

Odeon Theatres owned premises at 33 Charing Cross Road, London WC2, and Burlington Property owned the adjoining premises to the north at 35 Charing Cross Road (Figure 5).

The two properties had originally consisted of two terraced buildings separated by a longitudinally divided party wall. Number 33 retained its original form as a building occupying the whole of its site but the building formerly occupying the site of number 35 had been demolished and the site had been redeveloped. The whole thickness of the party wall remained standing and formed the northern flank wall to number 33.

Following its redevelopment, the northern portion of number 35 was occupied by a building known as Cameo Cinema. The southern portion, which adjoined the party wall, became Hunt's Court. This was a passageway which was dedicated as a public right of way. Four beams spanned Hunt's Court at first floor level, between the party wall and the southern wall of Cameo Cinema.

In 1937, Odeon Theatres decided to demolish the building at 33 Charing Cross Road in order to build the current Odeon Cinema,

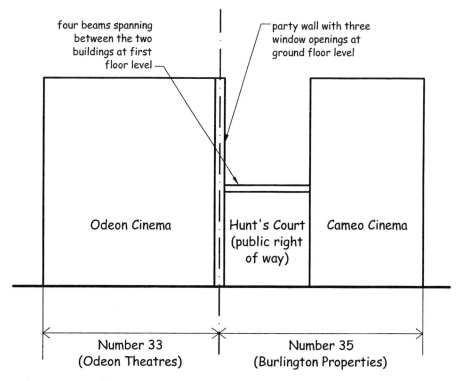

Figure 5: Burlington Property Company Limited v Odeon Theatres Limited

Leicester Square. This involved the demolition and rebuilding of the party wall between the two properties so they served the requisite party structure notice on Burlington Property under section 114(7) of the 1930 Act[1]. A difference arose under the Act and surveyors were appointed.

The surveyors' award provided that Odeon Theatres could make changes to the ground floor portion of the party wall when rebuilding it. The existing wall consisted of a solid wall containing three windows with fixed glass panes, which admitted light from Hunt's Court. The award purported to grant Odeon Theatres the right to enlarge the window openings to form a series of archways separated by pillars to enable the public to gain access from Hunt's Court into the new Odeon Cinema.

Appeal against surveyors' award

Burlington Property objected to the enlargement of the openings as half the thickness of the wall was owned by them. They therefore appealed to

---

[1] Broadly equivalent to section 2(2)(e) of the 1996 Act.

the county court under the Act. The county court dismissed the appeal and affirmed the award. The wall did not have to be constructed exactly as it had been before. Furthermore, as Burlington Property had no use for their half of the wall, the court considered that the surveyors had acted within their powers in providing for its rebuilding with arches.

Burlington Property appealed against this decision to the Court of Appeal. They maintained that the effect of enlarging the window openings would be to practically destroy the party wall at ground floor level. This would interfere with their rights in the wall in a number of ways.

It would deprive them of their entitlement to support for any premises which they subsequently decided to build over Hunt's Court. It would effectively dedicate parts of the wall within their ownership to the public without their consent and without providing them with any compensation in lieu. It would also deprive them of the light which was currently reflected into their property from the surface of the existing wall.

Interference with adjoining owner's property rights

The Court of Appeal allowed the appeal. One half of the wall was owned by Burlington Property and the surveyors had exceeded their jurisdiction in purporting to authorise these particular alterations to it.

> Slesser LJ: "In those circumstances, unless the respondents here can show some statutory right to interfere with what, on the face of it, is the absolute property of the appellants, it must necessarily be that an award which gives to them a power to do that which in law they have no power to do must be bad.
>
> Section 114(7) of the London Building Act 1930, seems to me to give them no such right. It confers a right 'to pull down any party structure which is of insufficient strength for any building intended to be built, and to rebuild the same of sufficient strength for the above purpose'. The whole complaint which the appellants make is that their moiety of the party structure has not been rebuilt as it formerly was, and certainly has not been rebuilt with any consideration of strength, seeing that the apertures in it are considerably wider than they were in the original building."

Whilst changes to the specification of the wall were therefore presumably acceptable to the extent that they were designed to increase the strength of the wall in accordance with section 114(7), an adjoining owner had a basic entitlement to be left with the same wall as that which he previously enjoyed.

> Greer LJ: "Moreover, instead of the three windows which existed in the party wall on the ground floor [the award] provides for the placing of entrances to, or exits from, the theatre which is to the south of Hunt's Court, thereby interfering with the rights of the appellants to have the party wall in existence in the same form in which it had been before for many years."

> **London Building Act 1930**
>
> PART IX: RIGHTS OF BUILDING AND ADJOINING OWNERS
>
> **114. Rights of building owner**
>
> The building owner shall have the following rights in relation to party structures (that is to say):
>
> (7) A right to pull down any party structure which is of insufficient strength for any building intended to be built and to rebuild the same of sufficient strength for the above purpose, upon condition of making good all damage occasioned thereby to the adjoining premises, or to the internal finishings and decorations thereof.
>
> **117. Settlement of differences between building and adjoining owners**
>
> (1) In all cases not specifically provided for by this Act, where a difference arises between a building owner and adjoining owner in respect of any matter arising with reference to any work to which any notice given under this Part of this Act relates, unless both parties concur in the appointment of one surveyor, they shall each appoint a surveyor, and the two surveyors so appointed shall select a third surveyor; and such one surveyor or three surveyors or any two of them shall settle any matter from time to time during the continuance of any work to which the notice relates in dispute between such building and adjoining owner, with power by his or their award to determine the right to do, and the time and manner of doing, any work, and generally any other matter arising out of or incidental to such difference; but any time so appointed for doing any work shall not, unless otherwise agreed, begin until after the expiration of the period by this Part of this Act prescribed for the notice in the particular case.

The legislation could not therefore be used as a pretext for undertaking general design changes to structures in joint ownership.

Appointed surveyors' jurisdiction

The judgment of Slesser LJ also emphasised the limited nature of the surveyors' jurisdiction:

"[The judge in the county court] refers to section 117 of the Act which provides that [the surveyors] shall 'settle any matter from time to time during the continuance of [the] work ... with power ... to determine the right to do, and the time and manner of doing, any work [and generally any other matter arising out of or incidental to such difference'] and he appears to be of the opinion that, by reason of that section, the award which the surveyors made, in so far as they purport to deal with the difference, had not exceeded their powers.

It seems to me entirely contrary to all recognised principles that arbitrators, not having differences at large submitted to them, but limited powers under a statute, can under pretext of the differences submitted to them adjudicate upon matters upon which the statute gives them no power to adjudicate.

In *Leadbetter v Marylebone Corporation* (1904) Lord Collins MR points out in terms that the jurisdiction which the surveyors have over such a dispute is a strictly limited jurisdiction ... If authority is necessary for the proposition that persons cannot make binding awards upon matters outside their jurisdiction, it is to be found in that case."

The decision

The Act contained no right for a building owner to permanently appropriate an adjoining owner's property. Odeon Theatres therefore had no right to form archways in the party wall which was partly owned by Burlington. The surveyors had exceeded their jurisdiction in purporting to grant such a right by their award. As a consequence, their award was invalid. Burlington's appeal was therefore allowed.

CARLISH v SALT (1906)
Chancery Division

In brief

- **Issues:** 1894 Act – surveyors' award – liabilities imposed on adjoining owner – change of ownership – whether award a material fact requiring disclosure to purchaser – purchaser's entitlement to rescind contract.

- **Facts:** Surveyors published an award for the repair of a party wall which required Carlish, as the adjoining owner, to contribute towards the costs. Before the work was undertaken he entered into a contract for the sale of his property to Salt. He failed to disclose the existence of the award. When Salt discovered this he refused to complete the purchase and demanded the return of his deposit.

- **Decision:** The existence of the award was a material fact which affected the value of the property. As it was also a latent defect in title there was also a duty to disclose it to a purchaser. Carlish was in breach of contract through failing to disclose the award and Salt was therefore entitled to rescind the contract. The court ordered Carlish to return the deposit.

- **Notes:** The following cases also deal with changes of ownership following statutory party wall procedures: *Re Stone and Hastie* (1903); *Mason v Fulham Corporation* (1910); *Selby v Whitbread & Co* (1917); *Observatory Hill Ltd v Camtel Investments SA* (1997).

The facts

Salt was the owner of 9 Portsmouth Street, Lincoln's Inn Fields, which was in a tumble-down condition. The party wall separating the property from number 10 Portsmouth Street was in a similar state and had recently been condemned by the District Surveyor.

In November 1903 the owner of number 10 served a party structure notice on Salt, under the 1894 Act, of his intention to demolish and rebuild the party wall. An award was made by surveyors which required Salt, as the owner of number 9, to contribute half the cost of the works. The award was made on 10 October 1904.

Two days later, on 12 October 1904, Salt entered into a contract to sell number 9 to Carlish. However, he made no mention of the award. The contractual completion date was 25 December 1904.

On 4 November, as a result of his earlier condemnation of the wall, the District Surveyor served a dangerous structure notice on Salt. Salt brought this to Carlish's attention and, at the same time, advised him about the existence of the award. Carlish demanded compensation from Salt as he maintained that he would not have entered into the contract had he known about the liability under the award.

Salt refused to pay compensation. Carlish therefore refused to complete the purchase on the basis of Salt's alleged breach of contract. On 6 January 1905 Carlish requested the return of the 10% deposit which he had paid on exchange of contracts. Salt refused to return this, claiming to be entitled to retain it following Carlish's own breach of contract in failing to complete on the contractual completion date. Subsequently, as he was now treating the contract as abandoned, Salt pulled down the house himself in order to comply with the dangerous structure notice.

Duty to disclose party wall award on exchange of contracts?

Carlish then issued proceedings against Salt to rescind the contract and to recover his deposit together with interest. He maintained that the award was a material fact which Salt had been under a contractual duty to disclose. In failing to disclose it he had committed a breach of contract which entitled him (Carlish) to rescind the contract and to recover his deposit.

Salt argued, firstly, that the existence of the award was not a material fact. Carlish had bought the property with the intention of demolishing and rebuilding it. The existence of the award did not therefore prejudice him. On the contrary, as the award also contained an obligation by the owner of number 10 to undertake the work, this was actually a benefit to Carlish.

He also argued that, even if the existence of the award was a material fact, it was not one which he was obliged to disclose. There could be no obligation to disclose even a material fact which was in the nature of a charge or outgoing on the property, rather than an impediment in the title. At most, like a mortgage, the award fell into this category and the proper course of action would have been to give him (Salt) an opportunity to clear it off the title. Instead, in refusing to complete, Carlish had ruled out this course of action and had instead, placed himself in breach of contract.

Party wall award as latent defect in title

The court rejected Salt's submission. It was of the view that the award was indeed a material fact which had affected the price that would be paid for the property. It acknowledged that non-disclosure of a material fact was not in itself a ground for rescission unless there was also some specific duty to disclose.

It considered the authorities on non-disclosure and concluded that there was a duty to disclose a latent defect in title. A *latent* defect is one which is not apparent on inspection. A defect in *title* is one which impacts directly on the owner's entitlement to enjoyment of the property rather than simply the quality of that enjoyment.

> Joyce J: "In the case of the sale of a chattel, the law as stated by Bramwell B in *Horsfall v Thomas* (1862) is that if there be a defect known to the manufacturer, and which cannot be discovered on inspection, he is bound to point it out.
>
> Upon consideration of the authorities, I am of opinion that the vendor of real estate is under a similar obligation with respect to a material defect in the title ... which defect is exclusively within his knowledge, and which the purchaser could not be expected to discover for himself with the care ordinarily used in such transactions.
>
> I hold that the party wall notice and the award constituted a material fact affecting the price to be paid, and in so far as they imposed a liability of uncertain amount at some future time on the owner of the premises, I am of the opinion that they constituted a latent defect not in the quality of, but in the title to, the property, and ought to have been disclosed."

The decision

The surveyors' award (and also the party structure notice) was a material fact which Salt was under a contractual duty to disclose. By failing to do so he had breached the contract and Carlish was entitled to rescind. The court therefore ordered Salt to repay the deposit to Carlish together with interest.

CHARTERED SOCIETY OF PHYSIOTHERAPY v SIMMONDS CHURCH SMILES (1995)

Official Referees' Court

In brief

- **Issues:** 1939 Act – nature of surveyors' awards – nature of appeals against awards – court's power to hear fresh evidence on appeal – court's power to overturn surveyors' findings of fact.

- **Facts:** The Society demolished and rebuilt the party wall separating their building from that owned by Simmonds. The work was undertaken under a surveyors' statutory award. Settlement damage subsequently occurred to Simmonds' building. A third surveyor's award attributed this to the Society's work and required them to make it good. The Society appealed against the award on the basis that the third surveyor's finding of fact about the cause of the damage was incorrect. Simmonds submitted that the court could not challenge surveyors' findings of fact and that appeals were restricted to matters of law.

- **Decision:** The surveyors' statutory role was more in the nature of an expert determination than a judicial process. Because of this it would be contrary to the interests of justice to restrict the nature of the enquiry to be undertaken on appeal. Appeals should therefore proceed by way of a complete rehearing of the issues considered by the surveyors. The court was entitled to hear fresh evidence and, if necessary, to overturn the surveyors' findings of fact.

- **Notes:** See also (on the nature of the surveyors' statutory role and on challenges to their awards): *Adams v Marylebone Borough Council* (1907); *Re Stone and Hastie* (1903); *Selby v Whitbread & Co* (1917); *Gyle-Thompson v Wall Street (Properties) Ltd* (1974).

The facts

Numbers 13 and 14 Jockey's Fields, London WC1, are two terraced properties separated by a party wall. Number 14 was owned by the Chartered Society of Physiotherapy (the Society) and number 13 was owned by a firm of solicitors, Simmonds Church Smiles (Simmonds).

The Society carried out some work to number 14 which involved demolishing and rebuilding the party wall between the two properties. They served the requisite party structure notice on Simmonds under the 1939 Act and, in due course, an award was made by the appointed surveyors.

After the works had been completed, Simmonds' surveyor alleged that the works had caused some settlement damage to his appointing owner's property which the Society should now make good. The Society's surveyor denied that any such damage had been caused and the matter was referred to the third surveyor.

The third surveyor made an award which found that the damage had indeed been caused by the works and that it should be made good by the Society. The Society appealed against the award on the ground that the evidence did not support the third surveyor's finding of fact about the cause of the damage. The appeal was subsequently transferred to the Official Referees' Court.

Court's power to overturn surveyors' finding of fact?

Simmonds made an interlocutory submission to the court that it had no power to overturn a finding of fact by the surveyors' tribunal. It could not therefore hear fresh evidence beyond that which had already been considered by the surveyors. The published law report is confined to the court's determination of this interlocutory submission.

Simmonds argued that the court's powers on appeal were akin to those of the Court of Appeal when hearing an appeal from a trial in the High Court. In this role the Court of Appeal could not generally receive fresh evidence which was not before the original court as its function was not to conduct a rehearing of the original issue. Simmonds argued that section 55(m)[1] of the 1939 Act made surveyors' awards conclusive and that they must therefore be subject to the same principle. As with appeals to the Court of Appeal, surveyors' awards could therefore only be challenged, on appeal, on a point of law.

The submission was considered by Judge Humphrey Lloyd QC and extracts from his judgment appear below. In reaching his decision he first considered the nature of surveyors' awards. He then went on to address the role of appeals against them under the Act.

The nature of surveyors' awards

The court considered whether the surveyors' statutory role should be regarded as a judicial function. A number of aspects supported such an interpretation:

> "In some respects the Act suggests that the difference is resolved by something in the nature of a statutory arbitration. The words 'difference', 'settle' and

---

[1] Equivalent to section 10(16) of the 1996 Act.

'award' and the provisions for the appointment of a sole 'agreed surveyor' and for decisions by two of the three surveyors or in default by the third surveyor are consistent with a judicial or quasi-judicial process."

But on the other hand:

"... the Act makes no provision for the parties to be heard or for the surveyor(s) to proceed as one might expect an arbitrator to act."

In particular the idea that the surveyors' award is an arbitration award was rejected:

"In the absence of authority I would not conclude from section 55(a) to (l)[2] ... that an award under the Act was an arbitration award ... The Act does not require the award to be a "speaking" award and there is no apparent obligation for the award to contain findings of fact or conclusions of law and, of course, awards are customarily and commendably direct and to the point. Furthermore, section 55(m)[3] plainly excludes the Arbitration Acts[4]."

The judgment concludes that the surveyors' role is not strictly judicial. Instead, the two surveyors perform a role which is more akin to that of professional advisors representing clients in negotiations to produce a negotiated settlement:

"... the Act envisages that if three surveyors are to be appointed, a party-appointed surveyor while no doubt retaining his professional independence is not obliged to act without regard to the interests of the party who appointed him. In practice matters in difference are regularly resolved by agreement between the two party-appointed surveyors without the need for the intervention of the third surveyor. Thus, the Act works well. The relevant owner

---

[2] Broadly equivalent to sections 10(1) to (15) of the 1996 Act.
[3] Equivalent to section 10(16) of the 1996 Act.
[4] Presumably because the section restricts rights of appeal to those contained in the Act and therefore by definition excludes the rights of appeal under the Arbitration Act(s) – an essential feature of an arbitration. Bickford-Smith and Sydenham (*Party Walls: The New Law*, Jordans, 1997, p. 52) point out that this is not necessarily correct. Section 94 of the Arbitration Act 1996 applies the (Arbitration) Act's provisions to statutory arbitrations to the extent that they are not 'inconsistent with the provisions of the enactment concerned'. Hence, if the procedures under the Party Wall etc. Act 1996 would otherwise constitute a statutory arbitration then the current section 10(16) will not prevent them from falling within this definition. The procedures could therefore theoretically be said to fall within the concept of a statutory arbitration which would be subject to the Arbitration Act to the extent that the Party Wall etc. Act 1996 had not varied this.

leaves it to the surveyor and has no need to prepare a case[5]. The facts are elicited informally by inspection and by perusal of proposals and counter-proposals..."

But, presumably because the surveyors are performing a statutory, rather than a contractual function, the court did not consider that this fully explained the nature of their role. The judgment simply concluded that: "an award under the Act is, in my judgment, *sui generis*[6] and is more in the nature of an expert determination".

The role of appeals against surveyors' awards

Having established the nature of a surveyors' award the court then turned its attention to the role of an appeal against such an award. In particular it had to consider whether an appeal meant a "true appeal" (along the lines of an appeal from the High Court to the Court of Appeal where the receipt of new evidence would be restricted[7]) or an appeal by way of rehearing (in which case new evidence would be admissible). The court decided that an appeal under the Act meant an appeal by way of rehearing.

The nature of the surveyors' role played a part in this decision. If the surveyors' role had been of a judicial or quasi-judicial nature then it would be more appropriate for the appeal to proceed by way of a "true appeal". However, as the nature of the process was not judicial, it would be contrary to the interests of justice to restrict the nature of the enquiry undertaken on appeal.

In particular the judgment notes that the surveyors' role is investigatory rather than adversarial and that there would be no adequate record of the facts before the original tribunal upon which to base a "true appeal". It would not therefore make sense for the appeal court to have to reconstruct the record before it could begin its "true appeal". It made more sense that the whole process should be opened afresh:

---

[5] So logically the award is simply an agreement negotiated on behalf of the parties, albeit one which is then given statutory authority. This view would certainly explain the reason why the award must then be served on the parties and why each of the parties ultimately has the final say via the machinery of an appeal. It would also support the view that an appeal should proceed by way of a rehearing. In other words if the surveyors had failed to properly carry out their functions, as delegated to them by their appointing owners, the owners have the right to re-open the whole matter. This would arguably not be the case if the surveyors were performing a judicial function.
[6] The only member of a species.
[7] Under the principles laid down in *Ladd v Marshall* (1954).

> **London Building Acts (Amendment) Act 1939**
>
> PART VI: RIGHTS ETC. OF BUILDING AND ADJOINING OWNERS
>
> DIFFERENCES BETWEEN OWNERS
>
> **55 Settlement of differences**
>
> Where a difference arises or is deemed to have arisen between a building owner and an adjoining owner in respect of any matter connected with any work to which this Part of this Act relates the following provisions shall have effect:
>
> (i) The agreed surveyor or as the case may be the three surveyors or any two of them shall settle by award any matter which before the commencement of any work to which a notice under this Part of this Act relates or from time to time during the continuance of such work may be in dispute between the building owner and the adjoining owner.
>
> (m) The award shall be conclusive and shall not except as provided by this section be questioned in any court.
>
> (n) Either of the parties to the difference may within fourteen days after the delivery of an award made under this section appeal to the county court against the award and the following provisions shall have effect:
>
>> (i) Subject as hereinafter in this paragraph provided the county court may rescind the award or modify it in such manner and make such order for costs as it thinks fit.

"All that may have happened is that the surveyor ('agreed' or 'third') visited the site and made a short non-speaking award which referred only to the building owner's proposals and to the counter-notice. The procedure would be investigatory and not necessarily 'adversarial'.

I do not believe that Parliament envisaged that, on an appeal, the court would first be asked to decide what evidence might be adduced to reconstruct the situation obtaining at the time of the award so as to provide the 'record' which would then set the scene for the appeal and for further argument as to what 'new' evidence might be admitted. Such a procedure would be unnecessarily costly and contrary to the statutory scheme which is self-evidently intended to provide a relatively inexpensive method of resolving disputes."

The court's powers in the legislation were also consistent with this interpretation:

> "Looking at the whole of section 55(n) and (o) it is, in my judgment, clear that the award is one which may be completely reopened if an appeal is duly made.
>
> Section 55(n)(i) provides that the county court may '... modify it in such manner and make such order as to costs as it thinks fit'. In my view, the words, 'as it thinks fit' plainly qualify 'modify in such manner' and are not limited to an 'order for costs' for otherwise 'such manner' is left hanging in the air.
>
> Thus, the court has, in my judgment, wide powers to alter any award and to do so must have the power to substitute its own finding or conclusion for any finding or conclusion that the surveyor(s) made or may be presumed to have made."

The judgment also noted that the Act contained no express restriction on the court's power to receive new evidence. In conclusion, the nature of an appeal against an award was described in the following terms:

> "Essentially the question which the court has to resolve is what award ought now to be made, taking into account all the facts established by admissible evidence, rather than the narrow question contended for by the respondent which is close to an investigation as to whether the award was made by a competent surveyor or surveyors."

The decision

Simmonds' interlocutory submission was dismissed. The appeal should proceed by way of a rehearing of all the original issues considered by the surveyors. This could include the consideration of fresh evidence and, where appropriate, the overturning of one of their findings of fact.

COWEN v PHILLIPS (1863)
Court of Chancery

In brief

- **Issues:** 1855 Act – entitlement to service of party structure notice – definition of "owner" – status of agreement for a lease for someone "in occupation" of premises – significance of equitable interests in land.

- **Facts:** Phillips commenced work on a party wall. He failed to serve notice on Cowen who occupied the adjoining property under an agreement for a three-year lease.

- **Decision:** Cowen's agreement for a lease took effect as an equitable lease for a three-year term. The Act was not confined to bare legal interests but also extended to equitable interests. As Cowen was in occupation of the premises by virtue of his three-year term he was an owner within the meaning of the Act and was therefore entitled to be served with notice. Judgment was therefore entered for Cowen.

- **Notes:** (1) See also: *Fillingham v Wood* (1891); *List v Tharp* (1897); *Orf v Payton* (1904); *Crosby v Alhambra Co Ltd* (1907); *Spiers & Son v Troup* (1915); *Solomons v Gertzenstein* (1954); *Lehmann v Herman* (1993); *Frances Holland School v Wassef* (2001). (2) The statutory definition of "owner" remained virtually unchanged from 1855 until the introduction of the 1996 Act. The definition in the 1996 Act omits the earlier references to persons "in occupation" of land. It is unclear whether this is intended to restrict the potential categories of owners or whether the term is now to be regarded as synonymous with the expression "in possession" which continues to appear. In any event a person under a contract for purchase or an agreement for a lease has now been separately included within the definition, whether or not they are also in occupation of the land.

The facts

Numbers 3 and 5 Bruton Street, London W1, were terraced properties which were separated by a party wall. Phillips was the freehold owner of number 5. Cowen was the occupier of self-contained shop premises on the basement and ground floors at number 3.

Cowen occupied his premises under a three-year lease at a yearly rent. However, although the lease was in writing it had not been executed as a deed, as required by the Real Property Act 1845.

Phillips wished to demolish and rebuild the party wall between the two properties[1]. He therefore served a party structure notice under the 1855 Act. He served this on Baker, who owned the freehold of number 3 but served no notice on Cowen.

Baker failed to respond to the notice and subsequently failed to appoint a surveyor. A surveyor was therefore appointed for him[2] and a surveyors' award was made which regulated the conduct of the works.

In due course Phillips' workmen commenced operations on site. They knocked holes in the party wall which separated number 5 from Cowen's shop and this caused disruption to his business.

Entitlement to service of party structure notice

Cowen commenced proceedings for damages and for an injunction to restrain Phillips' building operations. He argued that he should have been served with a party structure notice under the Act. Although he had no objection to the works taking place, he wished to ensure that his premises were adequately protected during the works and that he had an entitlement to have damage made good under a surveyors' award.

Cowen's entitlement to receive a notice[3] was dependent on him falling within the definition of "owner" in section 3 of the Act. He was certainly "in the occupation" of the premises within the terms of the section but the legal basis for this occupation was at issue between the parties.

If his 3-year lease had been made by deed he would have clearly fallen within the definition. However, as this was not the case, Phillips argued that the lease took effect only as an agreement for a tenancy from year to year or as a tenancy at will. On this basis, Cowen was not an owner within the Act.

Adjoining owner with agreement for a lease

The court rejected Phillips' argument. Cowen's lease was certainly valid as an agreement for a lease. However, although paying rent on a yearly basis, the agreement was not simply a legal agreement for a tenancy from year to year. It also took effect in equity as an agreement for a lease for the 3-year term. Cowen therefore had an equitable interest in the property for the 3-year term.

As this was an interest greater than from year to year the only remaining issue related to whether section 3 of the 1855 Act was intended to apply only to legal interests, as opposed to equitable interests.

---

[1] In pursuance of his right to do so under section 83(2) of the 1855 Act.
[2] Under section 85(9) of the 1855 Act.
[3] Under section 85(1) of the 1855 Act.

> ## Metropolitan Building Act 1855
>
> ### PRELIMINARY
>
> **3. Interpretation of certain terms in this Act**
>
> In the construction of this Act (if not inconsistent with the context) the following terms shall have the respective meanings hereinafter assigned to them; (that is to say):
>
> "Owner" shall apply to every person in possession or receipt either of the whole or of any part of the rents or profits of any land or tenement, or in the occupation of such land or tenement other than as tenant from year to year, or for any less term, or as a tenant at will.

Sir John Romilly MR: "The only question is whether the plaintiffs were the adjoining owners within the Metropolitan Building Act, which excepts tenants from year to year. If I am right as to the validity of this agreement here, then in equity the plaintiffs had an interest greater than from year to year, for it was an interest for three years, and unless this clause of the Act be confined to bare legal interests, the plaintiffs were clearly entitled to notice, and I have been referred to no case which says they are not.

If this Act were confined to bare legal interests, then in the case of a marriage settlement, where the legal estate is vested in trustees, although the husband and wife, the tenants for life, are in possession of their property, and the trustees abroad, the husband and wife are to have no notice of pulling down their house. I am satisfied that this is not the proper construction of the Act, nor could it be carried into effect."

The decision

The court held that, as a tenant in possession with an equitable agreement for a lease in excess of one year, Cowen was an adjoining owner within the Act. He should therefore have been served with notice of the works. Judgment was therefore entered for Cowen with an enquiry to be made into the level of damages which had been sustained by him.

CROFTS v HALDANE (1867)
Court of Queen's Bench

In brief

- **Issues:** 1855 Act – statutory purpose – right to raise a party wall – interference with easement of light – interference with adjoining owner's property rights – surveyors' jurisdiction.

- **Facts:** Haldane raised a party wall in accordance with a surveyors' award. The raised wall interfered with Crofts' easement of light.

- **Decision:** The statutory right to raise a party wall does not include a right to permanently deprive adjoining owners of their property rights. The Act therefore contains no right to interfere with an adjoining owner's easement of light. Because of this the surveyors had no jurisdiction to resolve disputes about easements of light. The court entered judgment for Crofts.

- **Notes:** (1) See also (on the interference with property rights): *Barry v Minturn* (1913); *Burlington Property Co Ltd v Odeon Theatres Ltd* (1938); *Gyle-Thompson v Wall Street (Properties) Ltd* (1974) – (on the surveyors' jurisdiction to adjudicate on matters of law): *Sims v The Estates Company* (1866); *Loost v Kremer* (1997) (2) Subsequent party wall legislation has expressly provided that rights granted by the Acts are not intended to interfere with rights of light or other easements. The relevant provision first appeared as section 101 of the 1894 Act and now appears in section 9 of the 1996 Act.

The facts

The case concerned two adjoining properties in Old Bond Street, London W1, which were owned respectively by Crofts and Haldane. The properties were separated by a party wall.

Haldane wished to demolish and rebuild his property and this involved the demolition and rebuilding of the party wall. He also wished to raise the height of the party wall when rebuilding it. He served the appropriate party structure notice on Crofts under the 1855 Act.

Surveyors were appointed by each party and an award was published which authorised the work. The work was then undertaken in accordance with the award. Unfortunately this had the effect of obstructing the light to windows in Crofts' property which he used as a picture gallery. Crofts had an easement of light in respect of these windows.

Crofts issued proceedings for infringement of his easement of light, claiming damages and an injunction.

Haldane argued, firstly, that he had not acted unlawfully because he had raised the party wall in accordance with the right contained in section 83(6) of the Metropolitan Building Act 1855.

Secondly, he argued that any dispute about the exercise of this right should be decided by the surveyors rather than the courts. The surveyors had approved the work, subject to Crofts' right of appeal to the county court. As Crofts had chosen not to exercise his right of appeal the right to raise the wall could not now be challenged by the courts.

No power to authorise infringement of right to light

The court held that the statutory right to raise a party wall does not authorise the interference with an easement of light. The rationale for this decision is explained in the judgment of Cockburn J:

> "The case seems to me quite clear. It turns entirely on the construction of [section 83(6)].
>
> Under that enactment [Haldane contends] that the building owner has the right to raise an external wall, although he interferes with the ancient lights in the neighbouring premises. In terms, there is certainly no such provision; and we ought not to imply it unless there are some terms from which such an inference might reasonably be drawn.
>
> But so far from there being any ground to imply this right, it appears to me that the language used leads directly to the opposite conclusion; when the statute gives the right to raise any party structure 'on condition of making good all damage to the neighbouring premises and the internal finishings', it clearly means that the building owner must restore the neighbouring premises to their condition previous to the erection; and cannot be forced into meaning, not only that he is to restore or make good any structural damage, but to give pecuniary compensation for destroying access to light ... the enactment has reference to structural damage, and not to paying damage for the invasion of a right.
>
> The second ground put forward, that [Crofts'] remedy was not by action, but by the arbitration of a surveyor ... also depends entirely on [section 83(6)]: because if the building owner has no right to raise a party wall so as to interfere with the right to light, there is no matter arising under the Act for the surveyors' arbitration.
>
> Both branches, therefore, of the argument on behalf of [Haldane] fail, the second depending on the first; and there must be judgement for the [Crofts]."

> ## Metropolitan Building Act 1855
>
> PART III: PARTY STRUCTURES
>
> RIGHTS OF BUILDING AND ADJOINING OWNERS
>
> **83. Rights of building owner**
>
> The Building Owner shall have the following rights in relation to party structures; that is to say:
>
> (6) A right to raise any party structure permitted by this Act to be raised, or any external wall built against such party structure, upon condition of making good all damage occasioned thereby to the adjoining premises or to the internal finishings and decorations thereof, and of carrying up to the requisite height all flues and chimney stacks belonging to the adjoining owner on or against such party structure or external wall.
>
> (7) A right to pull down any party structure that is of insufficient strength for any building intended to be built, and to rebuild the same of sufficient strength for the above purpose, upon condition of making good all damage occasioned thereby to the adjoining premises, or to the internal finishings and decorations thereof.

Effect of party wall legislation on private property rights

The court's rationale is echoed in the other judgments. These also provide a useful indication of the judicial approach to the party wall legislation in general. Although the legislation authorises the temporary interference with private property rights the limits of this are explained in the judgment of Blackburn J:

> "There is a series of enactments in Part 3 of the Act which do, for obviously good reasons, enable the building owner to deal with party structures in a way in which he would have no right at common law: but there is not one word in them to indicate an intention that he shall be at liberty to raise a wall so as to interfere with a neighbour's right to light any more than there is to take his land."

So, whilst the legislation certainly authorises acts which would otherwise amount to trespasses, it does not authorise the confiscation of an adjoining owner's property. This applies to incorporeal property, such as easements, in the same way that it applies to corporeal rights in the land itself[1].

A similar explanation is provided by Lush J who also makes reference to the underlying statutory purpose of the party wall legislation:

> "The argument [of Haldane] is founded on an entire misconception of the objects of the Building Act. There was no intention to interfere with rights such as easements of adjoining proprietors; the Act was simply to regulate the construction of buildings, and to prescribe the mode in which adjoining owners might proceed when dealing with party structures."

The decision

The Act did not authorise a permanent interference with property rights. The statutory right to raise a party wall did not therefore include a right to interfere with Haldane's easement of light. As the Act contained no such right there was nothing for the surveyors to adjudicate upon. Crofts had interfered with Haldane's easement and the court therefore entered judgment for Haldane.

---

[1] Although section 2(2)(m) of the 1996 Act does now authorise the appropriation of an adjoining owner's property through the permanent reduction in height of a party wall or party fence wall.

CROSBY v ALHAMBRA COMPANY LTD (1907)
Chancery Division

In brief

- **Issues:** 1894 Act – entitlement to service of party structure notice – definition of "owner" – definition of "adjoining owner" – whether more than one adjoining owner entitled to service – position of joint tenants and tenants in common.

- **Facts:** Alhambra wished to undertake work to a party wall. They served notice on the occupying tenant but failed to serve Crosby who had a reversionary interest in the property.

- **Decision:** Both the occupying tenant and Crosby were "owners" within the meaning of the Act. The reference to persons in possession "or" persons in occupation did not mean that only one of them was capable of falling within the definition. Furthermore, they both fell within the definition of "adjoining owner" and both were therefore entitled to service of a notice. The reference to "the owner or one of the owners" was not intended to restrict service to one adjoining owner where a number of people held separate legal estates in a property. The phrase related simply to co-owners (joint tenants or tenants in common) having a shared ownership of a single legal estate. Hence, whilst it was only necessary to serve notice on a single co-owner, all owners with a separate legal estate were entitled to service.

- **Notes:** (1) See also: *Cowen v Phillips* (1863); *Fillingham v Wood* (1891); *List v Tharp* (1897); *Orf v Payton* (1904); *Spiers & Son v Troup* (1915); *Solomons v Gertzenstein* (1954); *Lehmann v Herman* (1993); *Frances Holland School v Wassef* (2001). (2) The reference to "the owner or one of the owners" in the 1894 Act's definition of "adjoining owner" was removed from all subsequent legislation. Subsequent Acts, including the 1996 Act, reinforce the entitlement of all adjoining owners to receive notice by substituting the words "any owner".

The facts

Alhambra Company were the freehold owners of the Alhambra Theatre in Leicester Square. Crosby owned a 99-year lease in the adjoining property at number 28 Leicester Square. The two properties shared a party wall. In 1903 Crosby had sublet number 28 to the London County Council for a 21-year term.

Alhambra intended to carry out alterations to their theatre. These involved interfering with the fabric of the party wall and they therefore served a party structure notice on the Council under the 1894 Act. They served no notice on Crosby.

The Council consented to the works and Alhambra started the alterations in October 1906. In November Crosby became aware of the works and requested access to the theatre to inspect them. Alhambra refused him access so he issued proceedings against them. He requested an injunction to stop the work on the basis that he was entitled to be served with a party structure notice as an adjoining owner.

Service of party structure notice on more than one owner?

Alhambra submitted that, where there was more than one adjoining owner, it was only necessary to serve notice on one of them. As they had served notice on the Council, as the owner in occupation, they maintained that there was no requirement that they also serve on Crosby.

Fillingham v Wood (1891) had decided that, under the 1855 Act, it was necessary to serve notice on all adjoining owners. However, a different definition of adjoining owner had been inserted in the 1894 Act. This substituted the words "the owner or one of the owners" for the reference in the earlier Act to "the owner"[1]. Alhambra argued that the new form of words had been used with the specific intention of overruling *Fillingham* and avoiding the necessity of serving all adjoining owners.

The court's decision turned on the correct interpretation of sections 5(29) (definition of "owner") and 5(32) (definition of "adjoining owner") within the 1894 Act.

Definition of "owner"

Alhambra submitted that the definition of "owner" in section 5(29) could only refer to one interest in the property. It could include someone in occupation "or" someone in receipt of rents and profits. The use of the word "or" in the section clearly indicated that it was only intended to apply to one person from each of these two categories. It could not therefore apply to a succession of people with different tiers of leasehold interest in the same property, as in the present case.

The court did not accept this interpretation and held that every person who satisfied the definition was an "owner" within the meaning of the

---

[1] The phrase "any owner" has appeared in all subsequent legislation, including the 1996 Act.

Act. It found support for this view in the fact that the definition of owner in section 5(29) applied to the whole Act and not simply Part VIII which regulated the rights of neighbouring owners. For example, under section 179 (which dealt with the liability for expenses under the Act), the word "owner" was used in a way that clearly included parties with successive leasehold interests in the same property.

Definition of "adjoining owner"

Turning to section 5(32) the court noted that the limiting effect of the phrase "the owner or one of the owners" could mean one of two different things.

It could indeed mean, as submitted by Alhambra, just one of a number of persons who all owned different legal estates in the same property. Alternatively, it might simply mean just one of a number of persons who jointly owned the same legal estate in a property, whether as joint tenants or tenants in common.

In deliberating on the proper construction of this phrase the court made reference to the decision in *List v Tharp* (1897) which had been decided under the 1855 Act. In that case one Sir Charles Oppenheimer had been in receipt of rent from List under a building agreement in respect of the subject property. List was in occupation of the property under that agreement. An owner of adjoining property then served a party structure notice on Oppenheimer but not on List. List successfully sued the building owner on the basis that he was entitled to receive notice.

Crosby cited this case as authority for the proposition that people with different interests in the same property could simultaneously constitute an adjoining owner within the meaning of the legislation and that both were therefore entitled to receive service of a party structure notice. Alhambra favoured a different interpretation. They maintained that the case simply held that the tenant in possession (List) was entitled to receive a party structure notice and said nothing about the entitlement of more than one person to receive such notice.

The court rejected Alhambra's interpretation. There was nothing in the case to suggest that Oppenheimer was not entitled to notice as well. In the absence of this, if notice was only required to be served on one owner, then service on Oppenheimer should have been sufficient. On the contrary, in deciding that List was entitled to service the court must have been of the view that service on all adjoining owners was required.

> Neville J: "Chitty J held that the building owner could not proceed as against List notwithstanding the fact that he had served notice on Sir Charles Oppenheimer. In other words, it seems to me that the decision involves this, that notice both on Sir Charles Oppenheimer and on List

was necessary in order to enable the building owner to proceed under section 90 of the Act."

This convinced the court that the phrase "the owner or one of the owners" in section 5(32) was not intended to exclude persons falling within the definition of owner in section 5(29) from the definition of adjoining owner. All persons having an appropriate interest within the section 5(29) definition were adjoining owners and were therefore entitled to service of a party structure notice. The phrase simply referred to situations where a single estate was held jointly by more than one individual.

> Neville J: "I hold, therefore, that the true interpretation of the two subsections is that all the persons coming within the definition of 'owner' within section 5(2) must be served, except in the case where several persons hold, as tenants in common or as joint tenants, some particular interest in the land, in which case service on one of such tenants in common or joint tenants will be sufficient."

The decision

As the owner of a reversionary interest, Crosby was an owner within the meaning of the Act. The entitlement to receive notice was not restricted to owners who were in occupation of property and notice should therefore have been served on Crosby as well as his tenant. Accordingly, the court made a declaration that Alhambra were not permitted to proceed with their works without first serving a party structure notice upon him.

Metropolitan Building Act 1855

PRELIMINARY

3. Interpretation of certain terms in this Act

In the construction of this Act (if not inconsistent with the context) the following terms shall have the respective meanings hereinafter assigned to them; (that is to say):

"Owner" shall apply to every person in possession or receipt either of the whole or any part of the rents or profits of any land or tenement, or in the occupation of such land or tenement other than as tenant from year to year or for any less term, or as tenant at will.

PART III: PARTY STRUCTURES

PRELIMINARY

82. Definition of building owner and adjoining owner

In the construction of the following provisions relating to party structures, such one of the owners of the premises separated by or adjoining to any party structure as is desirous of executing any work in respect to such party structure shall be called the building owner, and the owner of the other premises shall be called the adjoining owner.

London Building Act 1894

PART I: INTRODUCTORY

5. Definitions

In this Act unless the context otherwise requires:

(29) The expression "owner" shall apply to every person in possession or receipt either of the whole or any part of the rents or profits of any land or tenement or in the occupation of any land or tenement otherwise than as a tenant from year to year or for any less term or as a tenant at will.

(32) The expression "adjoining owner" means the owner or one of the owners and "adjoining occupier" means the occupier or one of the occupiers of land buildings storeys or rooms adjoining those of the building owner.

London Building Act 1930

PART I: INTRODUCTORY

5. Definitions

In this Act, save as is otherwise expressly provided therein and unless the context otherwise requires, the following expressions have the meanings hereby respectively assigned to them (that is to say):

"owner" includes every person in possession or receipt either of the whole or of any part of the rents or profits of any land or tenement, or in the occupation of any land or tenement, otherwise than as a tenant from year to year, or for any less term, or as a tenant at will.

"adjoining owner" and "adjoining occupier" respectively mean any owner and any occupier of land, buildings, storeys or rooms adjoining those of the building owner.

CUBITT v PORTER (1828)
Court of King's Bench

In brief

- **Issues:** Wall subject to common acts of user – presumption that a party wall – trespass – status of party wall at common law – absence of clear evidence of ownership – presumption of tenancy in common – owner's entitlement to raise wall – owner's entitlement to demolish and rebuild wall.

- **Facts:** Cubitt and Porter's lands were separated by a freestanding wall which was subject to common acts of user. Porter demolished and then rebuilt the wall to an increased height. Cubitt alleged that both activities constituted a trespass against his property.

- **Decision:** The common acts of user meant that the wall was presumed to be a party wall. Furthermore, in the absence of clear evidence that each party had contributed an equal strip of land for its erection the wall was presumed to be owned by the parties as tenants in common. Tenants in common of a party wall had the right to raise the whole thickness of the wall and to demolish and rebuild it. No trespass had therefore been committed by Porter.

- **Notes:** (1) The lack of common law rights to undertake works to the whole thickness of a party wall is often assumed to have provided the rationale for enacting the various statutory codes, including the 1996 Act. This case demonstrates that these rights did exist at common law, at least in respect of pre-1926 party walls owned as tenants in common. Section 39 and Schedule 1, Part V of the Law of Property Act 1925 purported to preserve the existing rights in post-1926 party walls. It is an interesting question as to whether this also included the right to undertake works to the whole thickness of the wall. There are conflicting authorities on the point. For example, compare the decision in *Moss v Smith* (1977) with that in *Alcock v Wraith and Swinhoe* (1991). (2) Contrast the decision with that in *Watson v Gray* (1880) where a building owner was held to have no right to raise the height of a party wall. (3) See also: *Matts v Hawkins* (1813); *Wiltshire v Sidford* (1827); *Standard Bank of British South America v Stokes* (1878).

The facts

The case concerned a freestanding wall which separated plots of land owned respectively by Cubitt and Porter. There was evidence of "acts of user" of the wall by each of the parties although the nature of these is not described in the published law report. It may be that buildings, or other structures, had been erected against each side of the wall.

In about 1826 Porter demolished and rebuilt the wall to facilitate the construction of a cottage on his own land. He built the new wall to an increased height and connected the cottage to his side of the wall. Cubitt brought an action for trespass against him. He alleged that Porter had no right, either to demolish the wall or to rebuild it to a greater height.

At the trial, because of the evidence of common user, the jury decided that the wall was a party wall. The judge therefore found in Porter's favour. Because the wall was a party wall the parties must own it as tenants in common. As one of the tenants in common Porter was entitled to undertake work to it. No trespass can be committed in respect of common property by one tenant in common against another so Porter was not liable. Cubitt appealed against this decision on two grounds.

Legal status of party wall

The first ground of appeal related to the legal status of the party wall. In Cubitt's submission, the jury's finding that the wall was a party wall did not automatically lead to the conclusion that it was owned by tenants in common of the wall in undivided shares.

The other possibility, as occurred in *Matts v Hawkins* (1813), was that the wall was longitudinally divided with each party owning part of the thickness of the wall. He argued that this was the situation in the present case.

The court rejected this submission. Where there was clear evidence that a wall had been built astride a boundary then, as in *Matts v Hawkins*, the wall would be regarded as a longitudinally divided party wall. However, in the absence of clear evidence to this effect, the wall would be presumed to be owned by the parties as tenants in common under the principle in *Wiltshire v Sidford* (1827).

In the present case there was no evidence about the circumstances of the wall's original construction. In view of the common user of the wall the court therefore considered that the wall was more likely to have been built on the common land of the two owners rather than on portions belonging to each. The situation in *Matts v Hawkins* did not apply.

> Bayley J: "Where a wall is common property, it may happen either that a moiety of the land on which it is built may be one man's, and the other moiety another's, or the land may belong to the two persons in undivided moieties…
>
> …if the wall stood partly on one man's land, and partly on another's, either party would have a right to pare away the wall on his side, so as to weaken the wall on the other…

> ... It seems to me, the probability of the case is, that this was not a party wall according to the principle which was acted upon in the case of *Matts v Hawkins*, but that it was a wall built on the common property of the two, and that the wall was the common property of both...
>
> ... This case falls within the principle acted upon in *Wiltshire v Sidford*."

Ouster by demolition of party wall?

Under his second ground of appeal Cubitt argued that Porter's actions constituted a trespass as they had the effect of dispossessing (or 'ousting') him from his property.

It was accepted law that, in the absence of ouster, one tenant in common could not be liable to the other in trespass for an interference with what was the common property of both parties. However, Cubitt argued that both the demolition and the raising in height of the wall ousted him from his enjoyment of the wall. Both activities therefore constituted acts of trespass.

The court drew a distinction between the permanent destruction of a party wall and the temporary demolition of it as part of a combined demolition and rebuilding operation. Whilst the former was unlawful, the latter was not.

> Bayley J: "It has been contended that trespass is maintainable, on the ground that there was a destruction of the thing, and that if one tenant in common destroy that which is the subject of the tenancy in common, that is an actual ouster and exclusion by one of the other, and that the party so expelled may maintain an action of trespass for what has been done in that respect. Perhaps if one had entirely destroyed the wall, that might have been a foundation for an action of trespass.
>
> But I take it, that in the case of a wall, a temporary removal, with a view to improve part of the property on one side at least, and, perhaps, on both, is not such a destruction as will justify an action of trespass. There is no authority to show that one tenant in common can maintain an action against the other for a temporary removal of the subject matter of the tenancy in common, the party removing it having at the same time an intention of making a prompt restitution. It was not a destruction: the object of the party was not that there should be no wall there, but that there should be a wall there again as expeditiously as a wall could be made."

Ouster by raising a party wall?

The court was also of the view that Cubitt could not maintain an action in trespass on account of the raising of the wall.

Bayley J: "But then it is said the wall here is much higher than the wall was before. What is the consequence of that? One tenant in common has, upon that which is the subject matter of the tenancy in common, laid bricks and heightened the wall. If that be done further than it ought to have been done, what is the remedy of the other party?

He may remove it. That is the only remedy he can have. If there be land belonging to two as tenants in common, and one builds a wall on that land, the other cannot bring trespass, because he is excluded from the surface of that ground for a certain period of time, viz. for so long a period as that wall stands."

The decision

The court concluded that Porter, as a tenant in common of the party wall, was entitled to demolish and rebuild it, and to raise it. He had not therefore committed any trespass in undertaking these works. Cubitt's appeal was therefore dismissed.

DALTON v ANGUS (1881)
House of Lords

In brief

- **Issues:** Adjacent excavations – natural right of support – easement of support – acquisition by prescription – acquisition of increased right of support – whether enjoyed secretly – removal of support by contractor – whether building owner liable – non-delegable duties.

- **Facts:** Dalton was a contractor who carried out excavations on land owned by the Commissioners (a government agency). He failed to provide adequate shoring for Angus' adjacent building which collapsed as a result.

- **Decision:** A natural right of support could not exist for land which was subject to loads imposed by buildings. Such a right must be acquired as an easement. It was possible to acquire this right by prescription. A servient owner did not have to have precise knowledge of the loads being imposed by the building in order for this prescriptive right to be acquired. It was sufficient that the building had been openly erected without concealment of material facts. The same principles applied where an increased load was imposed by an existing building as a result of alterations. In these circumstances the increased load could not be said to be enjoyed secretly. Angus had acquired the necessary easement of support for his building by prescription and Dalton was liable for interfering with this. The Commissioners were also liable as Dalton's tort placed them in breach of a non-delegable duty to Angus.

- **Notes:** See also: *Bower v Peate* (1876); *Ray v Fairway Motors (Barnstable) Ltd* (1968); *Midland Bank plc v Bardgrove Property Services Ltd and John Willmott (WB) Ltd* (1992).

The facts

Two houses in Newcastle upon Tyne stood adjacent to each other. One was owned by Angus and the other was owned by a government agency, the Commissioners of Her Majesty's Works and Public Buildings.

Although they did not share a party wall each house stood on the extremity of its plot and was therefore in direct contact with the other. In view of their close proximity each house inevitably derived a degree of lateral support from the soil on which the other was built.

In 1849 Angus' house was converted into a coach factory. This involved the removal of its internal walls and the installation of iron girders to provide alternative support for the upstairs floors. The girders were built into a substantial brick pillar and chimney stack, which was also erected during the conversion. The effect of these alterations was to increase the lateral pressure on the soil beneath the Commissioners' adjacent building.

27 years later the Commissioners decided to demolish and rebuild their house as offices. They appointed Dalton, a contractor, to undertake the work and he, in turn, appointed subcontractors. The work included excavations for the construction of a new cellar to a depth several feet below that of the foundations to Angus' building. Both the main building contract and subcontracts contained requirements that shoring be provided to support the adjacent land.

When the excavations were undertaken the subcontractor failed to provide adequate shoring. As a result the pillar and chimney stack collapsed, pulling most of Angus' factory down with it.

Angus issued proceedings against Dalton and the Commissioners for damages. The history of the action was extremely complicated but Dalton and the Commissioners (the first and second defendants) eventually appealed to the House of Lords against the Court of Appeal's decision in Angus' favour.

The House of Lords had to consider whether Angus' factory building was entitled to a right of lateral support from the adjacent soil and whether Dalton and the Commissioners were liable for infringing it. This raised a number of separate issues.

Rights of support for land in non-natural state

The first issue related to whether the law recognised a right of lateral support for land against adjoining land. The court noted that natural rights of support exist in favour of land in its natural state but not where the land has been built upon.

However, in these circumstances a right of support could still exist, but it must first be acquired as an easement. Lord Selbourne LC explained the rationale for this in his judgment:

> "In the natural state of land, one part of it receives support from another, upper from lower strata, and soil from adjacent soil. This support is natural, and is necessary, as long as the status quo of the land is maintained; and, therefore, if one parcel of land be conveyed, so as to be divided in point of title from another contiguous to it, or (as in the case of mines) below it, the status quo of support passes with the property in the land, not as an easement held by a distinct title, but as an incident to the land itself, *sine quo res ipsa haberi non debet*[1] ...

---

[1] without which that thing cannot be had.

> ... [T]he doctrine laid down must, in my opinion be understood of land without reference to buildings. Support to that which is artificially imposed upon land cannot exist *ex jure naturae*[2], because the thing supported does not itself so exist; it must in each particular case be acquired by grant, or by some means equivalent in law to grant, in order to make it a burden upon the neighbour's land which (naturally) would be free from it."

However, this distinction related only to the manner in which the two categories of rights were acquired. The substance of the rights was identical in each case.

> Lord Selborne LC: "This distinction ... was pointed out by Willes J, in *Bonomi v Backhouse* (1858), where he said: 'The right to support of land and the right to support of buildings stand upon different footings, *as to the mode of acquiring them ... but the character of the rights, when acquired, is in each case the same*'."

Acquisition of right of support by prescription

Having decided that a right to lateral support for land which was built upon could be acquired as an easement, the court had to decide whether such a right could have been acquired in this case.

The issue before the court was whether the right could be acquired by prescription. Although their Lordships explained their reasoning in different ways, they unanimously decided that, potentially, it could.

Acquisition of increased right of support

The next issue before the court related to whether an easement for the additional support required by the factory conversion had actually been acquired by prescription over the 27 years since the work was undertaken.

In particular, the court had to consider the extent to which this was dependent on the Commissioners, as the servient owners, having specific knowledge of the additional forces being exerted on his soil by the conversion. Under the general law of easements, if the work had been undertaken secretly (*clam*), a right could not have been acquired by prescription.

---

[2] by virtue of a natural right.

Lord Selborne LC: "The enquiry on this part of the case is as to the nature and extent of the knowledge or means of knowledge which a man ought to be shown to possess, against whom a right of support for another man's building is claimed. He cannot resist or interrupt that of which he is wholly ignorant.

But there are some things of which all men ought to be presumed to have knowledge, and among them (I think) is the fact that, according to the laws of nature, a building cannot stand without vertical or (ordinarily) without lateral support.

When a new building is openly erected on one side of the dividing line between two properties, its general nature and character, its exterior and much of its interior structure, must be visible and ascertainable by the adjoining proprietor during the course of its erection.

When (as in the present case) a private dwellinghouse is pulled down, and a building of an entirely different character, such as a coach or carriage factory, with a large and massive brick pillar and chimney stack, is erected instead of it, the adjoining proprietor must have imputed to him knowledge that a new and enlarged easement of support (whatever may be its extent) is going to be acquired against him, unless he interrupts or prevents it.

The case is, in my opinion, substantially the same as if a new factory had been erected, where no building stood before. Having this knowledge, it is, in my judgment, by no means necessary that he should have particular information as to those details of the internal structure of the building on which the amount of incidence of its weight may more or less depend...

... if anything could be shown to have been done secretly or surreptitiously, in order to keep material facts from his knowledge, the case would be different. But here there was no evidence from which a jury could have been entitled to infer any of these things. Everything was honestly and (as far as it could be) openly done, without any deception or concealment ... I think, therefore, that in this case the kind and degree of knowledge which the adjoining proprietor must necessarily have had was sufficient; that nothing was done *clam*..."

The decision

The court was therefore satisfied that an easement of support was capable of being acquired in favour of land in its non-natural state; that it was capable of being acquired by prescription; and that an easement for increased support had indeed been acquired by Angus in this way. The only remaining issue was whether Dalton and the Commissioners could be held liable for an interference with this right of support by Dalton's independent subcontractor.

In this context the court approved the decision in *Bower v Peate* (1876). That case had decided that someone undertaking excavations adjacent to a house which had a right of support was subject to a non-delegable duty to the owner of that house. Both Dalton and the Commissioners were therefore liable for Angus' losses.

The House of Lords unanimously affirmed the judgment of the Court of Appeal and dismissed the appeal.

DODD v HOLME (1834)

Court of King's Bench

In brief

- **Issues:** Adjacent excavations – poor condition of adjoining owner's building – collapse of adjoining owner's building – absence of easement of support – negligence.
- **Facts:** Holme excavated adjacent to Dodd's building which was in a dilapidated condition. No shoring was provided and the building collapsed due to the excavations.
- **Decision:** Holme was liable in negligence.
- **Notes:** The basis of Holme's negligence is not stated. It is unclear whether this was due simply to his failure to provide shoring for Dodd's dilapidated building, or whether there was some other deficiency in his manner of working. See, for example, *Ray v Fairway Motors (Barnstable) Ltd* (1968). The decision should be treated with some caution as it predates the development of the modern law of negligence as well as traditional concepts relating to rights of support. See, for example, *Southwark & Vauxhall Water Company v Wandsworth District Board of Works* (1898). It seems unlikely, in the absence of an easement of support, that a building owner has a common law duty to provide shoring to his neighbour's building on the basis of its dilapidated condition alone. However, note the recent developments which extend the duties owned to adjoining owners in tort, for example: *Bradburn v Lindsay* (1983); *Holbeck Hall Hotel Ltd v Scarborough Borough Council* (2000); *Rees v Skerrett* (2001).

The facts

Dodd owned a very dilapidated house which, many years ago, had been built right up to the boundary with adjoining land owned by Holme.

Holme had recently demolished the warehouse situated on his land and was now in the process of redeveloping the site. This involved excavating to a depth of 6 ft below ground level to accommodate the foundations for a new building. The excavations did not encroach on to Dodd's land but were carried out within 4 ft of his dilapidated house. No shoring or other support was provided for the house during the excavations.

After the excavations had been undertaken the gable wall of the house started to bulge and, at that stage, Holme attempted to shore it up. This did not prevent further damage to the wall which "gave way in all directions" with the result that Dodd had to entirely rebuild it.

Negligence during adjacent excavations

Dodd issued proceedings for the cost of rebuilding. He alleged that the damage had been caused by Holme's negligence due to the manner in which he had carried out the excavations. In particular, he called expert witnesses who maintained that no damage would have occurred if the wall had been properly shored before the work commenced.

Holme denied any negligence. His expert witnesses attributed the damage to the condition of the wall and the way in which Dodd had been using it. The wall was said to have been in so rotten a state that shoring would not have prevented its collapse. Furthermore, the wall had inadequate foundations and was also under pressure from a great weight of rubbish which was being pressed against it from Dodd's side. As a result of these factors Holme maintained that, even if undisturbed, the wall would in any event have collapsed within six months.

Standard of care required?

On the evidence, the jury decided that Holme had indeed been guilty of negligence and entered a verdict for Dodd. However Holme moved for a retrial on the ground that the judge had misdirected the jury.

The judge's summing up had included the following words:

> "If I have a building on my own land, which I leave in the same state, and my neighbour digs in his land adjacent, so as to pull down my wall, he is liable to an action. If however, I have loaded my wall, so that it had more on it than it could well bear, he would not be liable."

In Holme's submission, this wrongly suggested that the mere act of digging near Dodd's house was an actionable negligence if this caused the wall to collapse. The real question for the jury was whether the work had been done in a negligent manner, or with as much care as the circumstances allowed.

Perhaps surprisingly, the court rejected this submission. Equally surprisingly, the court confined itself entirely to the question of negligence and appears not to have addressed the question of whether Dodd had a right of support which had been infringed.

Obligation to provide shoring where no right of support?

There is some ambiguity in the court's decision. In one sense it supports the principle that an act of excavation which causes a collapse is not of itself actionable, in the absence of negligence. In this context Taunton J observed that "If the building had fallen down merely in consequence of its infirm condition, that would not have been a damage by the act of the defendants...".

But this fails to properly address the common situation where a wall collapses, not simply because of its infirm condition, but due to some combination of this and a defendant's actions. In particular, if even the most careful excavation would result in a collapse due to the poor condition of the wall, is a potential defendant automatically subject to an obligation to provide shoring?

The court's decision suggests that the provision of shoring fell within Holme's duty of care. Although he would not have been liable if the wall had fallen simply as a result of its own infirmity, he should apparently have been aware of this infirmity in deciding on an appropriate methodology for his excavations. The logic of the court's decision suggests that this should have included appropriate shoring.

Holme could not therefore use the infirmity of the wall as a defence in circumstances where his own actions had undoubtedly contributed to the collapse. The only relevance of the wall's infirmity, in such circumstances, would be to reduce the damages for which he would be liable.

> Williams J: "The bad condition of the house would only affect the amount of damages. If it was true that the premises could have stood only six months, the plaintiff still had a cause of action against those who accelerated its fall: the state of the house might render more care necessary on the part of the defendant not to hasten its dissolution. There was evidence of an actual neglect in them; and, upon the whole, there is reason to think that the jury drew the proper inference."

Causation or breach of duty?

Unfortunately, apart from the failure to provide shoring, no further basis for the jury's finding of negligence is provided in the law report. Indeed, the court was content simply to accept the jury's finding at face value and made no further enquiry in this regard.

It was also unwilling to accept that there had been any misdirection by the judge which could have influenced the jury's finding. All the evidence considered by the jury had related directly to the question of whether there was some fault on Holme's part. It was therefore inconceivable that they should not have considered this issue in delivering their verdict.

> Taunton J: "... considering the length of time occupied by the cause, and the quantity of evidence gone into, it is impossible, even if the judge had been silent on the point, that the jury should have omitted to consider whether or not the act of the defendants was done by them negligently ... and ... I think we are

justified in saying that the minds of the jury were sufficiently directed to the question of how far the damage complained of arose from the improper act of the defendants."

In accepting the jury's findings there must be a suspicion that the court confused the separate issues of causation and breach of duty. There seems little doubt that Holme's actions, in combination with the infirmity of Dodd's house, directly caused the collapse of the wall. The fact that the damage occurred immediately after the excavations were undertaken must provide strong evidence for this.

However, unless the mere failure to provide adequate shoring is itself a breach of duty, it is difficult to identify any other breach capable of justifying a finding of negligence. The court appears to have based its decision largely on causation and this suspicion is supported by the judgment of Lord Denman CJ:

"The real point in the case was, the cause of the damage sustained by the plaintiff ... Upon that subject a great deal of evidence was given, and, no doubt, properly impressed upon the jury; and I think it was substantially left to them in charge of the learned judge, whether or not the result complained of was caused by the negligent act of the defendant. It being so left to them, I think, upon the balance of evidence, no other result could have been expected than the verdict they gave; the damage having occurred so soon after the act complained of."

The decision

The court were unwilling to challenge the jury's finding of negligence. Their verdict was therefore allowed to stand and Holme's request for a retrial was rejected.

DRURY v ARMY & NAVY AUXILIARY CO-OPERATIVE SUPPLY LTD (1896)

Queen's Bench Division

In brief

- **Issues:** 1894 Act – definition of type 'b' party wall – wall performing separating function for only part of its height – presumption that only party wall for part of its height – contrary statutory provision.

- **Facts:** Army & Navy proposed to install windows in external walls which performed a separating function at lower level. The District Surveyor sought to prevent this on the basis that the walls were party walls for the whole of their height. They must therefore comply with the technical requirements for the construction of party walls for their whole height.

- **Decision:** As a matter of fact the walls only performed a separating function up to first floor level. In the absence of an express statutory provision to the contrary the walls would therefore be presumed to be party walls up to this level and no higher. There was no such provision in the 1894 Act so the walls were not party walls above first floor level. Army & Navy were therefore free to install window openings in them above this level.

- **Notes:** (1) The courts have consistently held that a type 'b' party wall can exist for only part of a wall's height or length. The reference in the current statutory definition (1996 Act, section 20) to "so much of a wall" as separates buildings belonging to different owners now places this beyond doubt. (2) See also: *Weston v Arnold* (1873); *Knight v Pursell* (1879).

The facts

Army & Navy proposed to erect a warehouse building in Coburg Road in Haringey. The building was to be constructed to a height of five storeys with a light well in the centre to single-storey height (Figure 6).

They applied to the District Surveyor (Drury) for bye-law approval under the building control provisions in the 1894 Act[1]. They proposed to subdivide the building vertically with five walls to comply with the requirements to prevent the spread of fire in section 75 of that Act.

---

[1] Prior to 1986 building control functions in Inner London were undertaken by the District Surveyor in accordance with the London Building Acts.

As illustrated in Figure 6, two of these walls divided the five-storey sections of the building from the single-storey light well section. In each of these locations the walls would be carried upwards above the roof line of the single-storey section to form the external walls of the five-storey sections. Window openings were also to be included within the external wall portions of these walls to provide illumination from the light well.

Objections by District Surveyor

The District Surveyor objected to the proposal that window openings be installed in the external wall portions above the first floor roof level. As section 75 described the subdividing walls as 'party walls' he required them to be constructed in accordance with the Act's general requirements for the construction of such walls.

These included a requirement that party walls be carried up to a level, 3 ft above the roof of the highest building adjoining them[2] and a prohibition on the inclusion of window openings within them[3].

The walls certainly complied with the Act's requirements up to a height of 3 ft above the roof level of the single-storey section. However, according to the District Surveyor, they were party walls for their full five-storey height. They must therefore be constructed, as party walls, up to a height of 3 ft above the roof level of the five-storey sections. This necessarily prevented the inclusion of window openings in the external wall portions overlooking the light well.

Army & Navy appealed to the magistrate against the District Surveyor's decision and the matter was ultimately considered by the Queen's Bench Division.

Extent of a type 'b' party wall

Army & Navy cited *Weston v Arnold* (1873) and *Knight v Pursell* (1879) in support of their appeal. These cases demonstrated that, under earlier Building Acts, the courts had only treated walls as being party walls to the extent that they performed the function of dividing walls. They argued that the same interpretation should be applied to the 1894 Act and that the external wall portions of the walls were not therefore party walls within the meaning of that Act.

The District Surveyor sought to distinguish these earlier authorities, and in particular *Weston v Arnold*, from matters arising under the 1894 Act.

---

[2] Section 59(1).
[3] Sections 54(3) and 77(3).

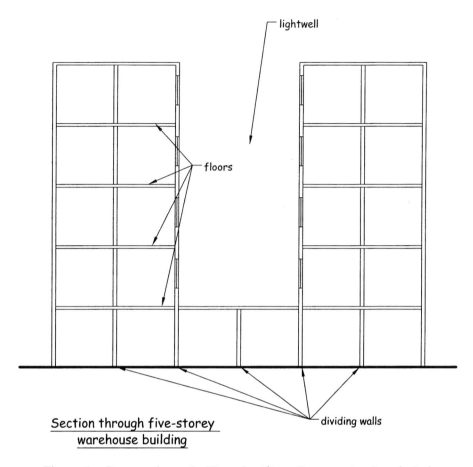

Figure 6: Drury v Army & Navy Auxiliary Co-operative Supply Ltd

In that case the upper part of a wall which was a party wall at a lower level had been held not to be a party wall. The court had held that walls are not party walls above the point at which they cease to be dividing walls unless there is some express statutory provision to the contrary. The Bristol Improvement Acts, under which that case had been decided, contained no such provision.

However, according to the District Surveyor, the necessary contrary provision was contained in section 59 of the 1894 Act. The present case could therefore be distinguished from the decision in *Weston* and the walls should be treated as party walls for the whole of their height.

Party walls and dividing walls

The court rejected the District Surveyor's argument. Firstly, it drew a distinction between the reference to "party walls" in section 75 and the technical use of the term as a wall which separates buildings in different ownerships. Whilst the definition of "party wall" in section 5(16) of the Act was consistent with the technical meaning, the court considered that its use in section 75 was synonymous with the term 'dividing wall'.

The isolated reference to "party walls" in section 75 therefore had no significance in the context of the other party wall provisions within the Act. To the extent that the wall performed the function of a dividing wall it could be regarded as a party wall but there was no requirement that it should be so regarded for its whole height.

> Wright J: "Obviously in [section 75] the words 'party wall' are used, not in their technical sense, but as a convenient phrase for dividing walls...
> The real ground of my decision is that section 75 has not the unnatural effect of making that which is not in the technical sense a party wall into a party wall at any point higher than 3 ft above the point at which it ceases to be a dividing wall.
> The walls in question here will be dividing walls up to the top of the first storey, and to that extent section 75 applies to them; but above the point where they will cease to be dividing walls I do not think section 75 or any other section makes them party walls in any sense, or attaches to them the obligation that they shall not be pierced by ordinary windows."

Party walls and external walls

Having decided that the walls were, as a matter of fact, only party walls up to first floor roof level, the court was equally unconvinced by the District Surveyor's argument that section 59 provided a rebuttal of this position in accordance with the rule in *Weston*.

> Collins J: "Counsel for the [District Surveyor] relied on section 59 as establishing that a wall, any part of which has been a party wall, is constituted a party wall right up to its full height. *Weston v Arnold* seems to me a decision that in the absence of any special enactment of the Legislature there is no sort of presumption that a wall which is a party wall in some part of it is a party wall throughout its whole height.
> In the present case section 59 is relied on as being such a special enactment, but when that section is examined it does not go so far as is contended. It simply provides that every party wall shall be carried up ... to such a height above the roof...
> So far as the wall is a party wall, no doubt it must conform to that provision; but so far as it is not a party wall it is not in any way fettered by the restrictions of the Act...
> For these reasons I am of the opinion that we are not bound to look upon

these walls above the height of the first storey as being anything but what they are – namely, external walls not used for the purpose of dividing either one house from another, or one portion of a warehouse from another. I, therefore, think that there is no reason why the windows should not be placed as they have been."

The decision

As a matter of fact the walls were only dividing walls up to first floor level. In the absence of any express statutory provision to the contrary they should be presumed to be party walls only up to this level. The Act contained no contrary express provision so the court held this to be the case. Army & Navy's appeal was therefore allowed and they were able to construct the warehouse and windows in accordance with their original proposals.

London Building Act 1894

PART I: INTRODUCTORY

5. Definitions

In this Act unless the context otherwise requires:

(16) The expression "party wall" means:

(a) A wall forming part of a building and used or constructed to be used for separation of adjoining buildings belonging to different owners or occupied or constructed or adapted to be occupied by different persons; or
(b) A wall forming part of a building and standing to a greater extent than the projection of the footings on the lands of different owners.

PART VI: CONSTRUCTION OF BUILDINGS

54. Rules as to recesses and openings

(3) An opening shall not be made in any party wall except in accordance with the provisions of this Act in relation thereto.

59. Height of party walls above roof

(1) Every party wall shall be carried up of a thickness in a building of the warehouse class equal to the thickness of such wall in the topmost storey and in any other building of eight and a half inches above the roof flat or gutter of the highest building adjoining thereto to such height as will give a distance (in a building of the warehouse class exceeding thirty feet in height) of at least three feet and (in any other building) of fifteen inches measured at right angles to the slope of the roof or fifteen inches above the highest part of any flat or gutter as the case may be.

75. Cubic extent of buildings

Except as in this section provided no building of the warehouse class shall extend to more than two hundred and fifty thousand cubic feet unless divided by party walls in such manner that no division thereof extend to more than two hundred and fifty thousand cubic feet.

77. Rules as to uniting buildings

(3) An opening shall not be made in any party wall or in two external walls dividing buildings which if taken together would extend to more than two hundred and fifty thousand cubic feet except under the following conditions:

- (c) Such openings shall have the floor jambs and head formed of brick stone or iron and be closed by two wrought iron doors each one fourth of an inch thick in the panel at a distance from each other of the full thickness of the wall fitted to rebated frames without woodwork of any kind or by wrought iron sliding doors or shutters properly constructed fitted into grooved or rebated iron frames.

EMMS v POLYA (1973)
Chancery Division

In brief

- **Issues:** 1939 Act – failure to serve party structure notice – nuisance – unlawful interference with enjoyment of land – noisy building operations – excessive hours of working – failure to properly supervise workforce – abnormally sensitive adjoining owner.

- **Facts:** Emms was a playwright who worked from home. He complained about prolonged and noisy building operations from the next door property which were being carried out by Polya. Polya, who worked elsewhere as a full-time hotelier, was converting the property to flats with casual labour. He was rarely on site and had not employed an architect or surveyor to supervise the works. He had also failed to serve the required notice under the 1939 Act.

- **Decision:** Polya was liable in nuisance. Although Emms had to put up with a certain level of inconvenience Polya was obliged to minimise this by exercising reasonable care and skill. He had failed to exercise proper control over his workforce and as a result the level of inconvenience had exceeded an acceptable level, according to the standards of reasonable people. It was irrelevant that Emms was abnormally sensitive to noise as any reasonable person would have found the level of disruption to be unacceptable. Damages were awarded to Emms for the nuisance suffered. No additional remedy was awarded on account of Polya's failure to comply with the 1939 Act.

- **Notes:** See also: *Matania v National Provincial Bank Ltd and The Elevenist Syndicate Ltd* (1936); *Andreae v Selfridge & Co* (1938); *Louis v Sadiq* (1996).

The facts

Mr Emms was the owner-occupier of a first floor flat at 1a Carlingford Road, Hampstead, London NW3. He was a playwright and worked from home.

The flat was part of a Victorian terrace and was separated from 3 Carlingford Road by a party wall. That property was owned by Mrs Polya who lived, with her husband, in Middlesex.

Mr Polya was, by trade, an hotelier. In 1972 he started converting number 3 into flats. Although he had very little knowledge of the building trade he decided to undertake the work himself by working from plans prepared on behalf of the previous owner and by directly employing casual labour. He undertook supervision of the work himself although he was not able to be present all the time because of his other commitments.

The work involved taking out chimney breasts, removing partitions, putting in steel joists, putting in new partitions, creating new rooms and new staircases, sanding floors and so on. It went on for over a year. On occasions it continued until seven or eight o'clock on weekday evenings and also sometimes during part of the weekend.

Noisy building operations

Mr Emms issued a writ against Mr Polya, claiming damages for nuisance caused by the noise of the construction operations. He claimed that the noise made life in his flat intolerable. The particulars of his claim are set out in the law report thus:

> "He complained of the noise of hammering, banging, drilling, falling plaster and rubble, thumping, shouting and swearing by the workmen – and incidentally, though not as giving him an independent cause of action, of the fact that although the work affected the party wall no party wall notice was ever served under the London Building Acts."[1]

Both parties accepted the established legal position that everyone must expect to put up with a certain amount of inconvenience arising out of the temporary building operations of his neighbour. However, Mr Emms claimed that the general effect of the noise over a prolonged period had exceeded what was reasonable. Mr Polya argued that no reasonable person would have complained about the amount of noise generated and that Mr Emms was an abnormally sensitive plaintiff[2].

Basis of liability in nuisance

The judge considered that two matters had to be considered in deciding the case.

---

[1] It is likely that no independent cause of action will lie simply for non-service of a notice under the Act.

[2] A plaintiff who is abnormally sensitive has no right to any greater freedom from interference than anyone else. See *Robinson v Kilvert* (1889); *Bridlington Relay Ltd v Yorkshire Electricity Board* (1965). Where an abnormally sensitive plaintiff nevertheless suffers interference which is unlawful by ordinary standards he is entitled to be fully compensated having regard to this abnormal sensitivity. See *McKinnon Industries Ltd v Walker* (1951).

Firstly he had to decide whether the noise was such that it materially interfered with the ordinary physical comfort of human existence in Mr Emms' flat according to "plain, simple and sober notions among English people", as opposed to it simply affecting Mr Emms due to a hypersensitivity to noise, as alleged by Mr Polya.

Secondly, if this *prima facie* case of nuisance was established, the judge then had to decide whether Mr Polya could nevertheless avoid liability on the basis that he had carried out his works with all proper care and skill so as to prevent a nuisance as far as possible in accordance with the rule in *Andreae v Selfridge & Co* (1938).

The decision

Regarding the first matter, the judge did consider that Mr Emms was abnormally sensitive to noise. However, he was satisfied that the degree of noise caused by the building operations was nevertheless such as to constitute a nuisance according to ordinary and reasonable standards.

Turning to the second matter, he held that Mr Polya had not carried out the work with the necessary care and skill that was required and in particular, he had failed to exercise proper supervision over his workforce:

> "Having no architect and no surveyor, he had no advice on how to keep the noise level down so as not to cause a nuisance. The truth of the matter was that he took no precautions whatever to mitigate the effect of his operations[3]."

Judgment was entered for Mr Emms who was awarded £350 in damages.

---

[3] If the work had been undertaken within the framework of the London Building Acts Mr Polya would have received the necessary guidance from the appointed surveyors and this would have protected him from this action in nuisance. See *Louis Sadiq* (1996).

FILLINGHAM v WOOD (1891)
Chancery Division

In brief

- **Issues:** 1855 Act – entitlement to service of party structure notice – definition of "owner" – requirement to serve on more than one owner – whether someone in occupation of only part of a building is an owner.

- **Facts:** Wood carried out work to a party wall which involved laying open adjoining premises occupied by Fillingham. Wood had served a party structure notice on the long leasehold owner of the adjoining building as the person in receipt of rents or profits. He argued that he was not also required to serve notice on Wood who only had a sub-tenancy of some rooms within the building for a three-year term.

- **Decision:** Notice should also have been served on Fillingham. He was in occupation of premises for a term which was longer than a tenancy from year to year so he was an owner within the Act. Despite the common practice of serving only on the person in receipt of rent or profits from the entire premises, all owners were entitled to receive notice. An owner, like Fillingham, who only occupied rooms within a building, was also entitled to be served with notice.

- **Notes:** (1) Changes have been made to the definition of "owner" in the 1996 Act (section 20). However, there is no suggestion that this will have any effect on this decision which reflects the current practice by party wall surveyors. (2) See also: *Cowen v Phillips* (1863); *List v Tharp* (1897); *Orf v Payton* (1904); *Crosby v Alhambra Co Ltd* (1907); *Spiers & Son v Troup* (1915); *Solomons v Gertzenstein* (1954); *Lehmann v Herman* (1993); *Frances Holland School v Wassef* (2001).

The facts

Wood was the freehold owner-occupier of 40 Margaret Street, off Cavendish Square, London W1. The property was separated from the adjoining premises at number 41 by a party wall.

The long leasehold estate in number 41 was owned by Smith who had sublet the whole of the property to two subtenants, including Fillingham. Fillingham had a three-year tenancy of various rooms on the ground and first floors and the remaining parts of the property were demised to the other subtenant.

In 1890 Wood served a party structure notice on Smith under the 1855 Act of his intention to demolish and rebuild his property, including the party wall with number 41. On the expiry of the notice he started demolishing number 40 and in the process laid open Fillingham's third floor bedroom, making it uninhabitable.

Fillingham issued proceedings against Wood for damages and an injunction in trespass. He maintained that Wood's interference with the party wall without the prior service of a party structure notice upon him was unlawful.

Service only on person in receipt of rents and profits?

Wood argued that, as an occupier of only part of the property, Fillingham was not an owner within the meaning of the 1855 Act. Although the definition of owner[1] included someone in possession or receipt of "the whole or any part" of the rents and profits from a tenement, the Act contained no equivalent reference to the occupation of only part of a tenement.

Wood submitted that the Act had never intended to put building owners to the trouble and expense of finding out every possible occupier. On this basis, he had "followed the invariable practice of London surveyors and architects, and served the person who is in receipt of the rents or profits". He made reference to the decision in *Hunt v Harris* (1865) in support of this practice.

In *Hunt* a building owner had sued an adjoining owner with a long leasehold estate in the adjoining property for a contribution towards the cost of repairing a party wall under the 1855 Act. The long leaseholder had unsuccessfully defended the claim on the basis that he had sublet his property to a number of subtenants and that they were therefore now liable for payment of the contribution. The court ruled that the building owner was entitled to sue the long leaseholder for the whole amount of the contribution. The rationale for this was explained in Chief Justice Erle's judgment:

> "According to the best interpretation I can put upon the statute, the building owner has a right to call upon the person who holds the entire premises for a long term, and is substantially in possession of all the rents and profits. I do not mean to say that the owner in fee simple may not also be liable. But I can see my way clearly to the conclusion that the owner of a long term, who is in receipt of the rents and profits, is liable."

Wood submitted that the decision in *Hunt* was directly applicable to the present case. Although it related to the responsibility to contribute towards the cost of repairs, by implication, it also addressed the definition

---
[1] In section 3.

of owner within the 1855 Act. In deciding that the long leaseholder was liable the court was clearly of the view that there could only be one owner within the Act and that this related to a person having an estate or interest in the whole property.

Service on more than one owner?

Fillingham argued, firstly, that there was nothing in the Act which limited the numbers of adjoining owners in respect of a single property. On the contrary, section 97(2), which provided for the sharing of contributions between adjoining owners, clearly anticipated the possibility of multiple adjoining owners. Furthermore, the identification of multiple adjoining owners by a building owner was not as onerous as had been suggested. The Act's requirements for service could easily be satisfied by placing a notice on a conspicuous part of the adjoining premises in accordance with section 98.

Secondly, he submitted that there was nothing to prevent someone who occupied only part of a building from being an owner within the meaning of the Act. He cited the decision in *Cowen v Phillips* (1863) in support of this.

In *Cowen* the plaintiff had been held to be entitled to receive notice as an adjoining owner on the basis of an equitable interest under an agreement for a lease. The decision had significance for the present case as the agreement related only to some self-contained shop premises on the basement and ground floors of the adjoining property. This, according to Fillingham, was authority that the statutory definition of owner extended to occupation of rooms within buildings, as well as to that of entire buildings.

Requirement to serve notice on all adjoining owners

The court considered that both the decision in *Hunt v Harris* and that in *Cowen v Phillips* were binding upon it. It therefore had to resolve the apparent contradiction between them. *Hunt* appeared to suggest that only one adjoining owner, namely the owner of the whole tenement, was an owner within the Act. In apparent contradiction to this, *Cowen* appeared to suggest that multiple adjoining owners, including an owner of part of the property, were entitled to receive service of a notice.

It considered that, in reality, the two decisions were not in conflict. The decision in *Cowen*, about the requirement for multiple notices, was perfectly clear. *Hunt* did not conflict with this. That case had addressed only the liability of the long leaseholder to contribute to the cost of the repairs and said nothing about any parallel liability of the various subtenants who had not been parties to the action.

This was explained by Chitty J in the following terms:

"Unquestionably, at first sight there seems to be some slight inconsistency in holding that under the construction of this statute the adjoining owner who is bound to contribute to the building owner the expense is not the same person as the adjoining owner entitled to notice before the work is commenced. But I think there is no inconsistency.

The point of law is that the court ... said that Harris at least is liable ... because he is one of the persons at any rate who under the definition clause is owner; and they held in effect that it was not necessary for the plaintiff Hunt to sue every owner; but such an interpretation cannot, it seems to me, be put on section 85.

Section 85 requires that notice should be given to the 'owner', which includes every person within the language of the interpretation section."

Requirement to serve notice on owner of part of property

There remained the outstanding question of whether an occupier of part of a property could properly be said to be an owner when this was not provided for by the wording in section 3. Why did that section anticipate an owner receiving part of the income from a land or tenement but not his occupation of part of the land or tenement itself?

The answer was to be found in the Act's use of the word "tenement" which refers to a legal title to land rather than to a physical building. Hence, as long as an owner occupied part of a building under a separate title, he occupied a whole tenement. It was therefore unnecessary for the Act to make provision for the occupation of part of a tenement as the law did not recognise such a concept. If the owner of a title shared the occupation of his premises with others this did not prevent him from being in occupation of the whole tenement. Similarly, if he owns his title jointly with these other occupiers, each joint owner is individually in occupation of the whole tenement.

> Chitty J: "I turn to section 3 ... the plaintiff is in possession and occupation of the rooms I have mentioned for a greater interest than the excepted interest. It is observable that in the definition of 'owner' the words 'whole, or any part of the rent or profits' are made use of, but that the clause does not go on to speak of the person being in occupation of any part of the land or tenement; and indeed it would have been difficult to put in such words with reference to land or tenement. I think, however, that the true interpretation of the last portion of that definition is to see whether the tenement is held under a title separately.
>
> ... I have said that in respect of his interest he is clearly an 'adjoining owner'[2]. The only question is whether he ought to be held not to be an adjoining owner because he does not hold the entirety of the house. I think,

---

[2] Defined in section 82.

however, on the true construction of the Act (which, I admit, is very difficult to construe), he does hold a tenement within the meaning of the interpretation clause."

The decision

A tenant of part of a property was an owner within the 1855 Act and was entitled to be served with notices.

> Chitty J: "... the plaintiff is the person who will suffer loss and inconvenience by the operation of the building owner; and I think it would be wrong, and a narrow interpretation of those parts of the Act with which I have specially to deal, if I held that he was not entitled to notice. Therefore, I think that substantially the case is concluded by the case of *Cowen v Phillips*."

The court therefore awarded an interim injunction to restrain any further work pending the final hearing of the suit.

Metropolitan Building Act 1855

PRELIMINARY

3. Interpretation of certain terms in this Act

In the construction of this Act (if not inconsistent with the context) the following terms shall have the respective meanings hereinafter assigned to them; (that is to say):

"Owner" shall apply to every person in possession or receipt either of the whole or of any part of the rents or profits of any land or tenement, or in the occupation of such land or tenement other than as tenant from year to year, or for any less term, or as a tenant at will.

PART III: PARTY STRUCTURES

PRELIMINARY

82. Definition of building owner and adjoining owner

In the construction of the following provisions relating to party structures, such one of the owners of the premises separated by or adjoining to any party structure as is desirous of executing any work in respect to such party structure shall be called the building owner, and the owner of the other premises shall be called the adjoining owner.

RIGHTS OF BUILDING AND ADJOINING OWNER

85. Rules as to exercise of rights by building and adjoining owners

The following rules shall be observed with respect to the exercise by building owners and adjoining owners of their respective rights:

(1) No building owner shall, except with the consent of the adjoining owner, or in cases where any party structure is dangerous, in which case the provisions hereby made as to dangerous structures shall apply, exercise any right hereby given in respect of any party structure, unless he has given at least three months previous notice to the adjoining owner by delivering the same to him personally, or by sending it by post in a registered letter addressed to such owner at his last known place of abode.

PART IV: MISCELLANEOUS PROVISIONS

97. Payment of expenses by owners

Where it is hereby declared that expenses are to be borne by the owner of any premises (including in the term "owner" the adjoining and building owner respectively) the following rules shall be observed with respect to the payment of such expenses:

(2) If there are more owners than one, every owner shall be liable to contribute to such expenses in proportion to his interest.

FRANCES HOLLAND SCHOOL v WASSEF (2001)
Central London County Court

In brief

- **Issues:** 1939 Act – definition of "owner" – whether statutory tenants are owners – statutory purpose – validity of surveyors' award – effect of estoppel by convention – surveyor's jurisdiction to make *ex parte* award – ten-day notice – refusal to act – neglect to act – validity of *ex parte* award.

- **Facts:** The School undertook party wall and adjacent excavation work in pursuance of a surveyors' award under the 1939 Act. The works caused damage to the Wassef's adjoining property and the third surveyor published an award requiring remedial work to be undertaken by the School. The two surveyors subsequently disagreed on other matters connected with the performance of the remedial work. The Wassef's surveyor (WS) considered that an addendum award was required to address these issues but the School's surveyor (SS) considered that this should wait until the work had been completed. WS served a ten-day notice on SS requiring him to enter into an addendum award, failing which he (WS) would publish an *ex parte* award. SS again reiterated his opinion that he did not consider an award was necessary. WS then published an *ex parte* award which cited SS's "refusal" to act. The School appealed against the award.

- **Decision:** Adjoining owners were entitled to be served with notice to protect their property rights. The Act did not contemplate that those without property rights in adjoining land would fall within the definition of "owner". The Wassefs were statutory tenants under the Rent Act 1977 and were not therefore owners. They had no right to receive notice and no right to appoint a surveyor. The surveyors' tribunal was therefore improperly constituted and had no jurisdiction to make any of the awards that had been made. However, both parties had proceeded on the basis of a shared misunderstanding and it would now be unconscionable for either party to withdraw from the arrangements that had been entered into. Under the doctrine of estoppel by convention the parties were therefore estopped from denying the validity of the earlier awards and the court was unwilling to declare them invalid. However, the *ex parte* award was invalid. There was no evidence that SS had either neglected or refused to act. Furthermore, the award cited SS's refusal to act but appeared to rely on his neglect following service of the ten-day notice. That award was therefore bad on its face and invalid.

- **Notes:** See also (on the effect of procedural irregularities on surveyors' awards): *Gyle-Thompson v Wall Street (Properties) Ltd* (1974) – (on the definition of "owner"): *Cowen v Phillips* (1863); *Fillingham v Wood* (1891); *List v Tharp* (1897); *Orf v Payton* (1904); *Crosby v Alhambra Co Ltd* (1907); *Spiers & Son v Troup* (1915); *Solomons v Gertzenstein* (1954); *Lehmann v Herman* (1993).

The facts

Frances Holland School owned a site at 21 and 23 Graham Terrace, London SW1. Number 21 was a building fronting on to Graham Terrace. Number 23 was a large garage building which lay directly to the rear of number 21 (Figure 7).

Access to number 23 was obtained from Graham Terrace via an external passageway at the ground floor level of number 21. This ran alongside the neighbouring terraced house at number 25. At first floor level the building at number 21 continued over the passageway and formed a party wall with number 25. Parts of number 23 were also separated from number 25 by a party wall (Figure 7).

Mr and Mrs Wassef occupied number 25 Graham Terrace as statutory tenants under the Rent Act 1977. Grosvenor Estates owned the freehold.

The School carried out work which involved demolishing both buildings on its site. A new building was erected on the site of the garage at number 23. The building at number 21 also had to be demolished to facilitate access for the construction work at number 23. This was then rebuilt in substantially its original form. The demolition and rebuilding work involved interference with the party wall with number 25. The ground level at number 23 was also lowered and this required parts of the party wall to be underpinned.

Primary, addendum and third surveyor's awards

Notices were served on the Wassefs under the 1939 Act and surveyors were appointed. The surveyors published awards dealing with the demolition and underpinning aspects of the work.

Unfortunately the underpinning work was poorly designed and executed and this resulted in substantial movement to the rear addition of number 25. The surveyors therefore published an addendum award which required the School to make good the damage by further remedial underpinning.

The two surveyors were subsequently unable to agree on aspects of the remedial underpinning work and these matters were referred to the third surveyor. The third surveyor's award (inter alia) required the under-

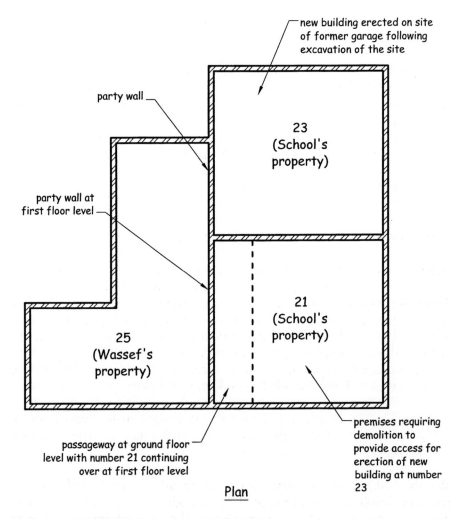

Figure 7: Frances Holland School v Wassef

pinning work to be protected by an insurance-backed guarantee in the parties' joint names.

Surveyors' failure to agree

Despite the presence of the third surveyor's award, the two surveyors were still unable to resolve a number of issues concerning the remedial underpinning work.

The Wassefs' surveyor (WS) wished to conclude an agreement on the extent of remedial work required and to agree arrangements for alternative accommodation for his appointing owners during the work. He also wished to address the fact that it had not been possible to obtain an insurance backed guarantee from insurers as required by the third surveyor's award.

The School's surveyor (SS) was unwilling to address these issues. He maintained that the provision of alternative accommodation was unnecessary and noted that no such arrangements had been made in respect of number 19 Graham Terrace which was affected by the works in a similar way to number 25. He also maintained that all outstanding matters concerning remedial works were best addressed together once the works at number 21 had been completed. He argued that to do otherwise would simply increase costs unnecessarily.

Adjoining owner's surveyor acting *ex parte*

After discussing the matter with his appointing owners, WS decided to proceed with an *ex parte* award in accordance with section 55(e)[1] of the 1939 Act. Having first obtained their authority to do so, he therefore wrote to SS requiring him to enter into an award on the outstanding issues within ten days. The letter noted that in the event of his failure to do so, the writer intended to act *ex parte* in accordance with section 55(e).

SS immediately responded by letter, reiterating his position in the earlier discussions. His letter included the following passage:

> "I am therefore somewhat surprised to see that you propose to act ex parte in this matter, particularly as there appears not to be a need for an award at this point between our respective parties."

WS then issued an *ex parte* award which addressed the various outstanding issues. The award recited "the building owner's refusal to act" and was accompanied by a covering letter. The letter referred to SS having "refused to consider the matter of relocation compensation and the matter of the insurance backed guarantee awarded by the third surveyor".

---

[1] The equivalent provision now appears as sections 10(6) and 10(7) of the 1996 Act. These sections refer to a surveyor's refusal or failure "to act effectively" rather than simply "to act". The other surveyor's power to act *ex parte* following a failure to comply with a ten-day request is also expressly limited to the subject matter of the request. It seems unlikely that these minor changes of wording will have any impact on the way that the courts interpret this provision, the substance of which appears to be unchanged. It should be noted however that the wording in the 1996 Act now authorises the other surveyor (or his appointing owner) to serve the ten-day request. Previously only the appointing owner had this authority which could only be exercised by the surveyor if he had been expressly authorised to do so. Although it remains good practice for the terms of a surveyor's formal appointment to include a general authority to serve notices, requests and appointments, this is now no longer strictly necessary in respect of this provision.

Appeal to the county court

The School appealed against the *ex parte* award to the Central London County Court. They argued that the award was made without jurisdiction as:

(1) the Wassefs were not "owners" within the meaning of the 1939 Act, and
(2) the Wassefs' surveyor had no grounds, under the provisions of the Act, for making an *ex parte* award.

The outcome of the appeal was based on the court's consideration of the following three issues.

Are statutory tenants "owners" within the Act?

The definition of "owner" applicable to the 1939 Act appeared in section 5 of the London Building Act 1930[2]. The definition included "every person ... in the occupation of any land or tenement otherwise than as tenant from year to year or for any less term or as a tenant at will."

The Wassefs' counsel submitted that, as statutory tenants, his clients had a superior interest to those of tenants from year to year or tenants at will. They must therefore fall within the definition. He also cited the case of *Brown v Minister of Housing & Local Government* (1953) in support of his contention. In that case the court held that a statutory tenant was an "occupier" within the Acquisition of Land (Authorised Procedure) Act 1946 and was therefore entitled to be served with notice of a compulsory purchase order.

The School's counsel replied that the issue in the *Brown* case related only to providing a statutory tenant with an opportunity to object to compulsory purchase proposals. The present case was entirely different as the 1939 Act was concerned with the regulation of property rights between owners. The term "owner" therefore had to be read in these terms.

He argued that the Act provided a building owner with rights that would not necessarily be available at common law. In so doing it also removed an adjoining owner's rights to certain common law remedies (for example, in trespass and nuisance) for the interferences sanctioned by these rights.

The legislative purpose of the service of notices, and the subsequent appointment of surveyors, was therefore to protect an adjoining owner's

---

[2] A similar provision now appears in section 20 of the 1996 Act.

property rights which had been superseded by the provisions of the statute. The Act was therefore limited to those who hold a legal interest in land of a greater duration than a yearly tenancy. Hence the excluded categories of monthly or weekly tenancies and tenancies at will.

The court agreed with this statement of the law and distinguished *Brown* from the present case, holding that the Wassefs did not fall within the definition of "owner". Judge Crawford Lindsay QC:

> "I have reached the conclusion that *Brown* can be distinguished upon the grounds relied on by [the School's counsel]. I also agree with his submission that the definition of owner in the 1930 Act is limited to those who hold legal interests in land of a duration greater than one year. Accordingly, I conclude that Mr and Mrs Wassef are not adjoining owners for the purposes of section 5 of the 1930 Act."

Effect of estoppel on procedural irregularities under the Act

The court then considered the role of estoppel by convention. This arises where parties make arrangements with each other on the basis of a shared misunderstanding, and where it would then be "unjust or unconscionable" for one of the parties to resile from these arrangements. In such circumstances the parties are estopped (prevented) from denying the existence of the situation forming the basis of the misunderstanding.

As both parties had proceeded on the understanding that the Wassefs were owners under the 1939 Act, their counsel now argued that the School was estopped from denying that this was not, in fact, the situation.

The court noted that estoppel by convention must be based on an assumption which had been agreed between the parties. The agreement could however be inferred from conduct, or even from silence[3]. Such an assumption clearly existed in this case. The court also noted that the estoppel need not be confined to agreed assumptions of fact but could also include those relating to law[4]. This included the parties' shared assumption that the Wassefs fell within the definition of "owner" in the 1939 Act.

Now that this understanding had been shown to be erroneous the court had to consider whether it would be "unjust or unconscionable" to allow the School to withdraw from the arrangements that they had entered into as a consequence of this. The court decided that it would indeed be "unconscionable" and was therefore unwilling to hold that the *ex parte* award was invalid simply on the basis that the Wassefs were not an "owner" under the Act.

---

[3] *India (Republic of) v India Steamship Co. Ltd (The Indian Endurance) (No. 2)* (1996).
[4] *Hiscox v Outhwaite (No. 1)* (1991).

> # London Building Act 1930
>
> PART I: INTRODUCTORY
>
> ## 5. Definitions
>
> In this Act, save as is otherwise expressly provided therein and unless the context otherwise requires, the following expressions have the meanings hereby respectively assigned to them (that is to say):
>
> "owner" includes every person in possession or receipt either of the whole or of any part of the rents or profits of any land or tenement, or in the occupation of any land or tenement, otherwise than as a tenant from year to year, or for any less term, or as a tenant at will.
>
> # London Building Acts (Amendment) Act 1939
>
> PART VI: RIGHTS ETC. OF BUILDING AND ADJOINING OWNERS
>
> DIFFERENCES BETWEEN OWNERS
>
> ## 55 Settlement of differences
>
> Where a difference arises or is deemed to have arisen between a building owner and an adjoining owner in respect of any matter connected with any work to which this Part of this Act relates the following provisions shall have effect:
>
> (e) If [a party-appointed surveyor] refuses or for ten days after a written request by either party neglects to act the surveyor of the other party may proceed ex parte and anything done by him shall be effectual as if he had been an agreed surveyor.

Whilst the estoppel would not apply to future dealings between the parties after the common assumption had been revealed to be erroneous, this was not the situation in the present case. The earlier awards addressed certain issues and reserved the possibility of further awards being made. The *ex parte* award addressed only the outstanding issues from these earlier awards and these had been entered into before the misunderstanding was discovered to be erroneous. The *ex parte* award was not therefore invalid on the basis that the Wassefs did not fall within the definition of "owner".

Surveyor's jurisdiction to make *ex parte* awards

A party-appointed surveyor has authority to make an *ex parte* award either where his opposite number has refused to act, or where he or she has neglected to do so for ten days after being requested so to do[5]. In view of the importance of the legal function performed by surveyors, the court placed great emphasis on WS's failure to clearly articulate the basis on which he claimed to be entitled to act *ex parte*.

The court considered that there was ambiguity as to whether he had relied on the other surveyor's refusal to act or his neglect to do so. The sending of a ten-day letter requesting SS to act, suggested that he relied on neglect. On the other hand, the award and its covering letter made reference to a failure to act.

The court found no evidence that SS had either refused or neglected to act. He had continued in correspondence throughout so there was no question of any neglect. The Wassefs' argument that the response to the ten-day letter was "deliberately ineffective" and therefore constituted a refusal to act could not be supported. It simply reflected the impasse that had been reached in the negotiations between the two surveyors.

The court's findings, which include some useful advice for surveyors, are summarised by Judge Crawford Lindsay QC:

> "I agree with [the School's counsel] that the Act has drastic consequences. The surveyor who proceeds *ex parte* can make an award that could lead to court proceedings. I conclude that any surveyor who wishes to avail himself of the provisions of section 55 must comply strictly with the provisions of the Act.
>
> This means that the surveyor can rely either upon a refusal, or upon a notice that complies with the provisions of the Act, or, where appropriate, upon both grounds. The relevant grounds must be expressed accurately in the *ex parte* award.
>
> In this case, there was no reference [in the *ex parte* award] to a neglect to act by [SS]. Accordingly, the *ex parte* award is inconsistent with the reference to the ten-day time limit in the [ten-day letter].
>
> The award accordingly refers to a ground, namely refusal, upon which [WS] did not rely. It does not refer to the ground upon which he purported to rely. In those circumstances, I consider that the *ex parte* award is bad on its face and invalid.

---

[5] 1939 Act, section 55(e); 1996 Act, sections 10(6) and 10(7).

Further, there is a factual difficulty, because there is no evidence to support either a refusal or a neglect to act by [SS] during the ten-day period. In these circumstances I find for the [School on this issue]."

The decision

But for the question of estoppel, the court would have found the award invalid for want of jurisdiction as the Wassefs did not constitute an "owner" within the 1939 Act. However, the court was prevented from drawing this conclusion because of its finding on the estoppel issue.

The court nevertheless held that the award was invalid on other grounds. As SS had not refused or neglected to act, WS had no jurisdiction to make an *ex parte* award and the award was therefore invalid on this basis.

However, the earlier awards remained effective (by virtue of the estoppel). The practical effects of this were explained by Judge Crawford Lindsay QC:

"... although I have ruled that the *ex parte* award is invalid, it nevertheless sets out in detailed form the remaining issues between the parties. I consider that, in relation to those remaining issues, the surveyors should continue to negotiate, and Mr and Mrs Wassef should be treated as adjoining owners in respect of those outstanding disputes."

FREDERICK BETTS LTD v PICKFORDS LTD (1906)
Chancery Division

In brief

- **Issues:** 1894 Act – incorporation of projecting fabric into building owner's building – type 'b' party wall – right to light – derogation from grant – scope of architect's authority – adjoining owner enclosing on external wall – trespass.

- **Facts:** Projecting fabric from Pickfords' building was incorporated into a new building on Betts' land. The external wall of this building thus became a type 'b' party wall and no window openings were therefore permitted within it. Pickfords had also enclosed on another portion of the building's external wall without consent. Betts sought the removal of all encroachments by Pickfords from their building.

- **Decision:** Betts' land had been demised to them by Pickfords for the erection of their building. Pickfords' failure to remove the projecting fabric deprived Betts of a right to light within the lease and was therefore a derogation from the grant. It was also a trespass. It was irrelevant that Betts' architect had agreed to the fabric remaining on their land as he had no authority to do so. Pickfords had also committed a trespass by enclosing on Betts' building. Pickfords were therefore required to remove all encroachments and an injunction was granted to this effect.

- **Notes:** See also: *Weston v Arnold* (1873); *Drury v Army & Navy Auxiliary Co-operative Supply Ltd* (1896); *London, Gloucester & North Hants Dairy Company v Morley & Lanceley* (1911).

The facts

Pickfords, the well-known removal company, owned a site in Long Lane, Bermondsey, which they used as a depot. A number of buildings had been erected on the site, including a large single-storey covered yard area or "cart shed" (Figure 8).

In October 1903 they granted a lease of the northern end of their site to Betts, a firm of builders' merchants. They retained the southern portion for their own use. Betts covenanted to erect a two-storey warehouse building on the demised land, in accordance with plans which Pickfords had already approved. These showed three windows in the southern wall of the warehouse, at first floor level, which overlooked Pickfords' retained land.

The parties also entered into a collateral contract[1] in which Pickfords agreed to clear the demised land of buildings. This included a requirement that they demolish those parts of the cart shed which were situated on the demised land.

Following the grant of the lease Pickfords demolished the northern end of the cart shed. However, in so doing, they left two iron stanchions, which had previously supported the roof, on the demised land. A number of roof timbers also projected over the boundary line on to this land. In total, the stanchions and timbers encroached on to Betts' land by approximately 3 in.

This encroachment subsequently presented difficulties for the contractor during the construction of Betts' new warehouse. These were resolved on site by the parties' architects who verbally agreed that the projecting roof beams and stanchions should be built into the south wall of the warehouse. Construction of the south wall proceeded on this basis although this was never brought to Betts' attention.

Service of notice by District Surveyor

On completion of the warehouse the District Surveyor served an enforcement notice on Betts under the 1894 Act. He alleged that the presence of window openings in the south wall placed it in breach of the statute's building control provisions. Because the projecting parts of Pickfords' cart shed had been incorporated into the wall, this had become a party wall within the meaning of the Act[2]. Under the Act, window openings were not permitted in party walls[3] so the District Surveyor required the three offending windows to be bricked up.

There was no doubt that the window openings were in breach of the Act. *Drury v Army & Navy Auxiliary Co-operative Supply Ltd* (1896) had established that a wall in single ownership could only be a party wall up to the level where it ceased to separate adjoining buildings. However, the same case had also confirmed that the construction requirements for party walls must continue to be observed for 3 ft above this level. Although, therefore, the south wall ceased to separate the two buildings at first floor level, window openings were not permitted within 3 ft above this[4]. The three window openings fell within 3 ft of the cart shed roof and were therefore in breach of the Act (Figure 8).

---

[1] A contract which is subsidiary to the main contract which induces a party to enter into the main contract.
[2] Because it had become "a wall forming part of a building and used ... for separation of adjoining buildings" within section 5(16) of the Act.
[3] Sections 54(3) and 77(3).
[4] Section 59(1).

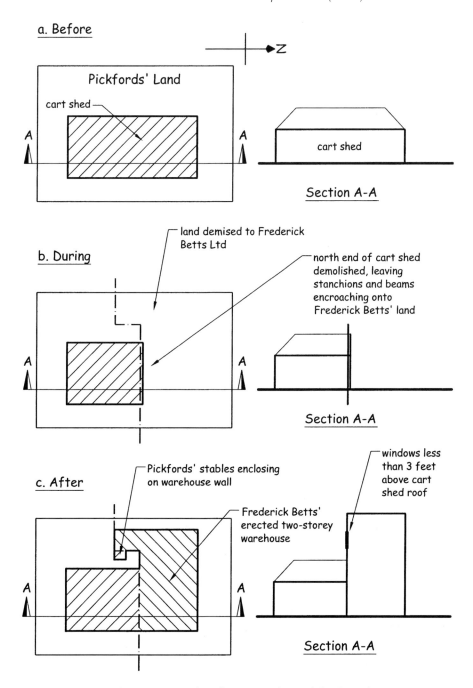

Figure 8: Frederick Betts Ltd v Pickfords Ltd

Betts made no challenge to the District Surveyor's findings. Instead they sought to prevent Pickfords from using the south wall as a party wall. By returning the wall to its intended use as an external wall of the warehouse there would be no impediment on the window openings remaining within it.

They therefore applied for an injunction, requiring Pickfords to remove the projecting parts of their building from the wall and to cease to use it, in any way, as a party wall. They based their claim on Pickfords' trespass and on derogation from their grant in the lease. The court gave detailed consideration to these issues.

Derogation from grant of right to light

Betts alleged that, in preventing them from using the windows as envisaged by the lease, Pickfords had derogated from the grant of the property which was contained in that lease. It is well established that where a landlord leases part of his property for a particular purpose, he cannot then use the remaining part in a way which prevents the leased portion being used for its intended purpose[5]. The court was satisfied that this principle applied to the present case.

The windows in the south wall were an essential part of the warehouse design which Pickfords had approved. The lease therefore impliedly granted an easement of light in respect of them. If Pickfords carried out some unlawful act on their retained land which prevented the use of the windows, this would amount to a derogation of this grant.

However, Pickfords raised three defences to these charges which had to be addressed by the court before it could determine any liability on their part.

Circumstances surrounding implied grant of easements

Firstly, Pickfords had cited *Birmingham, Dudley & District Banking Co v Ross* (1888) in support of the view that an implied easement must be related to what was actually contemplated by the parties and that this will depend on the surrounding circumstances.

They argued that Betts must have contemplated that the south wall would become a party wall when they entered into the lease. Because of its location on the plan it must have been obvious that it would be the boundary wall between the two properties and that, in all probability, because of the proximity of Pickfords' building, it would become a party wall.

---

[5] *Aldin v Latimer, Clark, Muirhead & Co* (1894).

On this basis, Betts' easement of light could only ever have been granted on the understanding that it was liable to being blocked up for non-compliance with the Act. As this is precisely what had happened there could be no derogation from grant on account of it.

The court acknowledged the principle laid down in the *Birmingham* case but felt that its application was more limited than was being suggested by Pickfords. On the facts it did not accept that Betts would have contemplated that their wall would become a party wall. Whilst they might have contemplated that Pickfords would rebuild their adjacent building it was not inevitable that the two buildings would share a single wall.

Indeed, even if they did contemplate the possibility of a shared wall, it was extremely unlikely that they would also contemplate the particular risk of the windows at first floor level being bricked up because of some breach of a particular provision in the 1894 Act. In effect, the court felt that the situation was too remote to have been in the parties' contemplation when the lease was entered into and therefore rejected Pickfords' first defence.

Acts capable of constituting a trespass/derogation from grant

Secondly, Pickfords contended that they had not personally carried out any acts which could be regarded as either a trespass or a derogation from their grant. In particular, they were not responsible for building the stanchions or roof beams into Betts' wall.

The court did not accept this argument. It found, as a matter of fact, that Pickfords were occupying Betts' property with their stanchions and roof beams and that they had no authority so to do. That was as much a trespass as if they had physically brought these items on to Betts' land and built them into the wall themselves. Pickfords' second defence therefore also failed.

Scope of architect's authority

Finally, Pickfords referred to the agreement reached between the two architects on site. They argued that Betts, acting through their architect as agent, had thereby given consent to the stanchions and beams being placed on their land. As these items were there by consent there could be no trespass or derogation from grant.

Betts denied that the agreement was within the scope of their architect's authority. He had been employed to prepare the plans and to supervise the works according to those plans. He had no authority to agree changes to those plans or to sanction unlawful invasions of their property.

The court considered the scope of the architect's authority. Its findings will be of general interest to those who undertake contract administration duties as well as those confronted by design changes in party wall matters.

> Kekewich J. "What is within the scope of an architect's authority I am not called upon to say; but I will say this much, that I apprehend his authority is strictly limited by the terms of his employment.
>
> So far as I know, in this case the architect was simply employed to superintend the erection of the buildings. That, I take it, he has done, but that cannot possibly give him any implied authority to allow these stanchions and roof beams to be built into the wall. He could not even have allowed, according to the constitution of his employment, the slightest deviation from the plans. His only duty was to superintend the erection according to the plans which had been approved.
>
> Of course, we know that in many cases the architect has certain powers given to him as to verifying the builder's work and certifying the result, and you may find that all that comes within his authority; and indeed there may be, probably are, many details which would be implied in those provisions; but beyond that it is impossible to say that an architect has any more authority than any other person who is employed to superintend work. His duty, to sum it up in a word, is to superintend and nothing more."

As the court had rejected each of Pickfords' three defences it found them liable for trespass and derogation from grant in respect of the encroaching stanchions and roof timbers.

Trespass to an external wall

The court also considered one final issue, which is also of general application in party wall matters. Because a portion of the demised land lapped round the western side of the retained land, part of the warehouse's east wall also formed the boundary wall between the two pieces of land. Subsequent to the issue of proceedings Betts discovered that, at basement level, Pickfords had enclosed on this wall by building stables against it. The court had no doubt that this also constituted a trespass.

> Kekewich J. "What right had the defendants to use that wall as a party wall? It is the external wall built by the plaintiffs on their own ground. If the defendants desire to use the adjoining premises for stables or for any other purpose they can erect their own containing wall; they have no business to use the wall of the plaintiffs, and it seems to me that they are doing that, and that there again they are using as a party wall what they have no business to use as a party wall."

Thus, although the legislation makes provision for party walls based on separation of adjoining buildings in addition to those based on shared ownership, this must generally arise by consent. The court's decision in this case confirms the notion that an unauthorised interference with the far side of a boundary wall, for example by enclosing upon it, will constitute a trespass.

The decision

The court therefore granted an injunction restraining Pickfords from committing any trespass on the walls of the warehouse, or from using any of the walls as party walls. They were also ordered to remove the stanchions and roof timbers from the demised land.

London Building Act 1894

PART I: INTRODUCTORY

5. Definitions

In this Act unless the context otherwise requires:

(16) The expression "party wall" means:

(a) A wall forming part of a building and used or constructed to be used for separation of adjoining buildings belonging to different owners or occupied or constructed or adapted to be occupied by different persons; or

(b) A wall forming part of a building and standing to a greater extent than the projection of the footings on the lands of different owners.

PART VI: CONSTRUCTION OF BUILDINGS

54. Rules as to recesses and openings

(3) An opening shall not be made in any party wall except in accordance with the provisions of this Act in relation thereto.

59. Height of party walls above roof

(1) Every party wall shall be carried up of a thickness in a building of the warehouse class equal to the thickness of such wall in the topmost storey and in any other building of eight and a half inches above the roof flat or gutter of the highest building adjoining thereto to such height as will give a distance (in a building of the warehouse class exceeding thirty feet in height) of at least three feet and (in any other building) of fifteen inches measured at right angles to the slope of the roof or fifteen inches above the highest part of any flat or gutter as the case may be.

77. Rules as to uniting buildings

(3) An opening shall not be made in any party wall or in two external walls dividing buildings which if taken together would extend to more than two hundred and fifty thousand cubic feet except under the following conditions:

- (c) Such openings shall have the floor jambs and head formed of brick stone or iron and be closed by two wrought iron doors each one fourth of an inch thick in the panel at a distance from each other of the full thickness of the wall fitted to rebated frames without woodwork of any kind or by wrought iron sliding doors or shutters properly constructed fitted into grooved or rebated iron frames.

GYLE-THOMPSON AND OTHERS v WALL STREET (PROPERTIES) LTD (1974)

Chancery Division

In brief

- **Issues:** 1939 Act – reduction in height of party fence wall – service of notices – appointment of surveyors – delivery of award – challenge to invalid award after expiry of 14-day appeal period.

- **Facts:** Wall Street wished to reduce the height of a party fence wall which separated their land from the plaintiffs' land. Party structure notices were served on a surveyor who had been retained to advise the plaintiffs. This surveyor later joined in the selection of a third surveyor. An award was signed jointly by the third surveyor and Wall Street's surveyor which authorised the works. This was posted to the plaintiffs' retained surveyor and Wall Street then started the works.

- **Decision:** There was no right to reduce the height of a party wall or party fence wall under the 1939 Act. The surveyors therefore had no jurisdiction to make an award which purported to authorise this. The award was further invalidated by procedural irregularities. The plaintiffs' surveyor had no authority to accept service of the notices or of the award so service of both documents was ineffective. Furthermore, this surveyor had never been formally appointed and therefore had no authority to join in the selection of the third surveyor. Consequently the third surveyor had no jurisdiction to join in the making of an award. The award was therefore invalid and the plaintiffs were entitled to challenge this, notwithstanding the expiry of the statutory 14-day appeal period. The court therefore granted an interlocutory injunction to restrain the works.

- **Notes:** (1) See also: *Burlington Property Co Ltd v Odeon Theatres Ltd* (1938); *Re Stone and Hastie* (1903); *Riley Gowler Ltd v National Heart Hospital Board of Governors* (1969). (2) Contrast with: *Whitefleet Properties Ltd v St Pancras Building Society* (1956). (3) Section 2(2)(m) of the 1996 Act now grants an express right to reduce the height of a party wall/party fence wall.

The facts

Wall Street (Properties) Ltd owned an old four-storey warehouse at 57–63 Old Church Street, Chelsea, London SW3. The western wall of the warehouse also formed the rear garden walls of some houses at 43–53

Paulton's Square. Gyle-Thompson and the two other plaintiffs in the case were the owners of three of these houses.

Wall Street planned to redevelop their property by demolishing the existing warehouse building and erecting houses, flats and an office building on the site. They intended to reduce the height of the western wall to 15 ft and for this to be retained as the boundary wall between the new development and the gardens of the houses in Paulton's Square.

The warehouse was substantially demolished, apart from the western wall. However, before this could be reduced in height, a plaque was discovered which indicated that it had been constructed astride the boundary line. Under the 1939 Act[1] the wall had therefore been a party wall and, following the demolition of the rest of the building, was now a party fence wall.

Appointment of surveyors and interim award

In December 1971 Wall Street formally appointed a surveyor (surveyor A) to act for them under the 1939 Act. Surveyor A then opened negotiations with the adjoining owners in Paulton's Square. In January 1972 three of them (the plaintiffs in this case) retained another surveyor (surveyor B), to represent their interests (although, at this stage, they did not formally appoint him under the Act).

On 22 February 1972 surveyor A served a party structure notice on all the owners of numbers 43–53 Paulton's Square in his capacity as Wall Street's agent. This notified them of Wall Street's intention to reduce the height of the western wall and to connect it to the new premises to be built on the site of the warehouse.

Surveyor B responded to the notices on behalf of his three clients. He refused to consent to the wall being reduced in height or to sign any award which would sanction the work. He maintained that the 1939 Act provided no such right and that the work would therefore be a trespass on his clients' property.

Nevertheless, as there was now a risk that the unsupported wall would collapse, the two surveyors signed an interim award on 20 June 1972. This permitted Wall Street to shore the wall but was expressed to be without prejudice to any future negotiations about the possibility of reducing its height.

The award also recorded the selection of a third surveyor and required Wall Street to pay the fees of surveyor B. Between July and September 1972 each of the three plaintiffs then signed a formal surveyor's

---

[1] The situation would have been the same under the 1996 Act. See the definitions of "party wall" and "party fence wall" in section 20 of that Act.

appointment under the 1939 Act in favour of surveyor B. This was intended to regularise the paperwork prior to payment of surveyors' fees under the award.

Role of third surveyor

It proved impossible to reach agreement regarding the reduction in the height of the wall so, in December 1972, surveyor A referred the matter to the third surveyor. The third surveyor advised the two surveyors that they no longer had any power to act under the notice. More than six months had elapsed since the date of service of the notice and this had therefore now expired[2].

By this stage Wall Street's architect had designed what he regarded as a compromise solution for the wall. This involved reducing the height of the wall to 19 ft rather than 15 ft and incorporating obscure glazing in its top half to maintain the level of illumination to the new development. Surveyor A therefore prepared new party structure notices on the basis of these revised proposals.

These were posted to surveyor B on 20 December 1972. He again refused to consent to the proposals or to sign any award which would authorise the work. His clients opposed any reduction in the height of the wall and he continued to maintain that the Act contained no right to reduce the height of a party wall.

Surveyor A therefore again referred the matter to the third surveyor. Surveyor A and the third surveyor then together published an award[3] which sanctioned the reduction in height of the wall on the basis of the revised proposals. This was posted to surveyor B on 2 March 1973.

On the morning of Saturday 17 March 1973 (as soon as the statutory 14-day appeal period had expired) Wall Street started demolishing the wall. The plaintiffs intervened personally and, with the help of the police, stopped the work progressing.

---

[2] 1939 Act, section 47(3). A similar provision now appears in section 3(2) of the 1996 Act although the six-month period has been extended to 12 months. The rewording of the new section has also unfortunately changed the sense of the provision. According to a strict reading of the new section both a 12-month delay and a failure to prosecute with due diligence are required before the notice becomes ineffective. It seems likely this is an error of draughtsmanship and that a failure to prosecute with due diligence, even where works have begun within the 12 months, will amount to the notice ceasing to have effect.

[3] Any of the three surveyors have jurisdiction to publish an award: 1939 Act, section 55(i); 1996 Act, section 10(10). However, the third surveyor should only join with one of the other surveyors to publish an award in the absence of disagreement between the two party appointed surveyors (for example, where matters have already been agreed but where one of the surveyors is unavailable to sign the final document). Where, as in the present case, there is disagreement between the two surveyors, the third surveyor alone should make the award: 1939 Act, section 55(j); 1996 Act, section 10(11).

On the following Monday the plaintiffs commenced proceedings for trespass and applied for an interlocutory injunction to restrain the work, pending a final hearing of the matter. The court considered the following issues in determining the application for interlocutory relief.

A right to reduce the height of a party (fence) wall?[4]

The plaintiffs based their application for relief on the ground that Wall Street had no right to reduce the height of a wall which they partly owned. Wall Street's proposals involved the demolition of the old western wall and its rebuilding to a reduced height (in effect, a reduction in the height of the wall).

Wall Street maintained that they had a right to do the work under the 1939 Act. Although there was no express right to reduce the height of a party wall or party fence wall, they argued that there was an implied right. This right was implicit in sections 46(1)(a) and (k)[5] which granted building owners various rights to demolish and rebuild party structures and party fence walls and to raise party fence walls and use them as party walls.

The court did not agree and concluded that in the absence of any express right within the legislation to reduce the height of a party fence wall the rights "to demolish and rebuild" required the wall to be reconstructed to the same height as previously. This decision was based on two considerations.

Firstly, the Court of Appeal decision in *Burlington Property Co Ltd v Odeon Theatres Ltd* (1938) supported this interpretation. In that case the surveyors' award had provided for a party wall to be demolished and rebuilt with enlarged openings, in place of windows, so as to facilitate access. The Court of Appeal held that the building owner had no right, in exercising the various rights under the legislation[6], to change the form of the party wall.

Secondly, this decision was simply an example of the general principle of statutory interpretation that, in the absence of a clear indication to the contrary, statutes are presumed not to expropriate private property.

---

[4] Section 2(2)(m) of the 1996 Act now includes an express right to reduce the height of a party wall or a party fence wall and this right to demolish and rebuild to a reduced height.
[5] See sections 2(2)(b) and (l) of the 1996 Act for equivalent provisions.
[6] In this case the right was being claimed under section 114(7) of the London Building Act 1930 which then appeared in substantially the same form as section 46(1)(f) of the 1939 Act and now, section 2(2)(e) of the 1996 Act.

Brightman J: "In my judgment the submission of counsel for the plaintiffs on this issue is correct. In the absence of an express right to lower a party fence wall it seems to me that the right to demolish and rebuild requires reconstruction to the same height. I think that this interpretation is in conformity with the decision in the *Burlington Property* case.

Further, if the building owner has the right to reduce the height of a wall which belongs to him and his neighbour without his neighbour's consent, such a right would be, in effect, a right for the building owner to expropriate the property of his neighbour. I would expect any such right of expropriation in an Act of Parliament to be conferred in clear terms, if it is conferred at all."

Brightman J also considered whether surveyors had a more general power to confer rights to undertake particular works, for example to reduce the height of a party wall, on a building owner. In this context he questioned the meaning of section 55(k)[7] which provided that the surveyors' award "may determine the right to execute … any work".

He was of the opinion that it did not confer a discretionary power on surveyors to confer rights which had not been anticipated by the Act. Rather it was a power to determine whether a particular fact or set of circumstances existed which were conditions precedent for the existence of various conditional rights conferred by the legislation:

"If it is asked what 'right' is within the contemplation of section 55(k) as appropriate to be determined by an award, an example applicable to section 46(1)(a)[8] would be the determination by the surveyors of the 'necessity' of the intended work on account of defect or want of repair; in the absence of such necessity the 'right' under that paragraph (for example) to underpin would not exist. In fact many of the 'rights' conferred by section 46(1) are conditional rights which are only exercisable on proof of some fact appropriate to be determined by the surveyors in their award. That, in my judgment, is the context in which section 55(k) enacts that the award may determine the 'right' to execute works."

There was therefore no right within the 1939 Act for a building owner to reduce the height of a party wall or party fence wall or to demolish one and rebuild it to a reduced height. The surveyors therefore had no power to make an award which purported to grant such a right.

---

[7] This provision now appears in substantially the same form in section 10(12) of the 1996 Act.
[8] The same provision is substantially repeated as section 2(2)(b) of the 1996 Act.

Challenges to awards after 14 days?

Wall Street argued that the award could no longer be challenged as the plaintiffs had failed to appeal to the county court within the statutory 14-day appeal period[9]. Section 55(m) provided that, apart from this right of appeal, surveyors' awards were conclusive and could not be challenged in any court[10]. The present challenge to the award in the High Court should therefore be rejected on this basis.

The court rejected this argument in situations where there was some legal impediment to the validity of the award:

> Brightman J: "In my judgement this submission is not correct in relation to an award which is ultra vires and therefore not a valid award. In the present case the defendants claimed a right which, in my judgement, they did not have, namely, a right to reduce the height of a party fence wall, and the two surveyors made an award which, in my judgement, they had no power to make. In my view the plaintiffs are entitled, in those circumstances, to come to this court to prevent a wrongful interference with their property."

Re Stone and Hastie (1903) was cited as authority for this view. In that case the surveyors awarded the adjoining owner a sum of money as payment for the building owner's extra use of the party wall. When the adjoining owner brought proceedings for the recovery of this sum the building owner asserted in defence that the relevant part of the award was ultra vires and therefore void. The adjoining owner submitted that it was not open to the building owner to raise this point as he had made no appeal to the county court within the 14-day period and that the award was therefore now conclusive and beyond challenge by the courts. The Court of Appeal rejected this submission and held that this part of the award was indeed void.

The significance of this decision was summarised by Brightman J in the following terms:

> "There are points of distinction between *Re Stone and Hastie* and the present case, but I do not think they are significant. The important point is that the building owner was not required to resort to the county court in order to free himself from an obligation imposed by the surveyors in excess of their jurisdiction."

---

[9] Under section 55(n) of the 1996 Act. The same right of appeal now appears in section 10(17) of the 1996 Act.
[10] The same provisions now appears as section 10(16) of the 1996 Act.

Effect of procedural irregularities?

The two issues considered above were sufficient to dispose of the case. However, the plaintiffs had also raised a number of procedural objections about the circumstances surrounding the making of the award. Whilst a court would not generally be sympathetic to procedural objections they should be taken more seriously in cases under this legislation because of its effect on property rights:

> Brightman J: "Section 46 *et seq* of the 1939 Act give a building owner a statutory right to interfere with the proprietary rights of the adjoining owner without his consent and despite his protests.
>
> The position of the adjoining owner, whose proprietary rights are being compulsorily affected, is intended to be safeguarded by the surveyors appointed pursuant to the procedure laid down by the Act. Those surveyors are in a quasi-judicial position with statutory powers and responsibilities. It therefore seems to me important that the steps laid down by the Act should be scrupulously followed throughout, and short cuts are not desirable.
>
> Having regard to the functions of surveyors under section 55 and their power to impose solutions of building problems on non-assenting parties, the approach of surveyors to those requirements ought not to be casual."

The following three procedural objections were identified. This meant that (irrespective of the award's substantial invalidity referred to above) the resulting award was procedurally invalid and Wall Street could not therefore rely on it.

Service of notices

Firstly, the party structure notices, which underpinned the surveyors' jurisdiction to make the award, had not been validly served. On 20 December 1972 these had been served on surveyor B rather than on the individual owners. According to the evidence he had no authority to accept service. The resulting award was therefore invalid.

Appointment of surveyors

Secondly, surveyor B had never been validly appointed under the Act. Section 55(h)[11] provided that all appointments must be in writing. Apart from the retrospective appointments during July and September 1972 in

---

[11] This requirement now appears in section 10(2) of the 1996 Act.

relation to the notices served on 23 February 1972 no such appointments had been made. These appointments were not effective in relation to the second set of notices purportedly served on 20 December 1972:

> Brightman J: "The validity of his appointment was vital to the validity of the award. If he were not validly appointed, he had no statutory authority to concur in the selection of ... the third surveyor. If [the third surveyor] were not validly selected ... then the award must be void."

The award was therefore invalid on this ground also.

Brightman J had the following advice for surveyors at the time of their original appointment:

> "It would be a wise precaution for the surveyor of the building owner and the surveyor of the adjoining owner to inspect each other's written appointment before they perform their statutory functions. Neither of them has power to concur in an award unless both of them have been duly appointed ...
>
> ... It would be a wise precaution for the third surveyor, on accepting office, to inspect the written appointments of those selecting him; unless they have been duly appointed, they have no power to select a third surveyor; if the third surveyor has not been validly selected in writing, he has no power to concur in an award."

Delivery of award

Thirdly, even had the award been valid, Wall Street would have had no authority to commence work on 17 March 1973, as the award had not, at that stage, been delivered to the plaintiffs[12].

Again, Brightman J's judgment contains some practical advice for surveyors:

> "It seems to me highly desirable that surveyors, who are performing their statutory duties under section 55, should bear in mind that important matters may turn on the date of the delivery of their award and I think they should take practical steps to ensure that there is no doubt what is the date of such delivery, and that the date is the same for the building owner as for the adjoining owner.

---

[12] The 1939 Act contained no clear requirements regarding delivery of the award. However, as the plaintiffs had neither received the award, nor been made aware of its existence before the work started, the court had no difficulty in deciding that the works had been started without authority. Once the award has been made, section 10(14) of the 1996 Act now requires the surveyors to serve it forthwith on the parties.

The problems which can arise if such practical steps are neglected are exemplified in *Riley Gowler Ltd v National Heart Hospital Board of Governors* (1969), to which I was referred in argument."

The decision

For the reasons discussed above, the surveyors' award was substantively and procedurally invalid. Wall Street therefore had no right to undertake the work which it purported to authorise. The court therefore granted the interlocutory injunction restraining them from continuing with the work.

London Building Acts (Amendment) Act 1939

PART VI: RIGHTS ETC. OF BUILDING AND ADJOINING OWNERS

RIGHTS ETC. OF OWNERS

46. Rights of owners of adjoining lands where junction line built on

(1) Where lands of different owners adjoin and at the line of junction the said lands are built on or a boundary wall being a party fence wall or the external wall of a building has been erected the building owner shall have the following rights:

(a) A right to make good underpin thicken or repair or demolish and rebuild a party structure or party fence wall in any case where such work is necessary on account of defect or want of repair of the party structure or party fence wall.

(k) A right to raise a party fence wall to raise and use as a party wall a fence wall or to demolish a party fence wall and rebuild it as a party fence wall or as a party wall.

47. Party structure notices

(3) A party structure notice shall not be effective unless the work to which the notice relates is begun within six months after the notice has been served and is prosecuted with due diligence.

DIFFERENCES BETWEEN OWNERS

55. Settlement of differences

Where a difference arises or is deemed to have arisen between a building owner and an adjoining owner in respect of any matter connected with any work to which this Part of this Act relates the following provisions shall have effect:

(a) Either:

(i) both parties shall concur in the appointment of one surveyor (in this section referred to as an "agreed surveyor"); or
(ii) each party shall appoint a surveyor and the two surveyors so appointed shall select a third surveyor (all of whom are in this section together referred to as "the three surveyors");

(h) All appointments and selections made under this section shall be in writing;

(i) The agreed surveyor or as the case may be the three surveyors or any two of them shall settle by award any matter which before the commencement of any work to which a notice under this Part of this Act relates or from time to time during the continuance of such work may be in dispute between the building owner and the adjoining owner;

(j) If no two of the three surveyors are in agreement the third surveyor selected in pursuance of this section shall make the award within 14 days after he is called upon to do so;

(k) The award may determine the right to execute and the time and manner of executing and work and generally any other matter arising out of or incidental to the difference;

(m) The award shall be conclusive and shall not except as provided by this section be questioned in any court;

(n) Either of the parties to the difference may within 14 days after the delivery of an award made under this section appeal to the county court against the award and the following provisions shall have effect:

(i) Subject as hereafter in this paragraph provided the county court may rescind the award or modify it in such manner and make such order as to costs as it thinks fit.

HOBBS, HART & Co v GROVER (1899)
Court of Appeal

In brief

- **Issues:** 1894 Act – party structure notice – required contents – meaning of "nature and particulars of the proposed work".
- **Facts:** Hobbs served a very general party structure notice on Grover which listed all the building owner's rights to undertake works to a party wall.
- **Decision:** The notice was invalid. Notices should contain sufficient information to enable recipients to understand what is being proposed and to enable them to decide how best to respond.
- **Notes:** See also: *Spiers & Son Ltd v Troup* (1915).

The facts

Hobbs owned 76 Cheapside in the City of London. Grover owned the adjoining building at number 75. The two properties were separated by a party wall.

Grover intended to demolish and rebuild his property and he therefore served a party structure notice on Hobbs' under the 1894 Act. The notice stated that he intended "to execute such of the following works to the said party structure as may on survey be found necessary or desirable". It then simply listed all the various rights which were granted to a building owner under section 88[1].

Hobbs claimed that the notice was invalid. They maintained that it failed to state "the nature and particulars of the proposed work" as expressly required by section 90(1) of the Act[2]. They therefore applied for an injunction to restrain Grover from interfering with the party wall.

Contents of party structure notice

Grover argued that it was not possible to give precise details about the nature of the work to be undertaken until further information was available regarding the condition of the wall. If notice could only be served

---

[1] Broadly equivalent to the rights listed in section 2(2) of the 1996 Act.
[2] The same requirement is now contained in section 3(1)(b) of the 1996 Act.

once full information was available it might not be possible to give the notice until the wall had been pulled down and then the site would have to be left vacant for a further two months following service of another notice. He argued that this could not have been the intention of the Act.

At first instance the judge was not satisfied that the notice was bad but felt that further particulars should be provided by Grover before he started raising the party wall. An order was therefore made in these terms. Hobbs' appealed to the Court of Appeal.

The court was not satisfied that the notice was valid and statements in the various judgments reflect this:

> Chitty LJ: "Should not the notice be such as will enable the adjoining owner to judge whether he shall consent or object to the proposed works?"

> Vaughan Williams LJ: "Ought not the notice to give such particulars of the proposed works as will enable the adjoining owners to judge whether it will be necessary to pull down the old wall?"

> Lindley MR: "How can the notice be sufficient unless it enables the adjoining owner to see what counter-notice he should give under section 89? Is it sufficient if the notice is expressed in such general terms that the adjoining owner cannot tell what works he should require to be executed under section 89?"

The decision

On the basis of these statements Grover gave an undertaking to amend his notice to include details of the kind and extent of the buildings he proposed to erect. The court made a consent order staying all further proceedings.

The view of the court was summarised by Lindley MR:

> "In my opinion the notice ought to be so clear and intelligible that the adjoining owner may be able to see what counter-notice he should give to the building owner under section 89. This is the key to the whole matter."

HOLBECK HALL HOTEL LTD v SCARBOROUGH BOROUGH COUNCIL (2000)

Court of Appeal

In brief

- **Issues:** Natural rights of support – degradation of servient land through natural causes – landslip causing damage to adjacent building – whether a duty to maintain support from servient land – measured duty of care – foreseeability of damage – misfeasance and non-feasance – latent and patent defects.

- **Facts:** The council owned land between the claimants' hotel and the sea. A series of landslips occurred over a number of years but the council took inadequate steps to prevent these reoccurring. In particular it failed to undertake geological investigations which would have revealed the risk of a sudden catastrophic collapse which would affect the claimants' hotel. A sudden catastrophic collapse then occurred which resulted in the total destruction of the hotel.

- **Decision:** The council owed a measured duty of care to the claimants to prevent loss of support notwithstanding the fact that the loss of support was due to natural causes. This required it to take reasonable steps to remove hazards to the claimants' adjoining land of which it was aware or ought to have been aware. The scope of the measured duty of care is more restricted than the normal duty of care in negligence. The court therefore held that the council could not be said to have been aware of the hazard. It could only have foreseen the risk of some damage occurring but not the sudden catastrophic failure which, in fact, occurred. Furthermore the council was not obliged to undertake the geological investigations which would have revealed the true nature of the hazard. The measured duty of care is only owned in respect of patent (visible) defects in the land. The geological fault was a latent (hidden) defect and the council were not expected to have knowledge of this type of defect. The council were not, therefore, in breach of their measured duty of care and judgment was entered in their favour.

- **Notes:** The court applied the principle in *Leakey v National Trust for Places of Historic Interest or Natural Beauty* (1980) in this case. The principle had previously been applied in a rights of support situation in *Bradburn v Lindsay* (1983) and has subsequently been applied in *Rees v Skerrett* (2001). Some of the earlier rights of support cases should be read in the context of these recent developments. See, for example, *Jones* *v Pritchard* (1908); *Sack v Jones* (1925); *Bond v Nottingham Corporation* (1940).

The facts

The claimants were the owners of the Holbeck Hall Hotel. This stood, in its own grounds, on a cliff overlooking the sea in Scarborough. The cliff was affected by coastal erosion and was inherently unstable. Scarborough Borough Council owned all the land between the hotel's grounds and the sea.

In 1982 a landslip occurred at the cliff edge. The council appointed engineers to recommend how the cliff could be stabilised. They reported in 1985, suggesting the installation of drainage but advising that further geological investigations should first be carried out to identify the location of the slip plane. Had this investigation been undertaken this would have revealed that there was a sudden risk of catastrophic movement which could cause the collapse of the hotel and its grounds. No investigation took place and in 1986 a further landslip occurred. This effectively doubled the size of the previous one and left the cliff edge only 30–35 metres from the edge of the hotel grounds.

By 1988 the council's chief engineer was expressing concern within the council that, if left unchecked, the coastal erosion could eventually affect land within the grounds of the hotel. In 1989 remedial works were eventually carried out but these were not to the design which had been recommended by the engineers. A further massive landslip occurred in 1993 during which the hotel's lawn disappeared and the ground underneath its seaward wing collapsed. The rest of the hotel was left in an unsafe condition and the whole building had to be demolished.

The claimants issued proceedings against the council for recovery of their losses on the basis that it had interfered with the natural right of support to their land. Amid much local publicity the trial judge gave judgment for the claimants. He held that the council had been aware, or ought to have been aware, of the hazard caused by the potential for a failure of support to the claimants' land. Under the principle established in *Leakey v National Trust for Places of Historic Interest or Natural Beauty* (1980), it was therefore under a measured duty of care to the claimants. In failing to carry out the further geological investigations recommended by the engineers it was in breach of that duty. The council appealed to the Court of Appeal and made three submissions in support of this.

A positive duty to maintain support to neighbouring land?

Firstly, the council submitted that, although the actual removal of the claimants' natural support would have been actionable, there could be no liability for simply allowing the existing support to deteriorate through natural degradation. In the council's submission there could be no duty on a servient owner to take positive steps to maintain the rights of support of neighbouring land.

The court noted that a number of first instance cases had indeed established that the owner of a servient tenement was only liable if he did something to withdraw support and that there was no positive duty to provide support. These included *Sack v Jones* (1925), *Macpherson v London Passenger Transport Board* (1946) and the dictum of Greene MR in *Bond v Nottingham Corporation* (1940).

It then traced the development of the principle in *Leakey* which was relied on by the claimants. This originated in *Sedleigh-Denfield v O'Callaghan* (1940) where a landowner was held liable for allowing a nuisance to continue on his land where the hazard had been caused by a trespasser. The Privy Council held in the Australian case of *Goldman v Hargrave* (1966) that hazards that arise by operation of nature are also subject to the same principle. The Court of Appeal then held in *Leakey* that this decision now represented the law of England.

The council had argued that the *Leakey* principle was entirely separate from the established law on rights of support. The Court of Appeal made no reference to the rights of support cases in *Leakey* so it could not have intended to modify them. Furthermore, this also appeared to be the view of the editor of *Gale on Easements* (16th edn, 1997) in paras 10.26 and 10.27. Far from challenging the existing law on rights of support, the decision in *Leakey* had been intended to apply only to encroachments or escapes from a defendant's land on to a plaintiff's land. The application of the *Leakey* principle in *Bradburn v Lindsay* (1983), where dry rot had spread from one property to another, was an illustration of this.

The court did not accept this argument and found that the *Leakey* principle had modified the previous law in relation to rights of support. The court made reference to the recent unreported case of *Bar Gur v Bruton* (1993) where the *Leakey* principle had also been applied by a county court in similar circumstances. The county court's decision had been upheld, on appeal, by Dillon LJ and had only eventually been reversed by the Court of Appeal (including Stuart-Smith LJ, who also sat in the present case) on the facts. The relevance of the *Leakey* principle to rights of support situations was therefore supported by this case, despite the eventual decision.

The council's first submission was therefore rejected. The owner of a servient tenement was under a duty to take positive steps to provide support for neighbouring land. The *Leakey* principle applied equally to a danger caused by loss of support or by any other hazard or nuisance.

> Stuart-Smith LJ: "[In *Bar Gur v Bruton*] ... the county court judge ... applied the decision in *Leakey v National Trust*. Dillon LJ upheld this deci-

sion. But I and Evans LJ allowed the plaintiff's appeal. We did so on ... the facts ... In the course of his judgment Dillon LJ, after citing the dictum of Greene MR in *Bond v Nottingham Corporation* said: 'The judge rightly recognised that in the light of Leakey's case that statement needs to be qualified and is no longer good law.' There is no further discussion of the matter. It is clear that both Evans LJ and I proceeded on the basis that *Leakey v National Trust* applied and had we thought that the wall constituted a nuisance, we would have dismissed the appeal...

... it is clear from Dillon LJ's judgment that we must have had in mind the law as stated in *Bond v Nottingham Corporation*. I cannot accept [the council's] submission that both Evans LJ and I might have distinguished *Leakey v National Trust*, but did not do so because we decided the case on the facts. That being so, the decision is binding upon us; [the council] does not submit it was *per incuriam*[1] ... In any event, for the reasons which I have given, I do not think there is any difference in principle between a danger due to lack of support and danger due to escape or encroachment of a noxious thing so far as the *Sedleigh-Denfield v O'Callaghan/Leakey v National Trust* principle is concerned. I therefore reject [the council's] first submission."

Knowledge of hazards to neighbour's land

Secondly, the council submitted that the judge was wrong in finding that it had been aware of the hazard to the claimants' land. Whilst it did foresee the likelihood of some damage occurring to the claimant's land, it could not, without further geological investigations, have foreseen the catastrophic damage which actually occurred.

The court accepted that this represented the state of the council's knowledge. In most situations where a duty of care existed, under the well-known rule in *Hughes v Lord Advocate* (1963), this would have been a sufficient basis for liability. However, this only applied in situations where there had been a breach of an existing duty of care (a misfeasance).

The present situation was a case of non-feasance. The council had done nothing to create the danger which had arisen by the operation of nature. In these situations the scope of the duty (a "measured duty of care") was much more restricted. It depended far more on the circumstances of the particular parties than the objective test which was applied where the normal duty of care had been breached. The court concluded that a defendant could only be held liable for damage which it could actually have reasonably foreseen.

---

[1] A decision made *per incuriam* (by carelessness) is made in ignorance of a relevant authority and is not therefore a binding precedent.

As the further geological investigations had not been undertaken the court accepted that the council could not have reasonably foreseen the damage which occurred. The court therefore accepted the council's second submission.

Presumed knowledge of hazards

Finally, the council submitted that the judge was also wrong in finding that it ought to have been aware of the hazard. In particular, it submitted that it had no obligation to undertake the further geological investigations which would have revealed the full nature of the risk.

The basis of the *Leakey* principle, on which the judge decided the case, is that liability arises in situations where a defendant "knew or ought to have known" about a potential hazard to neighbouring land. In the council's submission, the concept of "ought to have known" was confined to patent defects which can be observed by a reasonable landowner exercising reasonable care in the management of his land. There could be no obligation on a landowner to undertake investigations that would uncover latent defects in the land.

The court accepted this submission. It accepted that, what it described as "presumed knowledge" could only apply to patent defects and that there could be no obligation on a landowner to investigate the possibility of latent defects. The council had not therefore breached their measured duty of care by failing to undertake the geological investigations into a defect which was unforeseen.

Stuart-Smith LJ explained the nature of the duty, and the extent of presumed knowledge, in the following terms:

> "The duty arises when the defect is known and the hazard or danger to the claimants' land is reasonably foreseeable, that is to say it is a danger which a reasonable man with knowledge of the defect should have foreseen as likely to eventuate in the reasonably near future. It is the existence of the defect coupled with the danger that constitutes the nuisance; it is knowledge or presumed knowledge of the nuisance that involves liability for continuing it when it could reasonably be abated.
>
> [The council] submits that the defect must be patent and not latent: that is to say that it is a defect which can be observed. It is no answer for the landowner to say that he did not observe it, if a reasonable servant did so; or if as a reasonable landowner he, or the person to whom he entrusted the responsibility of looking after the land, should have seen it.
>
> But if the defect is latent, the landowner or occupier is not to be held liable simply because, if he had made further investigation, he would have discovered it. I accept [the council's] submission; in my judgement that is what is meant by the expression 'knowledge or presumed knowledge'."

The decision

In rejecting the council's first submission the court held that it was subject to a measured duty of care to the claimants, irrespective of the fact that the damage had occurred due to natural causes.

However, in accepting the council's second and third submissions, the court decided that it had not breached this duty. As the geological investigations had not been undertaken the council had no actual knowledge of the hazard to the claimants' land. Furthermore it was under no duty to investigate the possibility of latent defects and therefore had not breached a duty by failing to undertake the geological investigations.

The appeal was therefore allowed and judgment was entered for the council.

HUGHES v PERCIVAL (1883)
House of Lords

In brief

- **Issues:** 1855 Act – party wall – owner's entitlement to undertake work to wall at common law – reciprocal duties to adjoining owner at common law – damage to adjoining owner's property – damage caused by independent contractors – non-delegable duties.
- **Facts:** Hughes was undertaking substantial works to his property which affected the party walls on both sides. Towards the end of the works some independent contractors negligently damaged one of the party walls. This led to a major collapse which also damaged the other party wall with Percival's adjoining property.
- **Decision:** Hughes was liable for the damage suffered by Percival. Building owners owed a non-delegable duty to adjoining owners when undertaking work to party walls. They had a common law right to undertake work to a party wall and this necessarily involved a high risk of damage to the adjoining owner's property. In these circumstances the law imposed a reciprocal duty of care on them in favour of the adjoining owner. This duty was owed on a non-delegable basis. The duty was owed in respect of all work to the party wall and not simply in respect of those aspects which were particularly hazardous. The independent contractors had caused damage to Percival's property whilst undertaking work which had been entrusted to them by Hughes and this work affected the party wall. Hughes was therefore liable for the resulting damage.
- **Notes:** See also: *Cubitt v Porter* (1828); *Bower v Peate* (1876); *Alcock v Wraith and Swinhoe* (1991).

The facts

Hughes owned two houses at 1 and 2 Panton Street, London SW1. They stood at the corner of Panton Street and Haymarket. Percival owned 3 Panton Street. This was a new building which adjoined Hughes' property to the east and was separated from it by a party wall (Figure 9).

A third property was owned by one Baron. This was situated in Haymarket and adjoined Hughes' property to the south. It was also separated from it by a party wall. This party wall was in a poor structural condition. Its surface was uneven and the wall itself was out of plumb (Figure 9).

Hughes wished to demolish his existing buildings and to erect a new one

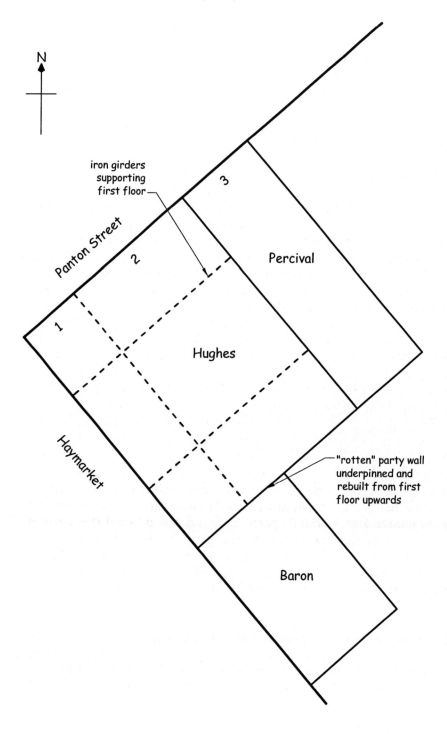

Figure 9: Hughes v Percival

in their place. This involved the demolition and rebuilding of the defective wall adjoining Baron's property. The work also included the placing of a connected series of steel joists into both party walls to provide the support for the first floor of the new building.

Party structure notices appear to have been served on the two adjoining owners under the 1855 Act[1]. Percival consented to the works but a difference appears to have arisen in respect of the other party wall adjoining Baron's property. Baron then appointed an architect[2] to negotiate with Hughes' architect about the proposed works which affected his property.

The two architects agreed on changes to the original specification. Instead of the whole wall being demolished and rebuilt they agreed that only the top half of the wall (affecting the first and second floors) should be renewed in this way. The lower half (affecting the basement and ground floor) was to be underpinned as part of the proposed excavation works to increase the depth of the basement to the new building. The rough surface finish to this part of the wall was also to be hacked off and levelled.

The works then proceeded without incident until they were virtually completed. At a very late stage of the project some subcontractors were employed to fit a wooden staircase leading from the ground floor at 1 Panton Street, down into the basement. During the course of this operation they cut a substantial channel into the old portion of the party wall to accommodate the new staircase. This was contrary to the plans and specification and was undertaken without the knowledge of Hughes' architect or of the main contractor.

The cutting in of the channel weakened the wall. Hughes' architect raised objections to the work as soon as he became aware of it and ordered that the damage be made good. Unfortunately, before there was time to undertake the remedial work, the wall collapsed. During the collapse the steel joists which were supported by this wall lost their bearing and became displaced. This, in turn, caused movement in the steel joists which were attached to Percival's party wall and this dragged this party wall over, causing cracking and other damage to Percival's house.

Percival sued Hughes for damages for negligence. At first instance the court found in his favour. Hughes appealed against the decision and the matter was ultimately considered by the House of Lords.

Liability for torts of independent contractors?

Hughes argued that he could not be held liable for damage caused by the workmen fitting the staircase. Because of their status as independent

---

[1] Although the published law report is a little unclear on this point.
[2] Presumably to act as appointed surveyor, although the published law report does not make this clear.

contractors he would only be liable for their torts if one of the various exceptions to the general rule excluding such liability could be shown to apply. For the following reasons he maintained that this was not the situation.

Firstly, he could not be said to have authorised the commission of the tort. The workmen had departed from all instructions and their act of negligence was therefore entirely collateral to the building work which he had authorised them to undertake.

Secondly, the exceptions recognised in the cases of *Dalton v Angus (1881)*, *Tarry v Ashton (1876)*, *Bower v Peate* (1876) and *Pickard v Smith* (1861) did not apply[3]. He acknowledged that the whole construction project was undoubtedly hazardous. However, the project was substantially complete when the damage occurred and all hazardous operations had already been undertaken. The fitting of the staircase which had actually caused the damage was not a hazardous act and the general rule excluding liability for the torts of an independent contractor must therefore apply.

Non-delegable duty

The court did not accept these arguments. Their Lordships were unanimously of the view that, in these circumstances, Hughes had been subject to a non-delegable duty of care to Percival which he had failed to discharge. He was therefore personally liable in negligence. The court described the nature of this duty in the following terms.

> Lord Watson: "... it was the duty of the appellant in carrying out his building operations to see that reasonable precautions were taken in order to protect from injury the eastern wall of his tenement, of which the respondent was part owner. The appellant does not deny that many of the operations which he contemplated, and which he had employed a contractor to execute, were such as would necessarily or possibly imperil the stability of the party-wall, if no precautions were used; nor does he dispute that it was incumbent upon him to see that these operations were safely carried out by the contractor."
>
> Lord Fitzgerald: "What was the defendant's duty? ... it was the duty of the defendant to have used every reasonable precaution that care and skill might suggest in the execution of his works, so as to protect his neighbours from injury, and that he cannot get rid of the responsibility thus cast upon him by transferring that duty to another. He is not in the actual position of being

---

[3] There is a certain amount of confusion in this case (and indeed in others as well) as to the distinction between these various exceptions. The distinctions that are often made by textbook writers today are often blurred into a single category of "hazardous acts".

responsible for injury, no matter how occasioned, but he must be vigilant and careful, for he is liable for injuries to his neighbour caused by any want of prudence or precaution, even though it may be *culpa levissima*[4]."

Exceptions to general rule?

Their Lordships specifically rejected each of Hughes' two submissions. Firstly, the workmen's negligence could not be described as entirely collateral to the works which he had authorised them to do.

> Lord Fitzgerald: "The act of the workmen may have been unauthorised and ill-judged but it was an act no doubt done in the execution of the work entrusted to them, and can in no case be said to have been entirely 'collateral', and it is not to be forgotten that the contract provides that 'complete copies of the drawings and specifications are to be kept on the buildings in charge of a competent foreman, who is to be constantly kept on the ground by the contractors, and to whom instructions can be given by the architects and who is of course to direct the workmen in their operations, and who ought to see that what they are doing is necessary and lawful, and carried out in the safest manner."

Secondly, contrary to Hughes' suggestion, the court did not accept that it was possible to separate the particular act which had caused the damage from the rest of the project.

> Lord Fitzgerald: "Then was the perilous portion of the work completed before the doing of the act which it is said led immediately to the fall of Barron's party-wall and the consequent injuries to the plaintiff's premises?
> ... The event shows that the danger was not over, and I should have thought that it could not be as long as anything was being done that could affect the stability of either party-wall.
> ... The conclusion I have reached is, that the defendant had undertaken a work which as a whole necessarily carried with it considerable peril to his neighbours."

Legal basis for non-delegable duty in respect of party walls?

The legal basis for the existence of Hughes' non-delegable duty is not altogether clear from the judgments of Lords Watson and Fitzgerald. Although they make general references to the perilous nature of the works

---

[4] A trivial fault.

they do not specifically explain the exception from the general rule in these terms.

Lord Blackburn's judgment is more helpful in this respect. He explains the basis of the duty in similar terms to those in *Dalton v Angus* and *Bower v Peate*. In those cases non-delegable duties were imposed on landowners when excavating their land which was subject to rights of support in favour of adjoining land. This was based on the premise that the act of excavation was in itself a tort (a nuisance) unless adequate protection was provided to prevent loss of support.

Lord Blackburn drew a parallel with the present situation. Prior to 1925 each owner's rights in a party wall were not expressed in terms of rights of support[5] as each owner was entitled to undertake work to the whole thickness of the wall. He suggested however that, because of this right to utilise the whole wall, there must be a reciprocal duty to safeguard the interests of the other owner when so exercising it. He therefore suggested that a non-delegable duty applied specifically to the case of party walls which is closely related to that in respect of rights of support laid down in *Dalton v Angus* and *Bower v Peate*.

> Lord Blackburn: "The first point to be considered is what was the relation in which the defendant stood to the plaintiff. It is admitted that they were owners of adjoining houses between which was a party wall and the property of both ... The defendant had a right to utilise the party wall, for it was his property as well as the plaintiff's; a stranger would not have had such a right.
>
> But I think that the law cast upon the defendant, when exercising this right, a duty towards the plaintiff. I do not think that duty went so far as to require him absolutely to provide that no damage should come to the plaintiff's wall from the use he thus made of it, but I think that the duty went as far as to require him to see that reasonable skill and care were exercised in those operations which involved a use of the party-wall, exposing it to this risk. If such a duty was cast upon the defendant he could not get rid of responsibility by delegating the performance of it to a third person."

The decision

Hughes owed a non-delegable duty of care to Percival in respect of the party wall which they both shared. In allowing independent contractors to cause damage to the wall he had breached this duty. The court therefore dismissed his appeal and entered judgment for Percival.

---

[5] Although they are now. See Law of Property Act 1925, sections 38 and 39 and Schedule I, Part V. The parallels between *Bower v Peate* and party walls are therefore now even closer than those suggested by Lord Blackburn.

J. JARVIS & SONS LTD v BAKER (1956)
Queen's Bench Division

In brief

- **Issues:** 1939 Act – repairs to a party wall – award requiring adjoining owner to contribute towards cost of works – appointment of contractor by building owner's surveyor – liability of building owner for entire cost of works – whether adjoining owner directly liable to contractor – effects of failure to serve account on adjoining owner within statutory time limit.

- **Facts:** Baker undertook repairs to a party wall in pursuance of an award under the 1939 Act. The award required the adjoining owner to contribute one half of the costs of the work. The building owner's surveyor failed to serve the contractor's account on the adjoining owner. Baker paid half of the account but refused to pay the other half on the ground that this was the adjoining owner's responsibility. The contractor sued Baker for the outstanding amount. Baker sought an indemnity for the sum claimed from the adjoining owner by bringing third party proceedings against him.

- **Decision:** Where costs are to be shared both parties to an award are jointly and severally liable for the contractor's account. As the contractor had chosen to bring proceedings against Baker rather than the adjoining owner judgment was awarded against her for the entire amount of the debt. However, in the third party proceedings, judgment was awarded against the adjoining owner in respect of the contribution which he was obliged to make under the award and under the terms of the 1939 Act. He could not evade this liability either on the grounds that he had not consented to the appointment of the contractor or that he had not been served with a copy of the account, as required by the Act.

- **Notes:** (1) The decision in the third party proceedings appears to contradict the finding in *Spiers & Son Ltd v Troup* (1915) which was decided under the 1894 Act. In that case the time for service of an account on the adjoining owner was held to be of the essence. If the rule in *Spiers* had been applied in the present case the adjoining owner would not have been held liable in the third party proceedings. (2) See also *Reading v Barnard* (1827).

The facts

Mrs Baker owned 22 Bishopsthorpe Road, Sydenham. Mr Benkert owned the adjoining terraced property at 20 Bishopsthorpe Road. Mrs

Baker wished to undertake repairs to the party wall between the two properties.

She therefore served a party structure notice on Mr Benkert under the 1939 Act and surveyors were appointed by each party. Mrs Baker appointed surveyor A who was also the superintending architect for the work. Mr Benkert appointed surveyor B. The published law report records that there were "a number of ridiculous disputes" between the two appointed surveyors, including one regarding the choice of builder for the works, J. Jarvis & Sons.

An award was signed in November 1949 which provided (inter alia) that Mr Benkert, as the adjoining owner, was to contribute one half of the cost of the work. At the time the award was signed the dispute about the builder remained outstanding.

However, the law report records that "the dispute over the builder died down by February 1950". The work was eventually started by Jarvis in May 1950 and completed by July 1950. Surveyor A never advised surveyor B that the work had commenced and appears to have deliberately tried to keep him away. Surveyor B therefore never had an opportunity to inspect the work until it was completed.

On completion of the works in July 1950 Jarvis delivered their bill for £592 to surveyor A, in his capacity as superintending architect. Unfortunately surveyor A ignored it and Jarvis sent a further copy in January 1951. Surveyor A again failed to respond and no action was taken on the matter for a further two years.

Eventually (presumably in or about January 1953) the bill seems to have been delivered to Mrs Baker who paid half of it on the basis that Mr Benkert was responsible for the other half under the terms of the award.

Liability for payment of expenses under party wall award

Jarvis than requested payment of the outstanding £296 from Mr Benkert. He refused to pay it on two grounds. Firstly, on the ground that he had never consented to the appointment of Jarvis and surveyor A therefore had no authority to appoint them to undertake the works. Secondly, he objected to the amount of the bill which he claimed was unreasonable for the work actually undertaken.

Jarvis therefore issued proceedings against Mrs Baker for the outstanding monies. Jarvis argued that their contract was with Mrs Baker so she was responsible for payment of the whole sum. Any arrangements for contribution by a third party were not their concern and were for Mrs Baker to pursue independently of her liability for the debt. Mrs Baker issued third party proceedings, in the same action, against Mr Benkert for the amount claimed by Jarvis on the basis of the award and/or his statutory obligation to contribute under the 1939 Act.

> **London Building Acts (Amendment) Act 1939**
>
> PART VI: RIGHTS ETC. OF BUILDING AND ADJOINING OWNERS
>
> EXPENSES
>
> **58. Account of expenses**
>
> (1) Within two months after the completion of any work executed by a building owner of which the expenses are to be wholly or partially defrayed by an adjoining owner in accordance with 56 (Expenses in respect of party structures) of this Act the building owner shall deliver to the adjoining owner an account in writing showing:
>
> (a) particulars and expenses of the work; and
> (b) any deductions to which the adjoining owner or any other person is entitled in respect of old materials or otherwise;
>
> and in preparing the account the work shall be estimated and valued at fair average rates and prices according to the nature of the work the locality and the cost of labour and materials prevailing at the time when the work is executed.
>
> (2) Within one month after delivery of the said account the adjoining owner may give notice in writing to the building owner stating any objection he may have thereto and thereupon a difference shall be deemed to have arisen between the parties.
>
> (3) If within the said month the adjoining owner does not give notice under subsection (2) of this section he shall be deemed to have no objection to the account.

The decision

The judge described the case as "a miserable little action, arising through the fault of one man". He gave judgment, in the main action, for Jarvis against Mrs Baker. He considered that surveyor A had authority, from both parties, to appoint Jarvis and that both of them thereby became jointly and severally liable for the whole amount.

Jarvis was therefore free to sue either one or both of the parties for the full amount and was not bound to look to each of them for half of the total. As he had chosen to issue proceedings against Mrs Baker he was entitled to recover the whole amount from her and judgment was entered for this amount.

In the third party proceedings the judge held that Mr Benkert was liable. Surveyor A had authority to appoint the builder and he was liable to contribute towards the bill under the terms of the award and the 1939 Act. He was not excused from liability merely because the account had not been served on him within the two-month statutory period required by section 58[1].

Nevertheless, the failure to serve the account on him meant that he had not had an opportunity to question the quantum. He must therefore now be provided with this. It was noted that he had agreed to pay £270 (a reduction of £26 on the amount claimed) and the court ordered that an enquiry should now be conducted to determine if this was an appropriate amount.

---

[1] See section 13 of the 1996 Act.

JOHNSON (T/A JOHNSON BUTCHERS) v BJW PROPERTY DEVELOPMENTS LTD (2002)

Technology and Construction Court

In brief

- **Issues:** Damage to adjoining property by fire – common law liability – Fires Prevention (Metropolis) Act 1774 – *Rylands v Fletcher* liability – negligence – nuisance – acts of independent contractor – vicarious liability – non-delegable duties.

- **Facts:** BJW's independent contractor negligently installed a fireplace in a sixteenth century timber party wall. As a consequence the adjoining property owned by Johnson was damaged by fire.

- **Decision:** BJW were not liable under *Rylands v Fletcher* because the use of a domestic fire grate was not a non-natural use of land within the rule. However, they were liable under the common law rules relating to the spread of fire. The fire had not escaped accidentally so there was no defence under the Fires Prevention (Metropolis) Act 1774. It had escaped because of the contractor's negligence. Unlike the modern law of negligence, vicarious liability for the torts of an independent contractor was still possible under the old common law rules in fire cases. BJW were therefore liable on this basis. BJW were also liable in nuisance and negligence on a separate basis. The installation of the fireplace related to work to a party wall which involved a special risk to the adjoining premises. In these circumstances a non-delegable duty of care was owed by BJW to Johnson. The contractor's negligent acts placed them in breach of this duty.

- **Notes:** This case provides a further example of the recent expansion of liability to adjoining owners in the torts of nuisance and negligence. See also: *Bradburn v Lindsay* (1983); *Holbeck Hall Hotel v Scarborough Council* (2000); *Rees v Skerrett* (2001).

The facts

Johnson owned and occupied a butcher's shop and living accommodation at 45/47 High Street, Newington, near Sittingbourne, Kent. BJW owned the adjoining public house at 49 High Street which was vacant and undergoing refurbishment. Both properties were timber framed buildings which had been erected in the sixteenth century. They were separated by a party wall. This was of timber frame construction with lath and daub infill.

As part of the refurbishment works to the pub BJW purchased a replacement fireplace from Young who was an antiques dealer. Young also installed it in the existing fire opening on BJW's side of the party wall. During the installation he made various alterations to the fire opening. These included the removal of a firebrick lining and fire curtain which had previously been installed to protect the timber party wall from fire.

Several weeks later the painting contractor lit a fire in the fireplace to keep warm whilst he was working. He extinguished the fire before he left the premises. Unfortunately, due to the earlier removal of the firebrick lining, the ends of some timbers which were embedded in the party wall had already started to smoulder.

During the night the smouldering spread along the entire length of these timbers which protruded into the adjoining premises. When the smouldering reached an open space under the floor it burst into flames. The resulting fire caused extensive damage to Johnson's property.

Liability for the spread of fire

Johnson issued proceedings against BJW for damages. He founded his claim on a number of alternative causes of action.

- Firstly, he relied on the traditional common law rules governing the spread of fire. These imposed liability on an occupier who allowed fire to spread from his property where this caused damage to an adjoining property. This liability was strict except where the fire was caused by a "stranger" over whom the occupier had no control.
- Secondly, he relied on the overlapping liability for the escape of dangerous things from land under the rule in *Rylands v Fletcher* (1868).
- Thirdly, he claimed that BJW were vicariously liable in negligence for Young's negligence in removing the fire protection from the fire opening.
- Finally, he also alleged vicarious liability in nuisance on the basis of the interference with his property by the spread of fire which had also been caused by Mr Young's negligent acts.

Common law liability

The court noted the modification of the strict liability common law rules by section 86 of the Fires Prevention (Metropolis) Act 1774. This provides that no-one can now be held liable for the spread of a fire which begins "accidentally". According to the judgment of Lord Denman CJ in *Filliter v Phippard* (1847) this will be the situation if a fire starts by mere chance or it is impossible to trace its cause. Where a fire is started, either deliberately or through some act of negligence, the occupier will be liable and he will not be able to rely on the immunity conferred by the Act.

The court then considered the situation, as in the present case, of a fire which is (deliberately) started in a domestic grate but which subsequently escapes and causes damage. Johnson argued that an occupier would always be liable in such circumstances as the fire was started deliberately rather than "accidentally". The court disagreed with this. It considered the legislative purpose of the 1774 Act and concluded that it was the spread of the fire, rather than its actual ignition, which was being addressed by the Act.

> Thornton J: "Before the Act, even if such a fire had escaped accidentally on to the adjoining occupier's land, the occupier of the land from which the fire had escaped could have been held strictly liable. Following the Act of 1774, fires that were started by an accidental escape were within the Act and an occupier of the land from which the fire had escaped could not be sued and held liable for the resulting damage. The critical question was whether the escape of fire was an accident, not whether the original fire had been started accidentally...
>
> ... Thus, in context, the Act of 1774 is intended to regulate the situation that arises once a fire has escaped from a domestic grate and has damaged the adjoining occupier's premises. If... that escape was accidental, the Act of 1774 provides a defence to the defendant occupier from whose domestic grate the fire has spread and escaped. If, on the other hand, the cause of the escape on to the adjoining occupier's land was the negligence of the defendant or for which he was responsible, the Act provides no defence.
>
> In [the present] case, the Act of 1774 has no relevance and is inapplicable to the original fire that was started deliberately in BJW's grate since the mere ignition of that fire gives rise to no liability, whether before or after the Act of 1774. Moreover, it provides the defendant with no defence to the claimant's claim in so far as that claim is based on the escape of fire on to the claimant's land as a result of negligence."

Johnson's claim was indeed based on the escape of fire on to his land as a result of negligence. Specifically he relied on Young's negligence in removing the fire protection from the fire opening and argued that BJW were vicariously liable for this. BJW admitted Young's negligence but argued that they could not be vicariously liable for the actions of an independent contractor. The court therefore had to consider whether occupiers were vicariously liable for the torts of independent contractors in the context of common law liability for the spread of fire.

Historically an occupier would certainly have been "vicariously" liable in these circumstances because liability was strict. It mattered not who actually caused the spread of the fire as (with the sole exception of a fire caused by a "stranger") the occupier would always be liable. However, these rules predated the development of the law of negligence and its related doctrine that vicarious liability did not apply to the torts of an independent contractor. The court therefore had to consider whether the old rules relating to fire had survived these later developments or whether, in common with the modern law of negligence, vicarious liability for the spread of fire now had a more restricted application.

The court considered the cases of *Balfour v Barty King* (1957) and *H & N Emanuel Ltd v Greater London Council* (1971) and concluded that occupiers continued to be vicariously liable for the torts of independent contractors in the context of liability for the spread of fire.

> Thornton J: "Thus an occupier's ancient vicarious liability for a fire spreading and escaping on to adjoining premises due to the non-accidental acts of anyone who is not a stranger survived ... the contrary trends in the general development in the law of negligence in the nineteenth and twentieth centuries against the imposition of vicarious liability for negligence.
>
> Indeed, the liability of an ... occupier for the damage caused by the escape of fire resulting from an independent contractor's negligence is the last vestige of the ancient strict liability for the escape of *ignis suus*."

In the present case BJW had authorised Young to install the fireplace and he had done so negligently. His negligence had caused the escape of fire which had damaged Johnson's premises. Such escape had therefore not been "accidental" and, for the reasons outlined above, BJW were vicariously liable for this.

> Thornton J: "It follows that the defendant[s] in this case [are] to be held vicariously liable for the damage caused by the escape of fire into and on to the claimant's premises since the fire only escaped and caused damage because of the negligent workmanship of the defendant's independent contractor for whom the defendant is to be held separately liable."

The court therefore found in Johnson's favour on the basis of his claim under the common law rules relating to the spread of fire. Although this was sufficient to dispose of the matter the court also addressed the other potential areas of liability.

Liability under Rylands v Fletcher

The court noted that strict liability for the spread of fire could potentially arise under the general rule in *Rylands v Fletcher* as well as under the traditional rules already discussed. Fire was clearly a "dangerous thing" within the meaning of the rule which could lead to liability if it was also a non-natural user of the land in question. Furthermore, a fire resulting from such non-natural use of land would be outside the protection of the

1774 Act as it would not be regarded as having been started accidentally: *Musgrove v Pandelis* (1919).

However, in the present context no liability could arise on this basis as the burning of a domestic fire in a domestic grate was an ordinary and natural use of the land. *J. Doltis Ltd v Isaac Braithwaite & Sons* (1957) provided the authority for this decision.

Liability in negligence and nuisance

The court noted recent developments in the law which were blurring the distinction between the torts of negligence and nuisance in the context of liabilities between neighbouring owners.

> Thornton J: "There is, in modern times, no significant distinction between liability in negligence and nuisance where the damage in question has been caused by an escape. Liability in nuisance for such an escape requires it to be established that the escape was caused by a failure by the defendant to exercise reasonable care in allowing the circumstances giving rise to an escape, or the escape itself, to start or by a failure to exercise reasonable care to control or to stop the escape once it has started...
>
> ... The assimilation of negligence and nuisance in escape cases is an aspect of the development of a general liability imposed on an owner or occupier of land that is based on a duty to take reasonable steps to prevent or remove a risk of damaging neighbouring property.
>
> This duty has been recognised and developed in a long line of cases starting with *Sedleigh-Denfield v O'Callaghan* (1940) and subsequently developed in *The Wagon Mound (No. 2)* (1967), *Goldman v Hargrave* (1966), *Leakey v National Trust for Places of Historic Interest or Natural Beauty* (1980), *Marchant v Capital & Counties Property Co Ltd* (1982), *Bradburn v Lindsay* (1983), *Alcock v Wraith* (1991), *Bar Gur v Bruton* (1993), *Holbeck Hall Hotel v Scarborough Borough Council* (2000) and *Rees v Skerrett* (2001).
>
> The law relating to liability for activities on land which cause damage to adjoining land or adjoining occupiers is still developing and the authorities have yet to finally clarify the extent to which the various types of liability giving rise to this more general liability have been assimilated."

It therefore considered BJW's potential liability under these two torts together. Once again, as BJW admitted Young's negligence, the main issue was whether they were vicariously liable for his actions. Once again, BJW argued that vicarious liability did not extend to the acts of an independent contractor.

Whilst acknowledging the general truth of BJW's argument, the court noted that vicarious liability can arise for the acts of an independent contractor in certain circumstances. The relevant test for these circumstances had been expressed in a variety of ways. Some cases had referred to activities giving rise to a "special danger", others to those of an "extra hazardous" kind and some made reference to a special risk of damage due to the difficulties inherent in carrying out a particular task. The court noted the particular significance of the party wall in these considerations:

> Thornton J: "The work need not have been on a party wall but if it was, any resulting escape and damage is more likely to give rise to vicarious liability. The authorities suggest that the work must have been extra hazardous to give rise to vicarious liability for damage to neighbouring land caused by work away from the boundary or party wall – *Matania v National Provincial Bank* (1936) and *Honeywill & Stein v Larkin* (1934) – whereas it need only create a special risk of damage if work causing such damage was carried out on or adjacent to the boundary or party wall – *Bower v Peate* (1876) and *Alcock v Wraith* (1991)."

In principle, therefore, an occupier would be vicariously liable for an independent contractor's negligence or nuisance if the offending work was to a party wall (or similar division between two properties), was authorised or approved by the occupier and involved some special risk or was from its very nature likely to cause damage if it was poorly performed. The court was satisfied that this was the situation in the present case.

> Thornton J: "... the work to be performed by Mr Young involved special risk to the party wall and to the claimant's property given its close association with and its proximity to the potentially inflammable party wall. The work was authorised by the defendant[s] given the rudimentary contract which merely required Mr Young to install the new fire surround. If that work was misperformed, resulting fire damage to the claimant's adjoining property would be a real possibility. Finally, Mr Young's work, although not being carried out directly on the party wall, would affect and was carried out adjacent to, and was associated with a party wall and with a structure that was common to both properties.
>
> I conclude that all the requirements that need to be shown to give rise to an occupier's vicarious liability for the negligence of his independent contractor have been made out. The relevant work was undertaken negligently on and adjacent to a party wall. It involved a special risk to, and by its very nature endangered, the claimant's adjoining premises. In consequence, the defendant[s] [are] liable for the fire damage that occurred."

The decision

BJW were not liable under the rule in *Rylands v Fletcher* as the use of a domestic fire grate was not a non-natural user of the land. However, they were liable for the spread of the fire under the common law rules as this had not occurred accidentally but due to the negligence of Mr Young. They were also vicariously liable in negligence and nuisance on the basis of Mr Young's negligence as their independent contractor. Judgment was therefore entered for Johnson with damages to be assessed.

JOHNSTON v MAYFAIR PROPERTY COMPANY (1893)

Chancery Division

In brief

- **Issues:** 1855 Act – type 'b' party wall – wall performing separating function for only part of its length – building owner's entitlement to enclose on adjoining owner's external wall – trespass.

- **Facts:** Mayfair demolished their house which shared a party wall with Johnston's house. They proposed to erect a new house against the whole length of the now exposed flank wall of Johnston's house. This involved enclosing on a portion of the wall which had not previously separated the two houses. Mayfair maintained that the flank wall was a party wall for the whole of its length. Johnston submitted that the portion which had not previously separated the houses was owned entirely by him and sought an injunction to restrain Mayfair from trespassing on it.

- **Decision:** Only those parts of the wall which had previously performed a separating function between the two houses were a party wall, both at common law and under the 1855 Act. The remainder was owned entirely by Johnston and Mayfair therefore had no right to enclose upon it. The court granted an injunction to prevent the work.

- **Notes:** (1) The definition of a party wall in the 1855 Act only included the type 'b' party wall. The separating function was therefore crucial to the existence of a statutory party wall under that Act. Nevertheless, it is thought that the decision would have been the same even if the additional type 'a' definition had been a possibility under the Act. There was no clear evidence that the line of junction ran down the centre of the wall serving only Johnston's property. Neither would this have been presumed by the court as this part of the wall was not subject to the common user of the parties. There would therefore have been no basis for finding that the wall was a type 'a' party wall. (2) See also: *Cubitt v Porter* (1828); *Moss v Smith* (1977).

The facts

The case concerned a terrace of three houses. The front wall of the middle house projected beyond those of the two end houses (Figure 10). Johnson owned the middle house. Mayfair owned one of the end houses under a building lease. Mayfair demolished their house prior to redeveloping the site. This involved leaving the entire flank wall between the two properties standing.

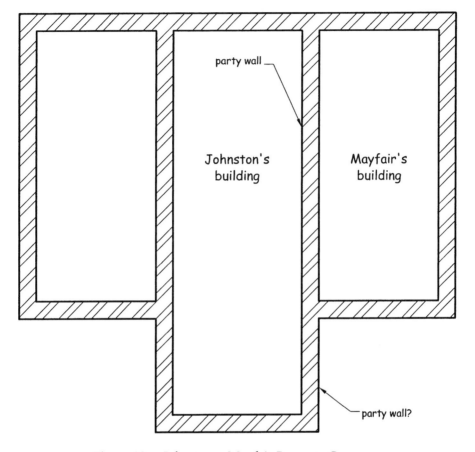

Figure 10: Johnston v Mayfair Property Company

The published law report does not provide a detailed account of the facts. Nevertheless, it appears that Mayfair intended to rebuild their house along the entire length of the flank wall, including that portion which protruded beyond the front wall of the demolished building. As shown in Figure 10, this portion was a direct continuation of the wall separating the two properties.

Mayfair appear to have served a party structure notice on Johnson, claiming a right to undertake the work under section 83(8) of the 1855 Act[1]. Johnson objected to the work to the extent that it affected the protruding front portion of the wall. He argued that this was owned entirely by him and was not a party wall. Mayfair therefore had no right to enclose upon it, either at common law or under the 1855 Act. On this

---

[1] This provided a right to cut into a party structure in similar terms to that contained in section 2(2)(f) of the 1996 Act.

> ## Metropolitan Building Act 1855
>
> ### PRELIMINARY
>
> **3. Interpretation of certain terms in this Act**
>
> In the construction of this Act (if not inconsistent with the context) the following terms shall have the respective meanings hereinafter assigned to them; (that is to say):
>
> "Party wall" shall apply to every wall used or built in order to be used as a separation of any building from any other building, with a view to the same being occupied by different persons.
>
> ### PART III: PARTY STRUCTURES
>
> ### RIGHTS OF BUILDING AND ADJOINING OWNER
>
> **83. Rights of building owner**
>
> The building owner shall have the following rights in relation to party structures; that is to say:
>
> (8) A right to cut into any party structure upon condition of making good all damage occasioned to the adjoining premises by such operation.

basis he applied for an injunction to restrain Mayfair from interfering with the front portion of the wall.

Ownership of flank wall

The court's decision presumably turned on the ownership of the front portion of the wall. Unfortunately, this aspect is not discussed in the published law report.

On the basis of *Cubitt v Porter* (1828), the portion of the wall separating the two houses was probably owned jointly by the parties as tenants in common. According to today's law[2], the boundary line would therefore

---

[2] Law of Property Act 1925, sections 38 and 39 and Schedule 1, Part V.

have run down the centre of this portion. It might be supposed that the front portion, as a continuation of this portion, would be owned on the same basis. This was presumably the rationale for Mayfair's assertion that the disputed wall was a party wall, on which they were entitled to enclose, either at common law or under the 1855 Act.

An opposing interpretation would have been presented by Johnson. He would have argued that, when the wall was built, there would have been no intention that the front portion should be a party wall as it served only the middle house. On this basis, it must therefore have been owned entirely by Johnson, notwithstanding the shared ownership of other parts of its length.

The decision

The court presumably accepted the latter interpretation as it held that Mayfair had no right to enclose on the front portion of the wall.

As the wall was owned entirely by Johnson, Mayfair had no right to interfere with it at common law and any attempt to do so would amount to a trespass.

The court also held that there was no right to interfere with the wall under the 1855 Act. The wall was not a "party wall" under the 1855 Act. The definition of "party wall" in that Act applied only to walls separating two buildings in separate ownership[3]. As the front portion of the wall served no separating function, it was not a party wall within the meaning of the Act. Mayfair therefore had no right to build against it and the court granted an injunction to restrain them from doing so.

---

[3] The definition is substantially the same as that for type 'b' party walls in section 20 of the 1996 Act.

JOLLIFFE v WOODHOUSE (1894)
Court of Appeal

In brief

- **Issues:** 1855 Act – entitlement to demolish and rebuild party wall – common law rights – statutory rights – relationship between common law rights and statutory rights – reciprocal duty of care at common law – meaning of "unnecessary inconvenience" – duty to perform work expeditiously – non-delegable duties.

- **Facts:** Woodhouse demolished and rebuilt the party wall between his property and that owned by Jolliffe. He took seven months to complete the work and Jolliffe was subjected to disruption for the whole of this period.

- **Decision:** Woodhouse was entitled to undertake the work, both at common law and under the 1855 Act. However, in either case, he owed a reciprocal duty to Jolliffe. A duty of care was owed at common law and, under the Act, he owed a duty not to cause unnecessary inconvenience. Both duties required him to complete the work expeditiously and he had failed to do this. It was no defence that Woodhouse had employed a competent architect and contractor. Both the statutory and common law duties were owed on a non-delegable basis. Woodhouse was therefore in breach of these duties and an order for damages was made against him.

- **Notes:** See also (on a building owner's common law right to undertake work to a party wall): *Cubitt v Porter* (1828); *Standard Bank of British South America v Stokes* (1878) – (on the reciprocal duties owed to adjoining owners): *Alcock v Wraith and Swinhoe* (1991); *Thompson v Hill* (1870) – (on unnecessary inconvenience): *Thompson v Hill* (1870); *Barry v Minturn* (1913) – (on non-delegable duties): *Bower v Peate* (1876); *Hughes v Percival* (1883).

The facts

Jolliffe was the tenant of 17 Coronet Street in Hoxton where he carried on business as a fish salesman. Woodhouse owned the adjoining terraced property at 19 Coronet Street. The two properties were separated by a party wall.

Woodhouse intended to demolish and rebuild his property. He undertook the demolition during the winter of 1891/1892. This revealed that the party wall between the two properties was not strong enough to support the new building which he intended to erect. He therefore served a party structure notice on Jolliffe, claiming his right under section 83(7) of the 1855 Act, to demolish and rebuild it.

An award was published which authorised the work and Woodhouse started demolishing the party wall in February 1892. The work continued for seven months and the rebuilding was completed by the end of September. During the course of the work Jolliffe suffered disruption to his property and issued proceedings against Woodhouse.

Unnecessary inconvenience

Jolliffe alleged that Woodhouse had been guilty of unreasonable delay in reinstating the demolished party wall and that he had therefore exceeded his statutory powers to demolish and rebuild. The court entered judgment in his favour and awarded him damages.

Woodhouse applied for a new trial. He maintained that he could not be held responsible for the delays. He had provided a competent architect and builder and was not therefore personally in breach of any obligation which he owed to Jolliffe.

The court noted that Woodhouse certainly had the right to demolish and rebuild the party wall. He had this right under section 83(7) of the 1855 Act and also, as a tenant in common of the wall, at common law, under the rule in *Cubitt v Porter* (1828). However, each of these rights was subject to a reciprocal duty.

The statutory right was subject to the duty, under section 85(3), not to cause unnecessary inconvenience to adjoining owners by the time or manner of its exercise. The common law right was subject to a duty, referred to in *Cubitt v Porter*, that the work be undertaken expeditiously and without delay.

The decision

Both the statutory and common law duties had been breached as there had been an unacceptable delay in reinstating the wall. On the basis of the decisions in *Bower v Peate* (1876) and *Hughes v Percival* (1883) the court held that both duties were non-delegable duties.

Woodhouse was therefore personally liable for the breaches. As the duties were non-delegable, it was no defence that he had employed a competent architect and contractor to undertake the work.

> Lord Justice Lopes: "*Cubitt v Porter* decided that an action would not lie against an adjoining owner for pulling down a party wall for the purpose of rebuilding it, but he was bound to do so without causing any unnecessary inconvenience to his neighbour, and without unreasonable delay.

> ## Metropolitan Building Act 1855
>
> ### PART III: PARTY STRUCTURES
>
> ### RIGHTS OF BUILDING AND ADJOINING OWNER
>
> **83. Rights of building owner**
>
> The building owner shall have the following rights in relation to party structures; that is to say:
>
> (7) A right to pull down any party structure that is of insufficient strength for any building intended to be built, and to rebuild the same of sufficient strength for the above purpose, upon condition of making good all damage occasioned thereby to the adjoining premises, or to the internal finishings and decorations thereof.
>
> **85. Rules as to exercise of rights by building and adjoining owners**
>
> The following rules shall be observed with respect to the exercise by building owners and adjoining owners of their respective rights:
>
> (3) No building owner shall exercise any right hereby given to him in such manner or at such time as to cause unnecessary inconvenience to the adjoining owner.

That also appeared from section 85(3) of the Metropolitan Building Act 1855, which provided that no building owner should exercise any right thereby given 'in such manner or at such time as to cause unnecessary inconvenience to the adjoining owner'.

An absolute duty was thrown upon the person knocking down the wall to do so within a reasonable time and in a reasonable manner, and he could not delegate that duty to a contractor."

Lord Justice Davey: "[Woodhouse's] argument could only be supported by confusing two classes of cases. Where a man owed a duty to his neighbour he could not free himself of that duty by delegating it to another person.

Where, on the other hand, a man employed a contractor to do work which, if properly and skilfully done, did not involve any interference with his neighbour's rights, but an injury had been caused by the negligence of the contractor, then it might be that the employer would not be liable if he employed a competent contractor.

Here [Woodhouse] had a statutory and common-law licence to take down this party wall, subject to the duty of using reasonable despatch, and when he employed a contractor to do the work he took upon himself the responsibility of seeing that that duty was adequately performed."

The court therefore rejected Woodhouse's application for a new trial and affirmed the earlier judgment in Jolliffe's favour.

JONES v PRITCHARD (1908)
Chancery Division

In brief

- **Issues:** Longitudinally divided party wall – cross easements to use flues – damage to servient flue – smoke escaping from dominant tenement into servient tenement – entitlement/obligation to repair subject matter of easement.

- **Facts:** A longitudinally divided party wall separated Jones' and Pritchard's houses. Each party had an easement to use the flues within the wall which served the fireplaces in their respective properties. Settlement to Pritchard's house caused damage to a part of his flue which was situated on Jones' side of the line of junction within the party wall. As a result, smoke escaped from the flue and caused damage to Jones' property.

- **Decision:** Both parties had an entitlement to repair the flue but neither was obliged to do so. A servient owner (Jones) has no obligation to repair the subject matter of an easement and can let it fall into disrepair. In exceptional circumstances a dominant owner (Pritchard) may have to repair the subject matter to prevent it causing damage to the servient property. Where, for example, the dominant owner lays water pipes over the servient property he must maintain them to prevent an escape of water. Pritchard owed no such duty in the present case. He was simply using the servient tenement in the way it was contemplated that he would use it when the easement was granted. In the absence of any fault on his part he could not therefore be held liable for damage which results from this. Jones' claim for an injunction was therefore dismissed and it was left for him to undertake the repairs at his own expense.

- **Notes:** It is unclear whether recent developments in the law of negligence and nuisance would have any impact on the decision in this case. For example, might Pritchard now be required to repair the flue to prevent damage to Jones' property even though it is situated on Jones' land? See: *Leakey v National Trust for Places of Historic Interest or Natural Beauty* (1980); *Bradburn v Lindsay* (1983); *Holbeck Hall Hotel Ltd v Scarborough Borough Council* (2000); *Rees v Skerrett* (2001).

The facts

The case concerned two large semi-detached houses in Beaumaris, Anglesea, which were separated by a longitudinally divided party wall (Figure 11).

Pre - 1886

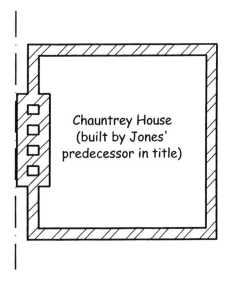

Post - 1900

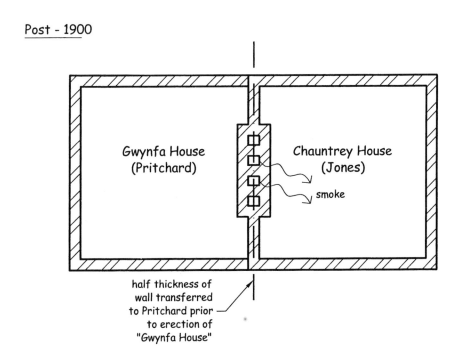

Figure 11: Jones v Pritchard

Jones' house, Chauntrey House, had been built as a detached house by Owen, the previous owner. Owen had built the house right up to the boundary with some adjacent vacant land owned by Pritchard. At the time he had anticipated that Pritchard might wish to enclose on to Chauntrey House at some stage in the future with a house to be built on his own land. He had therefore included additional flues in the construction of the gable wall facing on to Pritchard's land.

In 1886 Pritchard indeed decided to build a house on his land and before doing so he purchased half the thickness of the gable wall from Owen (Figure 11). He then immediately erected Gwynfa House against this wall and made use of the additional flues for fireplaces constructed in the property, including the fireplace in the dining room.

In 1900 Jones purchased Chauntrey House from Owen. The flue from Pritchard's dining room subsequently became defective due to cracks caused by settlement or subsidence of parts of Gwynfa House. As a result, smoke escaped through cracks and percolated into Chauntrey House, causing damage to its decorations and furniture.

Jones issued proceedings against Pritchard for an injunction to restrain the alleged nuisance. The case addressed the issue of whether the particular circumstances complained of amounted to an actionable nuisance. This involved a consideration of the law relating to longitudinally divided party walls and to aspects of the general law of easements.

Rights in longitudinally divided party walls

Firstly, Parker J described the property rights of the respective owners in this type of party wall:

> "Now if a man grant a divided moiety of an outside wall of his own house, with the intention of making such wall a party wall between such house and an adjoining house to be built by the grantee, the law will, I think, imply the grant and reservation in favour of the grantor and grantee respectively of such easements as may be necessary to carry out what was the common intention of the parties with regard to the user of the wall, the nature of those easements varying with the particular circumstances of each case.
>
> Subject, however, to the easements of which the grant or reservation will be so implied, the grantor and grantee, being respectively absolute owners of their respective divided moieties of the wall, may respectively deal with such moieties in such manner as they please."

In the present case each of the parties was therefore free to deal with their own half of the wall subject only to the implied easements of the other party. These easements related to the entitlement of each party to use the entire width of the flues from their own fireplaces, irrespective of the fact that one half of their width was situated within the thickness of the wall owned by the other party. Parker J described the primary obligation of each party in respect of the other's easement over the flues in the following terms:

> "In the first place, it appears to me that on principle the owner of a servient tenement cannot so deal with it as to render the easement over it incapable of being enjoyed or more difficult of enjoyment by the owner of the dominant tenement.
>   On this principle neither party can in the present case pull down his half of the wall or stop up his half of any flue used for the purpose of his neighbour's house."

However, there was no allegation, in the present case, that an easement was being interfered with. The issue before the court related to the responsibility for the repair of parts of a flue owned by Jones but used by Pritchard by virtue of an easement. This raised the question of whether each of the parties had an *entitlement* to undertake the repairs as well as whether they were *obliged* so to do.

Entitlement to undertake repairs to the subject matter of an easement

The court noted that each of the parties clearly had an entitlement to undertake the necessary repairs on Jones' side of the party wall, had they chosen to do so.

As already established, Jones, as the owner of this half of the wall, was entitled to deal with it as he pleased as long as he did not interfere with Pritchard's easement. This would clearly include an entitlement to undertake necessary repairs for the benefit of both parties.

Pritchard's entitlement to undertake the repairs arises out of his common law right to take reasonable steps to preserve the enjoyment of his easement as described by Parker J:

> "Once again, the grant of an easement is prima facie also the grant of such ancillary rights as are reasonably necessary to its exercise or enjoyment. Thus the grantee of an easement for a watercourse through his neighbour's land may, when reasonably necessary, enter his neighbour's land for the purpose of repairing, and may repair, such watercourse.
>   On this principle each party in the present case may do such acts on the property of the other as are reasonably necessary to the continued enjoyment of the easement; for example, each party would be entitled to repair the other's half of the wall in question so far as was reasonably necessary for the enjoyment of any easement impliedly granted or reserved."

Servient owner under no obligation to repair

Although both parties therefore had an *entitlement* to undertake the necessary repairs, this did not mean that either of them was necessarily *obliged* to do so. Firstly, Parker J had no doubt that Jones, as the servient owner, was under no obligation to repair:

> "… it appears to me that, apart from any special local custom or express contract, the owner of a servient tenement is not bound to execute any repairs necessary to ensure the enjoyment or convenient enjoyment of the easement by the owner of the dominant tenement. The grantor of a right of way over a bridge is not by common law liable, nor does he impliedly contract, to keep the bridge in repair for the convenience of the grantee.
>
> On this principle neither party is in the present case subject to any liability if, by reason of natural decay or other circumstances beyond his control, his half of the wall falls down or otherwise passes into such a condition that the easement thereover becomes impossible or difficult of exercise."

Dominant owner under any obligation to repair?

The more pertinent issue, in the present case, was whether Pritchard, as the dominant owner, was obliged to undertake the repairs necessary to prevent the encroachment of smoke into Jones' property. Parker J rejected the general notion that there was a common law rule that "he that hath the use of a thing ought to repair it". However, he did acknowledge that there were particular classes of easement where the dominant owner would be liable for losses resulting from his failure to repair the subject matter of the easement:

> "… there is undoubtedly a class of cases in which the nature of the easement is such that the owner of the dominant tenement not only has the right to repair the subject of the easement, but may be liable to the owner of the servient tenement for damages due to any want of repair.
>
> Thus, if the easement be to take water in pipes across another man's land and pipes are laid by the owner of the dominant tenement and fall into disrepair, so that water escapes on to the servient tenement, the owner of the dominant tenement will be liable for damage done by such water."

Despite this rule, Parker J did not consider that it was applicable to Pritchard's easement in respect of the flue:

> "In my opinion, however, the defendant's easement is in the present case of a different nature. It is, I think, the right to allow smoke from his fireplace to pass into the flue connected therewith, notwithstanding that at any point in its

passage up the flue it may pass from the defendant's moiety into the plaintiff's moiety of the flue.

If this be so, the principle I am considering can have no application, for, though there be cracks in the plaintiff's moiety of the flue, the defendant is none the less exercising his rights fairly and reasonably in the manner in which he was intended to exercise them when the easement was granted...

... if a grantor is doing, on land retained by him, only what it was at the time of the grant in the contemplation of the parties that he should do, and is guilty of no negligence or want of reasonable care or precaution, he cannot be liable for nuisance entailed upon the grantee.

And, of course, where the grantee is occasioning a nuisance to the grantor by doing on the land granted what it was at the time of the grant in the contemplation of the parties that he should do, and there is no negligence or want of care or precaution, the case is an a fortiori one."

The decision

There was no evidence that the settlement or subsidence in Pritchard's house was due to any negligence on his part, either in its construction or subsequently. In the absence of any such fault he could not therefore be liable for simply using the flue in the way contemplated by both parties.

Pritchard was not liable in nuisance. The action therefore failed and it was left for Jones, at his own expense, to undertake the necessary repairs to his own side of the party wall.

KNIGHT v PURSELL (1879)
Chancery Division

In brief

- **Issues:** 1855 Act – definition of type 'b' party wall – building owner's right to undertake work to wall in sole ownership of adjoining owner – wall performing separating function for only part of its length.

- **Facts:** A freestanding wall stood entirely on land owned by Knight. It marked the position of the boundary with Pursell's land. Both parties had erected lean-to sheds at various points along the length of the wall. Although the wall was owned entirely by Knight, Pursell proposed to demolish and rebuild it and served notice on Knight to this effect under the 1855 Act.

- **Decision:** Providing the wall was a party wall Pursell was entitled to undertake the work under the Act notwithstanding Knight's ownership of it. The wall was a party wall in respect of those portions where the parties had enclosed on both sides of it and where it therefore performed a separating function. Pursell was entitled to undertake the work to these parts of the wall but an injunction was granted to prevent any interference with the rest of the wall.

- **Notes:** See also: *Weston v Arnold* (1873); *Drury v Army & Navy Auxilliary Co-operative Supply Ltd* (1896).

The facts

Knight was the leasehold owner of land at 6 and 7 Surrey Row, off Blackfriars Road in Southwark. Pursell owned adjacent land at 172 Blackfriars Road.

The boundary between the two plots was marked by a freestanding wall. This was owned by Knight and built entirely on his land. At some stage in the past Pursell had constructed a substantial timber lean-to structure against his side of the wall. Subsequently, Knight had also erected a number of small lean-to sheds and closets against his side of the wall.

The wall was in poor condition and Pursell wished to demolish and rebuild it to an acceptable standard. He therefore served a party structure notice on Knight, under the 1855 Act, of his intention to do so. Knight objected to the proposed interference with his wall and applied for an injunction to restrain the alleged trespass.

Ownership of party wall

Pursell argued that the wall was a party wall within the meaning of the 1855 Act. Despite the fact that it was owned entirely by Knight, he therefore claimed to be entitled to do the work, subject only to compliance with the Act's procedural requirements.

Knight argued that a wall could not be a party wall within the Act unless both adjoining owners had some property in it. The Act was not intended to take control over a wall away from its proper owner.

The court did not agree with Knight's argument about the purpose of the legislation.

> Fry J: "[The plaintiff's counsel] ... contended that the Act could not be intended to take away the control over the wall from the proper owner of it, but Acts of Parliament often take away some control of owners over their property for public purposes...
>
> The object of the Act is to limit the acts of private owners for the general benefit of the public, to prevent the spread of fire, and for similar purposes."

The court considered the definition of "party wall" in section 3 of the Act. This made no reference to the ownership of the wall but simply to its function in separating buildings[1]. The court concluded that the legislation was concerned with the physical aspects of the wall, irrespective of the property rights within it:

> Fry J: "It appears to me, on reading the definition of a party wall contained in the third section of the Act, that the intention is to define a party wall, not by reference to the rights of ownership which the adjoining proprietors may have in any particular wall in dispute, but by reference to the mode in which the wall is used. It is a question, not of title, but of user.
>
> ... And therefore, in order to determine whether this wall is a party wall, it is not necessary to consider what rights of ownership the plaintiff and the defendant have, but what is the physical condition, position, and user of the wall."

Separation of buildings by party wall

The court considered that the structure on Pursell's side of the wall and the sheds and closets on Knight's side were all "buildings" within the meaning of the Act. Thus, for much of its length, the wall was indeed used for the

---

[1] In terms similar to the definition of a type 'b' party wall in section 20 of the 1996 Act.

> **Metropolitan Building Act 1855**
>
> PRELIMINARY
>
> **3. Interpretation of certain terms in this Act**
>
> In the construction of this Act (if not inconsistent with the context) the following terms shall have the respective meanings hereinafter assigned to them; (that is to say):
>
> "Party wall" shall apply to every wall used or built in order to be used as a separation of any building from any other building, with a view to the same being occupied by different persons.

separation of buildings occupied by different persons, as required by the Act's definition of "party wall" in section 3.

However, the court noted the case of *Weston v Arnold* (1873) which had held that a wall could be a party wall for only part of its height. It concluded that the same principles should be applied along the length of the wall in the present case.

> Fry J: "In the same way I hold this wall to be, laterally, a party wall for such distance as it is used by both plaintiff and defendant for their buildings and no further; the result is that for the distances that the plaintiff's closets extend, the wall is a party wall, and upon giving the proper notices under the Act, the defendant may deal with that part of the wall in accordance with section 88 of the Act."

The decision

The court upheld Pursell's right to undertake the works to the sections of the wall with buildings on either side. However, it also upheld the plaintiff's claim in respect of even the smallest lengths of wall where this was not the case:

> Fry J: "However small the extent of wall may be which is not a party wall, it is important to the plaintiff to have his rights over it protected, for, being a lessee, he is bound to deliver his holding up to his landlord in the same condition as when it was leased to him, and I cannot hold, therefore, that any trespass even

on a small part of the wall can be too trivial a nature to be remedied in this court."

The court therefore granted an injunction to restrain any interference, by Pursell, with those parts of the wall which were not a party wall within the Act.

LEADBETTER v MARYLEBONE CORPORATION [No. 1] (1904)

Court of Appeal

In brief

- **Issues:** 1894 Act – surveyors' jurisdiction – validity of surveyors' award – adjoining owner's right to undertake future work without service of party structure notice – meaning of "any other matter arising out of or incidental to the difference".

- **Facts:** Leadbetter demolished and rebuilt a party wall in pursuance of an award under the 1894 Act. The wall was rebuilt to an increased specification to facilitate future works by Marylebone to their adjoining property. The award also provided that they had the right, at any time in the future, to raise the wall. Two years later Marylebone started raising the wall without first serving a party structure notice on Leadbetter. Leadbetter issued proceedings for trespass.

- **Decision:** An injunction was granted to restrain the work. The new work was entirely separate from that previously undertaken. Marylebone were now acting in the capacity of building owner under the Act and were therefore required to serve a party structure notice before starting the work. Appointed surveyors had a strictly limited jurisdiction to adjudicate on matters referred to in a party structure notice. They had no jurisdiction to adjudicate on possible future works which had not yet been the subject of a party structure notice. The part of the surveyors' award which purported to authorise Marylebone's future work was therefore invalid. The surveyors' power to determine "any other matter arising out of or incidental to the difference" was not capable of clothing the surveyors with jurisdiction to decide matters which did not arise out of the particular works referred to in the party structure notice.

- **Notes:** See also: *Burlington Property Company Ltd v Odeon Theatres Ltd* (1938); *Woodhouse v Consolidated Property Corporation Ltd* (1993); *Leadbetter v Marylebone Corporation [No. 2]* (1905).

The facts

Numbers 33 and 34 John Street, Edgware Road, London, were adjoining properties, separated by a party wall. Leadbetter owned number 33 and Marylebone Corporation owned number 34.

In 1902 Leadbetter decided to demolish and rebuild his property and entered into a building contract with Ecclestone who agreed to undertake the work. Under the contract the property was to be leased to Ecclestone for the duration of the works and the lease was to be surrendered on completion.

The work involved the demolition and rebuilding of the party wall so Ecclestone (who was the building owner by virtue of his building lease) served a party structure notice on Marylebone under the 1894 Act. Marylebone wished to safeguard their ability to raise the wall in the future. They therefore served a counternotice on Ecclestone, requiring the wall to be built to a greater thickness than that originally proposed[1].

In due course the surveyors' award required the wall to be rebuilt to the thickness referred to in the counternotice. The award also provided that Marylebone were to have the right, at any time in the future, to raise the wall.

Neither party challenged the validity of the award. The building work was undertaken in accordance with its provisions and Ecclestone duly surrendered the building lease to Leadbetter upon completion.

The right to raise a party wall in the future?

In 1904 Marylebone started work to increase the height of the party wall. They served no prior notice on Leadbetter under the 1894 Act.

Leadbetter objected to the work being carried out without him receiving the Act's protection. He argued that, although Marylebone had been an "adjoining owner" when the earlier work was undertaken, they were now a "building owner" within the meaning of the Act. They must therefore serve a party structure notice on him in respect of their proposed works to the party wall.

Marylebone sought to rely on their entitlement to raise the wall which had been granted in the earlier award. Leadbetter argued that the surveyors had no jurisdiction to deal with future works which had not been referred to them and that this provision in their award was therefore ultra vires.

Leadbetter issued proceedings against Marylebone and obtained an interlocutory injunction to restrain the work. Marylebone appealed to the Court of Appeal.

New works require service of new party structure notice

The court held that when Marylebone became desirous of building upon the party wall they became 'building owners' within the meaning of the

---

[1] Under section 89. The equivalent provision now appears in section 4 of the 1996 Act.

Act and were therefore obliged to serve a party structure notice on Leadbetter before commencing the work.

Marylebone had argued that they retained their status as "adjoining owners" under the earlier award by analogy with section 95(2) of the Act[2]. That section provided for the payment of a proportion of the cost of work by an adjoining owner under an earlier award, if he subsequently makes use of work paid for by the building owner.

The court rejected this argument. It distinguished section 95(2), which simply reapportioned the cost of existing work, from the present situation where entirely new building operations were proposed.

> Collins MR: "... [section 95(2)] does not deal with an alteration of, or addition to, a party wall, but with claims to make a greater use of a party wall of dimensions already fixed, in which cases provision as to the mode in which the expenses are to be borne is made by the section.
>
> In other words, the section contemplates only the use to which the party wall may be applied, and has no application to building operations, to which different provisions are applicable; where fresh building operations are contemplated, there must be a building owner's notice."

Surveyors' jurisdiction

The court held that the surveyors (who they described as arbitrators) had acted beyond their jurisdiction in awarding that Marylebone were entitled to raise the wall in the future. This part of their award was therefore invalid.

In particular the surveyors' jurisdiction was limited to differences concerning work referred to in the party structure notice as this formed the basis for their appointment. The raising of the party wall was an entirely separate category of work from that referred to in Ecclestone's original notice and they had no jurisdiction to adjudicate upon it.

The judgments emphasise that the purpose of the legislation is to provide protection for adjoining owners. If the court had allowed the award to stand in its original form it would have deprived Leadbetter of this very protection.

> Collins MR: "According to the view put forward by the defendants, when once the jurisdiction of surveyors as to the dimensions of a party wall which the building owner is then proposing to erect has arisen, they may once and for all decide what is to be done as regards the wall when the position of the parties is reversed and the adjoining owner wants to build upon it.

---

[2] Equivalent to section 11(11) of the 1996 Act.

But, in my judgment, the code laid down in the Act necessarily negatives this contention, which would sweep away all the provisions contained in the code which were framed in contemplation of the adjoining owner himself becoming the building owner. I think, therefore, that the jurisdiction which the surveyors had over the dispute was a strictly limited jurisdiction, and that they have exceeded it in that part of their award which has been impugned."

Stirling L J: "I am of the same opinion and entirely adopt the view taken by my Lord. The opposite view would in my judgement deprive the adjoining owner of the benefit of the express statutory provisions passed for his protection...

Section 89[3] was passed for the benefit of adjoining owners, who are entitled to require various works to be done by a building owner who is building or altering a party wall, one of which (I mention it by way of illustration) is to build the requisite chimneys...

... if the award is good, the defendants would in this respect be emancipated from the obligations of a building owner, and it will be possible for them to raise the party wall to any height without building corresponding chimneys. This seems to me to be impossible."

Surveyors' powers

Marylebone had also argued that the wording of section 91[4] was wide enough to give validity to the disputed parts of the surveyors' award. This invested the surveyors with power to determine a number of matters by their award, including "any other matter arising out of or incidental to" the original difference. The court was unimpressed by this argument and, once again, emphasised that surveyors were appointed on the basis of work referred to in the original notice. The Act had to be understood in these terms.

Collins MR: "Reliance is placed by the defendants upon section 91, which gives the surveyors 'power by his or their award to determine the right to do and the time and manner of doing any work, and generally any other matter arising out of or incidental to such difference'.

It is contended that the section clothes the surveyors, who are deciding as to the conditions under and the mode in which a party wall shall be built, with jurisdiction to deal with what is to be done after the wall has been built, and with jurisdiction to allow the adjoining owner at some future date to build upon it without giving the notice or resorting to the machinery prescribed by the Act.

But, in the first place, section 91 is expressly and in terms limited to matters

---

[3] The entitlement to serve a counternotice.
[4] Equivalent to section 10(12)(c) of the 1996 Act.

referred to in the notice; it does not give the surveyors general jurisdiction over every dispute 'in respect of any matter arising with reference to any work to which any notice given under this part of the Act relates'; it cannot oust the fresh jurisdiction of a fresh surveyor in a case where the person who was originally an adjoining owner has turned himself into a building owner.

I think, therefore, that the surveyors had no jurisdiction to decide any question as to the right of the defendants to build on the party wall; that that part of their award was, therefore, made without jurisdiction, and that the parties are left to the code provided by the statute. The appeal must, therefore, be dismissed."

The decision

The court dismissed Marylebone's appeal. They had no right to raise the party wall without first complying with the 1894 Act's procedural requirements. The injunction therefore remained in force against them.

London Building Act 1894

Part VIII: RIGHTS OF BUILDING AND ADJOINING OWNERS

89. Rights of adjoining owner

(1) Where a building owner proposes to exercise any of the foregoing rights with respect to party structures the adjoining owner may by notice require the building owner to build on any such party structure such chimney copings jambs or breasts or flues or such piers or recesses or any other like works as may fairly be required for the convenience of such adjoining owner and may be specified in the notice and it shall be the duty of the building owner to comply with such requisition in all cases where the execution of the required works will not be injurious to the building owner or cause to him unnecessary delay in the exercise of his right.

90. Rules as to exercise of rights by building and adjoining owners

(1) A building owner shall not except with the consent in writing of the adjoining owner and of the adjoining occupiers or in cases where any wall or party structure is dangerous (in which cases the provisions of Part IX of this

Act shall apply) exercise any of his rights under this Act in respect of any party fence wall unless at least one month or exercise any of his rights under this Act in relation to any party wall or party structure other than a party fence wall unless at least two months before doing so he has served on the adjoining owner a party wall or party structure notice stating the nature and particulars of the proposed work and the time at which the work is proposed to be commenced.

91. Settlement of differences between building and adjoining owners

(1) In all cases (not specifically provided for by this Act) where a difference arises between a building owner and adjoining owner in respect of any matter arising with reference to any work to which any notice given under this Part of this Act relates unless both parties concur in the appointment of one surveyor they shall each appoint a surveyor and the two surveyors so appointed shall select a third surveyor and such one surveyor or three surveyors or any two of them shall settle any matter from time to time during the continuance of any work to which the notice relates in dispute between such building and adjoining owner with power by his or their award to determine the right to do and the time and manner of doing any work and generally any other matter arising out of or incidental to such difference but any time so appointed for doing any work shall not unless otherwise agreed commence until after the expiration of the period by this Part of this Act prescribed for the notice in the particular case.

95. Rules as to expenses in respect of party structures

(2) As to expenses to be borne by the building owner:

If at any time the adjoining owner make use of any party structure or external wall (or any part thereof) raised or underpinned as aforesaid or of any party fence wall pulled down and rebuilt as a party wall (or any part thereof) beyond the use thereof made by him before the alteration there shall be borne by the adjoining owner from time to time a due proportion of the expenses (having regard to the use that the adjoining owner may make thereof):

(i) Of raising or underpinning such party structure or external wall and of making good all such damage occasioned thereby to the adjoining owner and of carrying up to the requisite height all such flues and chimney stacks belonging to the adjoining owner on or against any such party structure or external wall as are by this Part of this Act required to be made good and carried up;

(ii) Of pulling down and building such party fence wall as a party wall.

LEADBETTER v MARYLEBONE CORPORATION [No. 2] (1905)

Court of Appeal

In brief

- **Issues:** 1894 Act – compliance with injunction – statutory time limit for continued effectiveness of party structure notice – effect of time limit on surveyors' award.

- **Facts:** Marylebone had previously started raising a party wall separating their property from that owned by Leadbetter. An injunction had then been granted restraining any further works until regularised in accordance with the 1894 Act. They subsequently served a party structure notice. An award had still not been made six months later. Leadbetter alleged that the statutory time limit for commencing the work had therefore expired. He sought the removal of the earlier works to the party wall on the basis that these placed Marylebone in breach of the injunction.

- **Decision:** Marylebone were not in breach of the injunction as they were actively seeking to comply with the Act's requirements. The six-month time limit for commencement of works only applied where an adjoining owner consented to the works and the building owner therefore proceeded under the authority of the original notice. Once a difference arose in respect of a notice the statutory time limit had no further relevance. In these circumstances the building owner could only proceed under the authority of an award and this might take time to finalise. The statutory time limit had no effect on the proceedings and Marylebone were therefore free to proceed with the works as soon as an award was published.

- **Notes:** (1) Because of this decision it is good practice for surveyors to include a time limit for commencement of the works in their award. Awards commonly require works to commence within 12 months. The statutory time limit for proceeding on the authority of a party structure notice with an adjoining owner's consent has also been increased to 12 months by section 3(2)(b) of the 1996 Act. (2) See also: *Leadbetter v Marylebone Corporation [No. 1]* (1904).

The facts

This was the second of two cases between Leadbetter and Marylebone Corporation concerning their respective properties at 33 and 34 John

Street. Marylebone wished to raise the party wall between the two properties. In the previous case the Court of Appeal had upheld an injunction restraining them from undertaking the work without first serving a party structure notice on Leadbetter.

The party structure notice was then served on 11 August 1904 but Leadbetter failed to respond to it. A difference was deemed to arise under the Act but he then failed to appoint a surveyor. Eventually, he grudgingly appointed a surveyor after Marylebone had served a ten-day notice on him under section 91(3) of the 1894 Act[1]. The two party-appointed surveyors then began negotiations about the terms of an award.

By 11 February 1905, six months after service of the party structure notice, an award had still not been published. Leadbetter therefore advised his appointed surveyor that, in accordance with section 90(4)[2], his powers to make an award under the Act had now lapsed. His surveyor then declined to take any further steps to progress the award.

Leadbetter then applied to the county court for an order directing that Marylebone remove those parts of their new building which had been erected prior to commencement of the earlier proceedings. He maintained that the presence of these works amounted to a breach of the earlier injunction as Marylebone had failed to regularise the situation under the Act. The court at first instance refused to grant the order. Leadbetter appealed to the Court of Appeal.

Expiry of party structure notice?

The Court of Appeal did not accept that the party structure notice had expired under section 90(4). The six-month time limit imposed by that section only applied to circumstances where an adjoining owner consents to a notice. Where, as in the present case, a difference arises and surveyors are appointed to make an award, it had no application. It would be impractical to impose a time limit for the making of an award and this was not the Act's intention.

> Mathew LJ: "[Section 90(4)] appears to me to provide for cases in which a building owner's notice is given with respect to a party wall, and, the adjoining owner consenting thereto, no difference arises between the building owner and the adjoining owner.
>
> But another and most important section in reference to this subject is section 91, which contemplates the case in which, a building owner's notice

---

[1] Equivalent to section 10(4) of the 1996 Act.
[2] Similar provisions appear in sections 3(2)(b) and 6(8) of the 1996 Act although the time limit is now 12 months.

> **London Building Act 1894**
>
> Part VIII: RIGHTS OF BUILDING AND ADJOINING OWNERS
>
> **90. Rules as to exercise of rights by building and adjoining owners**
>
> (4) A party wall or structure notice shall not be available for the exercise of any right unless the work to which the notice relates is begun within six months after the service thereof and is prosecuted with due diligence.
>
> **91. Settlement of differences between building and adjoining owners**
>
> (3) If either party to the difference make default in appointing a surveyor for ten days after notice has been served on him by the other party to make such appointment the party giving the notice may make the appointment in the place of the party so making default.

having been served, the differences arise, in which case there is to be an arbitration by surveyors for the settlement of those differences.

I cannot see any indication in the Act that the limit of six months given by section 90(4) is to apply to such a case. Under section 91 there may be an appeal to the county court from an award of the surveyors, or in some cases to the High Court, and a point of law may be raised by means of a case stated for the opinion of the High Court. It is obvious that in such a case a period exceeding six months may necessarily elapse before the work can be begun."

Rationale for statutory time limit

As well as discussing the practical difficulties of commencing works within six months, Cozens-Hardy LJ also referred to the possible reason for the provision as providing protection for a consenting adjoining owner:

> "Certain rights are exercisable by a building owner, who gives a party-wall notice under section 90, as against an adjoining owner who consents to that notice.
>
> It is a very reasonable provision that the building owner who has served such a notice, to which the adjoining owner has consented, should not be at liberty to exercise those rights after allowing a long period of time, say, for instance, two years, to elapse without acting on the notice.

But different considerations appear to me to apply to a case where, differences having arisen, further proceedings have to be taken, and surveyors appointed to settle those differences under section 91. In that case the rights of the parties under the Act would appear to depend not so much on the party-wall notice as on the award made by the surveyors.

I cannot think that the intention was that in that case the whole of the proceedings should become nugatory, unless the award could be perfected, which might involve the stating and decision of a case for the High Court on a legal point, within the period of six months from the giving of the party wall notice. In my view that is not the true effect of the Act."

The decision

As the party structure notice had not expired Marylebone were acting in accordance with the court's earlier decision in attempting to secure an award to authorise the work. They were not, therefore, in breach of the injunction against carrying out the work without proper authority. The court therefore dismissed Leadbetter's appeal.

LEAKEY v NATIONAL TRUST FOR PLACES OF HISTORIC INTEREST OR NATURAL BEAUTY (1980)

Court of Appeal

In brief

- **Issues:** Damage to neighbouring land from natural causes – building owner's liability to adjoining owner in tort – negligence – nuisance – relationship between negligence and nuisance.

- **Facts:** A steep mound of earth on the Trust's land collapsed and caused damage to the plaintiffs' adjoining land. Although the mound collapsed due to natural causes the Trust had been aware that it was unstable and had taken no action to stabilise it.

- **Decision:** The Trust were liable for the damage. The law imposed a duty of care on occupiers to prevent or minimise the risk to adjoining owners and their property. The duty is owed in respect of hazards emanating from the occupier's property, whether arising from man-made or natural causes. The scope of the duty is less onerous than the duty of care in negligence. A subjective test of reasonableness is applied which takes account of the parties' respective resources and ability to remove the hazard. The Trust were in breach of the duty and therefore liable. It was unclear whether liability arose in nuisance or negligence. Previous authorities suggested that liability arose in negligence. It was no bar to the plaintiffs' claim that they had pleaded their case in nuisance. Liability probably arose in nuisance but if this was wrong it made no difference to the success of the plaintiffs' claim.

- **Notes:** See also: *Bradburn v Lindsay* (1983); *Holbeck Hall Hotel Ltd v Scarborough Borough Council* (2000); *Rees v Skerrett* (2001); *Johnson (T/A Johnson Butchers) v BJW Property Developments Ltd* (2002).

The facts

The plaintiffs were the owners of two houses at the base of a steep hill known as Burrow Mump. Burrow Mump was owned by the National Trust and was in an unstable condition. Its soil was composed of keaper marl which made it prone to cracking and slipping. As a result of the natural weathering of the land there had been numerous landslides into the plaintiffs' property over the preceding years.

In 1976 a prolonged summer drought was followed by an unusually wet winter. This led to a large crack opening up in the hill immediately above one of the plaintiffs' houses. The plaintiffs brought this to the Trust's attention and warned them of the risk of collapse.

The Trust advised the plaintiffs that they had no responsibility for the situation because it was simply a result of the natural movement of the earth. They refused to take any action and a few weeks later a large landslide occurred. The debris filled the plaintiffs' gardens and threatened to cause damage to their houses if not immediately removed. The Trust refused to remove the debris or to take any protective measures to prevent further landslides.

The plaintiffs issued proceedings against the Trust for damages and for an injunction. They founded their claim in the law of nuisance. At first instance the court held that the Trust were liable. An injunction was therefore granted and damages were awarded. The Trust appealed to the Court of Appeal.

Liability for damage to adjoining land from natural causes?

Under their first ground of appeal, the Trust maintained that no liability could arise in nuisance where damage to adjoining land arose from entirely natural causes. This required the court to consider the significance of fault on the part of a defendant in actions for nuisance.

Some degree of fault is almost always necessary for liability in nuisance to arise. A defendant will therefore be liable where he directly causes the nuisance, or where he fails to take some action which would have prevented it from arising in the first place. But what of the situation, as in the present case, where a nuisance arises due to no fault on the part of the defendant? Can he become liable, in these circumstances, for simply allowing it to continue?

There are a number of early authorities which support the view that there is no liability for a nuisance which arises from natural causes[1]. However, in *Sedleigh-Denfield v O'Callaghan* (1940), an occupier of land was found liable for a nuisance caused by a trespasser which he had subsequently allowed to continue. Viscount Maughan explained in that case that an occupier will be liable for continuing a nuisance "if, with knowledge or presumed knowledge of its existence, he failed to take any reasonable means to bring it to an end".

In the Australian case *of Goldman v Hargrave* (1966) the Privy Council subsequently applied the *Sedleigh-Denfield* principle to a nuisance which had arisen from entirely natural causes. In that case a landowner was held liable for damage caused to neighbouring properties by fire. This had started when lightning struck a tall tree on his land but he had failed to take adequate steps to extinguish it.

The question which arose in the present case was whether the decision

---

[1] *Hodgson v York Corporation* (1873); *Giles v Walker* (1890).

in *Goldman v Hargrave* accurately stated the law of England, or whether there was a distinction between nuisances caused by trespassers, where liability would arise under the *Sedleigh-Denfield* principle, and those arising from natural causes, where it would not.

Nature of duty to adjoining owners

The court held that no such distinction should be made and that the decision in *Goldman v Hargrave* also represented the legal position in England. There was a general duty of care imposed on occupiers in relation to hazards occurring on their land, whether natural or man-made. The nature of this duty was to do all that is reasonable in the circumstances to prevent or minimise the risk to neighbours and their property. The scope of this duty was explored by Megaw LJ:

> "This leads me on to the question of the scope of the duty. This is discussed, and the nature and extent of the duty is explained, in the judgment in *Goldman v Hargrave*. The duty is a duty to do that which is reasonable in all the circumstances, and no more than what, if anything, is reasonable, to prevent or minimise the known risk of damage or injury to one's neighbour or to his property.
>
> The considerations with which the law is familiar are all to be taken into account in deciding whether there has been a breach of duty, and, if so, what that breach is, and whether it is causative of the damage in respect of which the claim is made... What... are the chances that anything untoward will happen or that any damage will be caused? What is to be foreseen as to the possible extent of the damage if the risk becomes a reality? Is it practicable to prevent, or to minimise, the happening of any damage? If it is practicable, how simple or how difficult are the measures which could be taken ... and what is the probable cost of such works?"

A subjective test of reasonableness

Despite the familiarity of these considerations, Megaw LJ emphasised that the law did not require the imposition of an objective "reasonable man" test in these circumstances. To avoid imposing an excessive burden on occupiers where a hazard has arisen naturally on their land, the law would take account of the resources of the particular defendant in deciding what is reasonable:

> "The defendant's duty is to do that which it is reasonable for him to do.
> The criteria of reasonableness include, in respect of a duty of this nature, the factor of what the particular man, not the average man, can be expected to do,

having regard, amongst other things, where a serious expenditure of money is required to eliminate or reduce the danger, to his means.

Just as, where physical effort is required to avert an immediate danger, the defendant's age and physical condition may be relevant in deciding what is reasonable, so also logic and good sense require that, where the expenditure of money is required, the defendant's capacity to find the money is relevant.

But this can only be in the way of a broad, and not a detailed, assessment; and, in arriving at a judgment on reasonableness, a similar broad assessment may be relevant in some cases as to the neighbour's capacity to protect himself from damage, whether by way of some form of barrier on his own land or by way of providing funds for expenditure on agreed works on the land of the defendant."

This raises the question as to whether the duty of care to prevent the continuance of a nuisance which has already arisen forms part of the law of negligence or of nuisance. The more subjective concept of reasonableness has more in common with that from the law of nuisance. However, the concept of a distinct duty of care clearly owes more to the law of negligence than of nuisance.

A duty in negligence or nuisance?

The Trust's second ground of appeal focused on this issue. They argued that, if the duty described in *Goldman v Hargrave* did apply to the present situation, it must arise within the law of negligence. As the plaintiffs had pleaded their claim in nuisance, rather than negligence, the action must fail.

The Privy Council's decision in *Goldman* provides some support for the Trust's position, although it also leaves the relationship between the two torts unresolved. It placed the duty firmly within the tort of negligence but sidestepped the issue of whether it also fell within the boundaries of the law of nuisance.

> Lord Wilberforce (in *Goldman v Hargrave*): "Their Lordships propose to deal with the issues as stated without attempting to answer the disputed question whether if responsibility is established it should be brought under the heading of nuisance or put in a separate category.
>
> As this board recently explained in *The Wagon Mound (No 2)*, the tort of nuisance, uncertain in its boundary, may comprise a wide variety of situations in some of which negligence plays no part, in others of which it is decisive. The present case is one where liability, if it exists, rests upon negligence and nothing else; whether it falls within or overlaps the boundaries of nuisance is a question of classification which need not be resolved here."

The Court of Appeal, in the present case, was also unwilling to articulate the precise relationship between the two torts. However, they were unsympathetic to the Trust's submission and rejected this second ground of appeal. In contrast to the decision in *Goldman*, the court was of the view that the duty arose within the law of nuisance. However, even if this was incorrect, and it actually arose in negligence, it would not have mattered that the claim had been framed in nuisance.

> Megaw LJ: "The plaintiff's claim is expressed in the pleadings to be founded in nuisance. There is no express reference to negligence in the statement of claim. But there is an allegation of a breach of duty, and the duty asserted is, in effect, a duty to take reasonable care to prevent part of the defendants' land from falling on to the plaintiff's property.
>
> I should, for myself, regard that as being properly described as a claim in nuisance. But even if that were, technically, wrong, I do not think that the point could or should avail the defendants in this case. If it were to do so, it would be a regrettable modern instance of the forms of action successfully clanking their spectral chains; for there would be no conceivable prejudice to the defendants in this case that the word 'negligence' had not been expressly set out in the statement of claim."

The decision

The court dismissed the Trust's appeal. They were in breach of their duty to prevent the continuance of the nuisance which had arisen on their land and were therefore liable, in nuisance, for the losses suffered by the plaintiffs.

LEHMANN v HERMAN (1993)
Chancery Division

In brief

- **Issues:** 1939 Act – definition of "owner" – definition of "building owner" – position of joint tenants and tenants in common – service of party structure notice – whether service by a single joint tenant is valid.

- **Facts:** Mr and Mrs Lehmann and Mr and Mrs Herman owned adjoining properties. Mr Herman wished to repair the party fence wall between their respective gardens. He served a party structure notice on Mr and Mrs Lehman in respect of this.

- **Decision:** Service of the party structure notice was invalid. The Act required service by the "building owner". Mr and Mrs Herman owned their property as joint tenants so they together constituted the building owner. Both joint tenants must join in the service of the notice as it commenced a process which involved the adjustment of property rights. Mr Herman, as a single joint tenant, therefore had no authority to serve the notice alone.

- **Notes:** (1) This case addresses the service of party structure notices *by* co-owners. Contrast this with service *on* co-owners, where service on a single owner is acceptable: *Crosby v Alhambra Company Ltd* (1907). (2) See also: *Solomons v R. Gertzenstein Ltd* (1954).

The facts

Mr and Mrs Lehmann were owner-occupiers of 35 Reddington Road, a terraced house in London NW3. They owned the freehold estate in the property as joint tenants.

The adjoining terraced property at number 33 was divided into a ground floor floor flat with garden (flat A) and a first floor flat (flat B) (Figure 12). Mr and Mrs Herman owned a long lease in flat A as joint tenants.

The rear gardens of numbers 33 and 35 were separated from each other by a party fence wall. This was in a poor state of repair. Mr Herman decided to repair the wall and he appointed a number of consultants to deal with the matter on his behalf. His surveyors, as his agents, notified the Lehmanns about the intended work by serving a party structure notice on them under the 1939 Act.

There had been a long history of strained relations between the Lehmanns and the Hermans. Upon receipt of the notice the Lehmanns

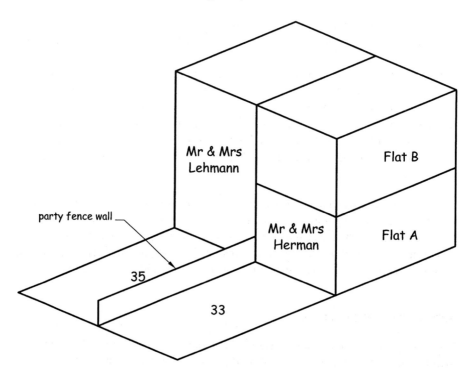

Figure 12: Lehmann v Herman

argued that it was invalid. It should have been served on behalf of Mr and Mrs Herman rather than simply Mr Herman. They requested that Mrs Herman should undertake to treat herself as bound by the notice in order to regularise the situation.

Mrs Herman refused to provide the undertaking. Mr Herman argued that he alone was the building owner within the Act. He was the person who proposed to carry out the works. He had appointed the consultants who were responsible for the work and, in due course, he would pay for the works by cheques drawn on his account. His wife had no involvement in the works and there was no intention that she should have any, "save perhaps to the extent of offering mugs of tea to the building contractors from time to time".

Mr and Mrs Lehmann then applied to the court for a declaration that the party structure notice was invalid as it had not been served by both joint tenants.

Service on co-owners

Mr Herman argued that the Act was not intended to raise complex issues of property law and that only an owner who actually desired to carry out work was required to serve a party structure notice. He argued that each joint tenant was, by himself, a building owner within the meaning of the Act.

In support of this argument he cited the case *of Crosby v Alhambra Co Ltd* (1907) which had considered whether there was a requirement to serve notice on more than one owner. That case had been decided under the 1894 Act and had considered the combined effects of sections 5(29) (definition of "owner") and 5(32) (definition of "adjoining owner"). Mr Herman referred, in particular, to the following extract from the Neville J's judgment:

> "I hold, therefore, that the true interpretation of the two subsections is that all the persons coming within the definition of 'owner' in subsection 29 must be served, except in the case where several persons hold, as tenants in common or as joint tenants, some particular interest in the land, in which case service on one of such tenants in common or joint tenants would be sufficient."

Service by joint owners

If service of a notice *on* one of two joint tenants was effective, Mr Herman argued that, by analogy, a service *by* one of two joint tenants must be equally valid.

The court did not agree. It considered that there was a clear distinction between the service *on* a single joint tenant (which was valid) and the service *by* a single joint tenant (which was not).

In the case of service *on* a single joint tenant the relevant issue is simply whether bringing something to the attention of one tenant constitutes doing the same in respect of all of them. The same practical considerations apply to this as to service of other legal documents.

Service *by* a single joint tenant is however an entirely different matter. This is concerned, not with simply bringing something to someone's attention, but with the instigation of a course of action that has the potential to interfere with existing property rights. Because of this, property law issues are of greater significance. This point is emphasised in the judgment:

> "Simply in practical terms it would be very odd if the statute provided for one of two joint owners to deal with an adjoining owner without the other joint owner being involved. In real property law terms the concept of one joint owner being able to deal with the property without the other being party to the transaction has been foreign to English law since the 1925 property legislation."

Therefore, the reference in the Act's definition of "building owner" to "such one of the owners of adjoining land" did not mean that any one of a number of joint owners could, by themselves, be a building owner within the Act. These words must be construed as referring to any one owner of a qualifying estate or interest in land rather than to any one of a number of joint owners of the same estate or interest.

Definition of "owner"

The reason for the court's rejection of Mr Herman's submission is explained in its deliberations on the statutory definition of "owner" in section 5. In common with the definitions in earlier Acts, the 1939 Act defined "owner", either in terms of the possession or receipt of rents from property, or, in terms of its occupation.

The judgment explains why, within the meaning of the Act, Mr Herman was not in possession of his flat, nor in receipt of rents from it:

> "In order for a person to be a building owner for the purposes of the Act, he must first be an 'owner'. The definition of 'owner' is not exhaustive – it begins 'owner includes'. It then continues: '... every person in possession or receipt either of the whole or of any part of the rent or profits of any land'.
>
> It is not suggested that Mr Herman alone could be regarded as being in possession of his flat, nor that he is in receipt of any rents or profits. In any event this phrase itself must be read as being limited in some sense. It does not include an agent who is in receipt of rents: see *Solomons v R. Gertzenstein Ltd* (1954).
>
> Similarly, in the ordinary case, neither joint tenant of premises let at a rent could be said, when taken alone, to be in receipt of any part of the rent. Technically the whole rent would be payable to them jointly as trustees (even if received by one as agent for both) and the money, so far as available for distribution paid out as trust income, distributable to each beneficiary according to his entitlement."

The judgment then explains why, in a legal sense, Mr Herman was not in the occupation of the flat either. In this context a distinction is drawn between simply being "in occupation" of premises (a practical concept) and being "in the occupation" of premises (a legal concept):

> "The remaining words of the definition are: 'or in the occupation of any land or tenement otherwise than as tenant from year to year or for any less term or as a tenant at will'.
>
> Here 'in the occupation' must be construed restrictively. This is clear from the following words; the Act cannot be intended to be construed so that a tenant for 11 months certain is not an owner, but a mere licensee is.
>
> The clue lies in the words 'the occupation'. The Act is concerned with persons who are in *the* occupation, not persons who are occupying. Where, as in

the present case, there are joint tenants occupying premises, it cannot be said that either alone is in *the* occupation of the premises.

Nor can the definition of owner properly be stretched as to include one of the occupiers on the basis that the definition is not exhaustive."

The decision

The court concluded that a party structure notice which is served by, or on behalf of, only one of two joint tenants has not been served by a building owner as required by the Act. Such service is therefore invalid. The court made the declaration requested by the Lehmanns to this effect.

London Building Acts (Amendment) Act 1939

PART I: INTRODUCTORY

1. Short title construction and citation

This Act may be cited as the London Building Acts (Amendment) Act 1939 and shall be read and construed as one with the London Building Acts 1930 and 1935 and may be cited with those Acts as the London Building Acts 1930 to 1939.

PART VI: RIGHTS ETC. OF BUILDING AND ADJOINING OWNERS

RIGHTS ETC. OF OWNERS

47. Party structure notices

(1) Before exercising any right conferred on him by section 46 (Rights of owners of adjoining lands where junction line built on) of this Act a building owner shall serve on the adjoining owner notice in writing (in this Act referred to as a 'party structure notice') stating the nature and particulars of the proposed work the time at which it will be begun and those particulars shall where the building owner proposes to construct special foundations include plans sections and details of construction of the special foundations with reasonable particulars of the loads to be carried thereby.

London Building Act 1930

PART I: INTRODUCTORY

5. Definitions

In this Act, save as is otherwise expressly provided therein and unless the context otherwise requires, the following expressions have the meanings hereby respectively assigned to them; (that is to say):

"building owner" means such one of the owners of adjoining land, as is desirous of building or such one of the owners of buildings, storeys, or rooms separated from one another by a party wall or party structure, as does or is desirous of doing a work affecting that party wall or party structure.

"owner" includes every person in possession or receipt either of the whole or of any part of the rents or profits of any land or tenement, or in the occupation of any land or tenement, otherwise than as a tenant from year to year, or for any less term, or as a tenant at will.

LEMAITRE v DAVIS (1881)
Chancery Division

In brief

- **Issues:** Independent contractor – demolition of building owner's property – damage to adjoining owner's property – easement of support – prescription – acquisition of easement by underground vault – whether support enjoyed secretly – non-delegable duties.

- **Facts:** Davis' independent contractor demolished his property. He failed to provide adequate support for Lemaitre's adjacent building. As a consequence damage was caused to Lemaitre's underground wine vault which was deprived of support.

- **Decision:** Lemaitre's property had acquired an easement of support from Davis' property by prescription. Such rights were capable of acquisition, either against neighbouring land, or against adjacent buildings. Both parties were aware of the existence of each other's cellars/vaults. The acquisition of an easement in favour of the vault could not therefore be defeated on the basis that it had been enjoyed secretly. The contractor had interfered with Lemaitre's easement of support. Duties to neighbouring owners during building operations are owed on a non-delegable basis. Davis was therefore liable for the damage caused by his contractor.

- **Notes:** See also *Bower v Peate* (1876); *Dalton v Angus* (1881); *Hughes v Percival* (1883).

The facts

Lemaitre owned the Mitre Tavern in Fish Street, London EC3. Davis owned adjoining premises at 1 Monument Yard. Both properties had been erected shortly after the Great Fire of London. They did not share a party wall but their neighbouring external walls were in contact with each other and relied on each other for support.

Davis wished to demolish and rebuild his property and engaged Crabb, a contractor, to undertake the work. The contract required Crabb to make adequate provision to protect Lemaitre's property.

Crabb then demolished the building at 1 Monument Yard and this included digging out the old foundations and cellar at the property. Unfortunately, in breach of his contract with Davis, he failed to provide any shoring or other support for the neighbouring property. In consequence, the removal of the western cellar wall caused serious damage to the eastern wall of Lemaitre's wine vault, which partially collapsed.

Lemaitre issued proceedings against Crabb and Davis for an injunction and for damages. He claimed that they had interfered with an easement of support for his property which had been acquired by prescription. At a preliminary hearing Crabb gave an undertaking to make good all the damage and to provide all necessary support whilst the works were completed.

By the time of the full hearing, Crabb had complied with his undertaking. All the substantive issues had therefore been resolved, leaving only the question of costs to be decided by the court. Davis submitted that he should not be subject to an order for costs as he was not responsible for Lemaitre's losses. He raised a number of arguments in support of his submission.

Easement of support from neighbouring building

Firstly, Davis argued that the law did not recognise a prescriptive easement of support against other buildings.

In the absence of an express or implied grant of an easement, Lemaitre had based his claim on an easement acquired by prescription. He had relied on *Dalton v Angus* (1881) as establishing that a building could acquire an easement of support in this way.

Davis attempted to distinguish *Dalton v Angus* from the present case. The decision in *Dalton* related to the support of buildings by adjacent land. In the present case Lemaitre was claiming an easement of support for a building from an adjacent building. There was no authority for the grant of such a right by prescription.

The court considered that Davis was drawing an artificial distinction between the two types of support and rejected his submission.

> Hall VC: "For the defendant Davis it was contended that such support can be claimed from land only, and that the decision in *Dalton v Angus* determined that alone; but if an owner has a claim of support against land, can he not enforce such a claim against a building upon it? There is nothing in that case which conclusively settles the point...
>
> ... if there be a right of support against land, why should there not be against a building? It may be an inconvenient right, but that may be said against all servitudes, as, for instance, that of light – all are attended with some inconvenience. Still it is considered better that they should exist than that property should be injured, and not be capable of being protected.
>
> Having that case to guide me, and seeing no distinction drawn between the two rights of support from land and from a building, I consider the law to be as applicable to support from an adjoining building as it is to support from land."

Easement of support for cellar wall

Secondly, Davis argued that Lemaitre could not have acquired a prescriptive right of support for his vault, which was the only part of the Mitre Tavern which had suffered damage.

A right can only be acquired as an easement by prescription if it has been enjoyed *as of right* for the requisite period. The right must therefore have been enjoyed *nec per vim, nec clam, nec precario* (without force, without secrecy and without permission).

Davis argued that, as the vault wall was hidden inside Lemaitre's building, the enjoyment of support from the wall of his own cellar had taken place secretly. As Lemaitre had made use of his wall secretly, rather than as of right, no easement could have been granted by prescription.

Whilst acknowledging the need for the support to have been enjoyed openly the court found, on the facts, that this had been the situation in the present case.

> Hall VC: "In a state of things, where the origin of these two buildings goes so far back, it is very difficult to deal with the case, it being almost impossible to prove anything, on the one hand or the other, affirmatively.
>
> Therefore the conclusion which I come to is that the enjoyment would not be of right if it was *clam*, but I think the evidence shows that the right was open – that each proprietor of the two tenements knew of the existence of his neighbour's cellar; therefore, as a matter of fact, I hold, so far as it may be necessary, that the enjoyment of the right has not been *clam*, or otherwise than open."

Non-delegable duties to adjoining owners

Thirdly, Davis argued that he could not be held liable for Crabb's wrongful acts.

A client is not generally liable for the torts of his independent contractor unless he has specifically authorised these. Davis had made proper provision in his contract with Crabb for the protection of the adjoining property by shoring. The damage to Lemaitre's property had been caused entirely by Crabb's failures in this regard. Davis therefore argued that, as he was not personally at fault, he could not be held liable.

The court rejected this submission and held that Davis owed a personal non-delegable duty to Lemaitre. The rationale for this was explained by Hall VC:

> "The defendant Davis doing the works with the assistance of a contractor was doing his own works, and it was his duty to see that the works were properly done and that certain precautions were taken, either by himself or his agent, by

shoring, to prevent any injury to his neighbour's vault. Such precautions were not taken.

The defendant Davis cannot shift the responsibility from himself by saying that he employed a contractor and that it was his wrongful act. It would be a strange thing if principals should be allowed to escape from liability when altering their premises, and erecting new buildings, by saying that they employed contractors under the specifications which were drawn up for their guidance, and that the contractors only were liable for any injury which might happen."

As with other cases involving non-delegable duties, the rationale is not entirely clear. The suggestion that such duties will be imposed in any situation involving the alteration of premises or erection of new buildings appears rather broad in the light of later cases. Nevertheless, it is consistent with the earlier decision in *Bower v Peate* where a non-delegable duty was also imposed in a situation involving the removal of support which was the subject of an easement.

The decision

The court concluded that Crabb and Davis were jointly liable for Lemaitre's losses. Judgment was therefore entered for Lemaitre with the costs to be paid by both Crabb and Davis.

LEWIS & SOLOME v CHARING CROSS, EUSTON & HAMPSTEAD RAILWAY COMPANY (1906)

Chancery Division

In brief

- **Issues:** 1894 Act – demolition of building owner's building – work undertaken pursuant to statutory powers – exemption from 1894 Act's party wall requirements – interpretation of empowering statute – reliance on other rights to undertake work – requirement to serve party structure notice if not relying on statutory rights – relationship between statutory and common law rights – requirement to serve party structure notice when exposing a party wall.

- **Facts:** The Company wished to demolish their building and to expose the party wall with Lewis & Solome's adjoining building. No other interference with the party wall was involved. The work was to be undertaken as part of the development of a station under powers conferred on the Company by statute.

- **Decision:** As the demolition work did not involve an interference with the party wall it was not necessary for the Company to serve a party structure notice on Lewis & Salome. However, notice would have been required if work had been undertaken to the fabric of the party wall itself. When the empowering statute was read in the context of the 1894 Act it was clear that there was no exemption from compliance with that Act's party wall requirements. Furthermore, the Company could not have avoided these requirements by claiming to proceed with work under some other authority. The right to undertake work to a party wall could only exist under the Act and its procedures must be complied with.

- **Notes:** (1) *Standard Bank of British South America v Stokes* (1878) provided the authority for the finding that rights to undertake work to a party wall can only exist under the Act. (2) *Major v Park Lane Co* (1866) also held that service of a party structure notice is not required for the demolition of a building owner's building which exposes the party wall. A notice is now required in these circumstances due to the inclusion of section 2(2)(n) in the 1996 Act.

The facts

The defendant company in this case had been incorporated by statute (the Charing Cross, Euston, and Hampstead Railway Act 1893) to develop the

Northern Line of the London Underground. It purchased a house at number 21 Cranbourn Street, London WC2 by agreement with the owner of that property, under powers contained in the statute.

The company then started demolishing the property as it wished to use the site for the construction of the present Leicester Square underground station. Lewis & Salome owned the adjoining house at 22 Cranbourn Street which was separated from number 21 by a party wall.

They objected to number 21 being demolished without the prior service of a party structure notice upon them, and issued proceedings against the company. They sought damages for wrongful interference with their possession of number 22 and an injunction to restrain any further demolition work which would interfere with the party wall. The company maintained that it had authority to proceed with the works without first serving a party structure notice. It raised two separate defences to the action.

Exemption from party wall legislation?

Firstly, the company argued that section 31 of the Charing Cross, Euston, and Hampstead Railway Act 1893 exempted them from the provisions of the 1894 Act in respect of party walls. After investing the company with certain powers to acquire land by agreement, that section provided that:

> "Any buildings erected on any land acquired under this section (except such building or parts of buildings as may be used for the purposes of a station) shall be subject to the provisions of the Acts relating to buildings in the metropolis."

The company argued that the exception relating to the erection of stations must apply equally to the prior demolition of existing buildings from the sites of stations. As the demolition of number 21 fell within this exemption, Lewis and Salome's claim must fail.

The court rejected this submission. The exemption in section 31 should be understood in its proper context. Section 201 of the 1894 Act already exempted certain railway works, including work to stations, from the building control requirements in Parts VI and VII of that Act.

Section 31 was intended to remove most of this exemption in respect of the works which were to be carried out in developing the Northern Line. The only exception to this was to be work to stations. The specific reference to stations in section 31 therefore simply preserved the 1894 Act situation in respect of these buildings, rather than granting any additional exemption.

The station exemption in section 31 must therefore apply only to the building control provisions in Parts VI and VII of the 1894 Act, rather than to the Act as a whole. Specifically, there was no exemption from Part VIII of the 1894 Act, which regulated the rights of adjoining owners.

> Warrington J: "... the buildings erected for the purpose of the railway would, under section 201 of the London Building Act, have been exempted from Parts VI and VII of that Act, and the effect of section 31 seems to me to be to remove that exemption and to render the buildings used for the purpose of a railway, but not for the purpose of a station, subject to the provisions of the Act.
>
> In my judgement this section has no reference to the provisions as to adjoining owners in the London Building Act of 1894, but was intended to refer to buildings to be erected on land to be acquired by the company, and not to, what is a totally different matter, the interference by the defendant company with party structures.
>
> The whole tenor of that section shows that it was intended to exempt buildings used as stations from the provisions of the London Building Act 1894, relating to buildings, and nothing more. It seems to me therefore, there is nothing in section 31 of the special Act to exempt the defendant company from the liabilities imposed by the London Building Act of 1894 in reference to party structures."

Reliance on existing common law rights?

Secondly, the company argued that, as it did not wish to rely on the rights contained in the 1894 Act, there was no requirement for it to comply with its provisions in respect of party walls.

Those provisions enabled a building owner to go beyond his ordinary common law rights in certain situations, subject to compliance with the Act's procedural requirements. Therefore, if a building owner had no need of the additional statutory rights, but was content with his existing (usually common law) rights, then he had no need to comply with the Act's provisions.

The company had authority to do certain acts, including demolishing buildings, under section 16 of the Railway Clauses Consolidation Act 1845. It argued that, as it had this authority, there was no need to rely on authority under the 1894 Act. It could not therefore be required to serve a party structure notice before undertaking the demolition.

This submission was also rejected by the court. Warrington J cited the observations of Sir George Jessel in *Standard Bank of British South America v Stokes* (1878) to rebut the argument that it was possible to proceed with party wall works outside the provisions of the London Building Acts.

> Warrington J: "In my judgment the result of that decision is this, that, whatever at common law may be the rights of an adjoining owner in relation to a party structure, his rights in relation to a party structure within the area comprised by the Building Act depend upon that Act. He has the rights which that Act gives him and no others."

Although section 16 of the Railway Clauses Consolidation Act 1845 may well have given the company authority to undertake demolition works, the 1894 Act was the sole source of such authority when undertaking works which affected the rights of an adjoining owner in a party wall.

> Warrington J: "In my judgment the fallacy of [the company's] argument lies in this. I will assume that the special Act authorised the defendant company to demolish the building. It does not, however, say anything about interfering with or affecting a party structure, and it does not necessarily follow that in demolishing a building they would demolish or affect the party structure.
> ... It seems to me, therefore, that their statutory enactments cannot be read so as to authorise them to deal with a house on land which they acquire in such a manner as to affect the party structure without giving the necessary notice.
> ... If, in doing the work, the defendant company are parties so desirous of doing a work affecting the party wall or structure, they must, it seems to me, like any other adjoining owner, give the notice required by the Act."

The decision

Both of the company's submissions on the law had been rejected. However, on the facts, the court found in their favour.

After hearing evidence and viewing the premises, Warrington J found that the demolition works had not actually interfered with the party wall. Because of this they fell outside the 1894 Act and a party structure notice was not therefore required.

The action was dismissed although, having regard to "the position taken by the defendant company in the correspondence", no order was made against Lewis & Salome for costs.

London Building Act 1894

APPLICATION OF ACT

201. Buildings exempt from parts of Act

The following buildings and works shall be exempt from the operation of Parts VI and VII of this Act:

(8) ... Any building or structure situate upon the railway or within the railway or station premises and used for the purposes of or in connection with the traffic of a railway company.

Railways Clauses Consolidation Act 1845

16. Subject to the provisions and restrictions in this and the special Act, and any Act incorporated therewith, it shall be lawful for the company, for the purposes of constructing the railway, or the accommodation works connected therewith, hereinafter mentioned, to execute any of the following works; (that is to say,) ... They may do all other acts necessary for making, maintaining, altering, or repairing, and using the railway ...

45. ... it shall be lawful for the company, in addition to the lands authorized to be compulsorily taken by them under the powers of this or the special Act, to contract with any party willing to sell the same for the purchase of any land adjoining or near to the railway, not exceeding in the whole the prescribed number of acres for extraordinary purposes; (that is to say,) for the purpose of making and providing additional stations, yards, wharfs, and places for the accommodation of passengers, and for receiving, depositing, and loading or unloading goods or cattle to be conveyed upon the railway, and for the erection of weighing machines, toll houses, offices, warehouses, and other buildings and conveniences...

Charing Cross, Euston, and Hampstead Railway Act 1893

5. Subject to the provisions of this Act, the company may make and maintain, in the lines and according to the levels shown on the deposited plans and sections, the railways and other works hereinafter described, with all necessary and proper stations, platforms, approaches, passages, bridges, stairs, subways, tunnels, sidings, shafts, lifts, buildings, apparatus, generating plant, depots, machinery, appliances, works and conveniences connected therewith, and may subject as aforesaid enter upon, take and use such of the lands delineated on the said plans and described in the deposited book of reference as may be required for those purposes...

31. The company may take by agreement for the extraordinary purposes mentioned in the Railways Clauses Consolidation Act 1845, any quantity of land not exceeding in the whole five acres, but nothing in this Act shall exonerate the company from any action, indictment or other proceeding for nuisance in the event of any nuisance being caused or permitted by them upon any land taken under the powers of this section: provided always that for the purposes of this section extraordinary purposes shall not without the consent of the council include the erection of buildings or works for generating electricity or the provision of yards, wharves and places for receiving, depositing and loading or unloading goods or cattle. Any buildings erected on any land acquired under this section (except such buildings or parts of buildings as may be used for the purposes of a station) shall be subject to the provisions of the Acts relating to buildings in the metropolis.

LIST v THARP (1897)
Chancery Division

In brief

- **Issues:** 1894 Act – entitlement to service of party structure notice – definition of "owner" – possession of land – status of building agreement – requirement for qualifying interest in land – nature of building agreement – equitable interest.
- **Facts:** Tharp obtained an award under the 1894 Act. He then raised the party wall separating his building from the adjoining site. He had first served notice on the freehold owner of the adjoining plot but not on List who occupied the site under a building agreement.
- **Decision:** List was an owner within the Act and was entitled to be served with notice. His agreement for a lease stated that he was entitled to "possession" of the site and this was sufficient to bring him within the Act's definition. The basis on which he was entitled to be "in occupation" of the land was therefore of no relevance. However, his building agreement also operated as "an equitable interest of great value in the land and the buildings which he had thereon erected".
- **Notes:** See also: *Cowen v Phillips* (1863); *Fillingham v Wood* (1891); *Orf v Payton* (1904); *Crosby v Alhambra Co Ltd* (1907); *Spiers & Son v Troup* (1915); *Solomons v Gertzenstein* (1954); *Lehmann v Herman* (1993); *Frances Holland School v Wassef* (2001).

The facts

Oppenheimer owned an empty plot at 36 St James' Place. He had previously demolished the house which had stood on the plot, leaving the party wall with number 37 intact. Number 37 was owned by Tharp.

Oppenheimer entered into a building agreement with List for the redevelopment of the plot at number 36. The agreement provided (inter alia):

- (Clause 6) List should forthwith, upon the signing of the agreement, be entitled to possession of the land.
- He should erect certain buildings on the land in accordance with the terms of the agreement.
- (Clause 7) Oppenheimer and his surveyor should have the right to enter the land at all times during the building work for the purpose of inspecting the same.

- Upon satisfactory completion of the work Oppenheimer would grant a 90-year lease of the property to List.
- (Clause 18) Nothing in the agreement should be construed as a demise at law so as to vest any estate in List.

Definition of "owner"

Whilst the work was proceeding Tharp decided to raise the party wall for the benefit of his own property at number 37. He therefore served a party structure notice on Oppenheimer under the 1894 Act, in February 1896. He served no notice on List as he did not consider that he was an owner within the meaning of the 1894 Act. An award was published and Tharp started work shortly afterwards.

List issued proceedings against Tharp to restrain him from raising the wall on the basis that he had not been served with a party structure notice. The court therefore had to decide whether List was entitled to receive notice. This depended on whether he was an "owner" within section 5(29) of the 1894 Act.

This section defined an owner as a person in possession (including being in receipt of rents and profits) or the occupation of land, otherwise than as a tenant from year to year or for any less term, or as a tenant at will[1].

Possession of land

The court decided that List was an owner as he was in possession of the land on the basis of his building agreement.

> Chitty LJ: "Now, regard being had to the statutory definition of 'owner', the first enquiry is whether the plaintiff was, in February 1896, in possession of the land. It appears to me that the only answer must be that he was. Clause [6] of the agreement is precise; he was entitled to the possession, and, having entered upon the land, he was lawfully in possession accordingly.
>
> The reservation in the 7th clause of a right of entry for the surveyor and others during the erection of the buildings shows that the term 'possession' is used in the 6th clause in its ordinary sense: the power is reserved as against the plaintiff's possession.
>
> It was argued that he was in possession only for the purpose of erecting the buildings, and, consequently, that as his right to possession was thus limited, he was not in possession within the meaning of the definition in the Act.

---

[1] Under section 20 0f the 1996 Act, it is now sufficient to be a purchaser of an interest in land under contract without the need to prove possession or occupation.

> ## London Building Act 1894
>
> PART I: INTRODUCTORY
>
> ### 5. Definitions
>
> In this Act unless the context otherwise requires:
>
> (29) The expression "owner" shall apply to every person in possession or receipt either of the whole or of any part of the rents or profits of any land or tenement or in the occupation of any land or tenement otherwise than as a tenant from year to year or for any less term or as a tenant at will.

> But I cannot accept this argument. A man is not the less in possession because as between himself and the person from whom he receives it, he is under a contractual obligation to use the property for some particular purpose; as, for instance, when he is under a covenant to use a house for the purpose of a private dwelling-house only."

Occupation of land

Having established that List was in possession within the meaning of the definition's first leg, there was nothing about his situation that made it possible to "cut down" his rights within the context of the second leg of the definition.

> Chitty LJ: "The next enquiries are whether the plaintiff was in occupation of the property 'as a tenant from year to year, or for any less term, or as a tenant at will'. Here, again, the agreement supplies the answer: nothing in the agreement was to be construed into a demise at law or to vest any estate in the plaintiff. Consequently he was not tenant from year to year, or for any lesser term; nor was he tenant at will.
>
> The provision in clause 18 of the agreement that he was only to have the right of entry for the purpose of performing the agreement, I have already dealt with incidentally. It cannot, in my opinion, be made use of for the purpose of cutting down the right to the possession which was previously imposed by express terms.
>
> After his entry on the land he was in possession as between himself and Sir Charles Oppenheimer, though he was bound by the contract as to the user of the land. And as against strangers he had all the ordinary rights and remedies incidental to possession."

The decision

The precise nature of List's interest in the land was described by Chitty J in the following terms:

> "The plaintiff in February 1896 had, by virtue of the building agreement, an equitable interest of great value in the land and the buildings which he had thereon erected."

The court held that, as an owner within the Act, List should have been served with a party structure notice. As Tharp had failed to serve this upon him he had no authority to raise the party wall. An injunction was therefore granted to restrain the work.

LONDON & MANCHESTER ASSURANCE COMPANY LTD v O & H CONSTRUCTION LTD (1989)

Chancery Division

In brief

- **Issues:** 1939 Act – failure to serve party structure notice – demolition of party fence wall – erection of new wall – encroachment on adjoining owner's land – crane oversailing into adjoining owner's airspace – trespass – interlocutory injunction – order to remove offending structures.

- **Facts:** The parties owned adjoining plots of land which were separated by a party fence wall. O & H demolished the party fence wall and began erecting a new building on their plot without planning permission and without first serving a party structure notice. The new building occupied part of the space formerly occupied by the party fence wall and also encroached by 5 in on to London & Manchester's land. O & H also allowed a crane boom to oversail London & Manchester's land without their consent. They did not deny that their actions amounted to a trespass but refused London & Manchester's various requests that these should cease.

- **Decision:** The encroachment on to London & Manchester's land was a clear case of trespass. The flagrant defiance of the 1939 Act was equally serious as this too involved an interference with London & Manchester's property rights. The court issued a mandatory injunction for the removal of the offending structures. Although this would cause hardship to O & H they had brought it entirely on themselves. London & Manchester had behaved reasonably throughout but O & H had rushed ahead to get their building up regardless of anyone else's rights. In these circumstances there was no question of the court exercising a discretion. It was obliged to recognise London & Manchester's right to regain possession of their property.

- **Notes:** See also (on the consequences of failing to serve a party structure notice): *Leadbetter v Marylebone Corporation [No. 1]* (1904); *Adams v Marylebone Borough Council* (1907); *Emms v Polya* (1973); *Louis v Sadiq* (1996).

The facts

The case concerned an area of land on the south side of the Thames, between Battersea and Albert Bridges in London. The land had formerly consisted of a number of wharves but since the early 1980s had been used predominantly for warehouse storage.

London & Manchester owned an area of this land to the west, known as Albion Wharf. O & H owned some adjoining land, to the east, which was known as Albert Wharf. The two plots were separated by a substantial, 2 ft thick, wall. There was clear evidence that it was a party wall and that it had been treated as such by the respective owners of the two plots for many years[1].

O & H wished to demolish the warehouse on their land and to redevelop the site with new buildings. They had been in negotiations with the local planning authority for the erection of new buildings on the site. However, at the time of the present case, these had proved unsuccessful and no planning permission had been granted.

On September 15 1987 O & H appointed a surveyor to act for them in connection with the party wall. However, towards the end of September, without informing him or London & Manchester, they demolished the party wall. No procedural steps had been commenced under the 1939 Act at that stage.

Encroachments on to adjoining land

They then immediately, and without planning consent, commenced the erection of a new building in the proximity of the boundary. In fact, this building encroached on the site of the former party wall and also, on parts of London & Manchester's own land. O & H also erected a crane whose boom encroached into the airspace above London & Manchester's land.

London & Manchester became aware of O & H's actions on 15 October 1987 and the next day sent a letter of complaint to them. O & H continued with the trespass, so in December 1987 London & Manchester's solicitors formally demanded that this should cease. O & H again failed to respond so London & Manchester issued proceedings against them for trespass on 25 February 1988.

London & Manchester maintained that O & H had no authority to place the new building on their land, nor to swing the crane boom over their land without their consent. They further maintained that they had no authority to demolish the party wall or to build upon its site unless they either obtained their consent or alternatively, a proper authority under the 1939 Act. The present proceedings arose out of London & Manchester's application for interlocutory relief to restrain the various encroachments on to their property.

---

[1] Although the wall is referred to as a party wall throughout the published law report, it was strictly a party fence wall within the 1939 Act.

On the evidence the court was completely satisfied that the encroachments complained of had taken place and O & H produced no evidence to contradict this. The court noted O & H's "high-handed expropriation of rights in defiance of the law" and granted the interlocutory relief requested. In view of the clear cut nature of the trespass, this included unusual interlocutory relief in the form of a demolition order.

Trespass by crane oversailing

Harman J noted that there was now established authority[2] that cranes could not be swung over a neighbour's land without their consent:

> "It is, of course, notorious that the use of a crane swinging round from the useful position upon a neighbouring site is liable to intrude into the airspace of other persons and thereby to commit a trespass. The matter has been the subject of a good deal of litigation. It is, in my view, beyond any possible question on the authorities and the law that a party is not entitled to swing his crane over neighbouring land without the consent of the neighbouring owner."

Trespass to land

The more significant part of the court's decision relates to the encroachments on to the land and the interference with the party wall. Harman J explains his decision to order mandatory interlocutory relief in the following terms:

> "The more difficult questions arise as to whether I should make mandatory orders for removal of structures (a) on what is expressly admitted by the defendants to be a trespass on the plaintiffs' undoubted land ... (though only for a width of perhaps 5 in) and (b) erected in ... breach of the London Building Acts by placing the new wall over and along the line of the former party wall...
>
> On that basis, it seems to me, there really is a case for warranting the very unusual course of the court in ordering interlocutory and without more ado, without trial or cross-examination, that these structures should be removed.
>
> I cannot see that any court could ever say that a trespass by the building of a structure on a neighbour's land, in circumstances where the neighbour has in no way encouraged but has at all times protested against the building, where the neighbour would have had great difficulty in establishing the boundary

---

[2] *Anchor Brewhouse Developments Ltd v Berkley House (Docklands Developments) Ltd* (1987).

after the works had been started, and where the neighbour had behaved in an entirely reasonable way, should not be rectified by removal of the structure.

In my judgment, no court could conceivably refuse to that neighbour the right to possession of his own land and the right to removal of structures wrongly placed on it. As it seems to me, there is not a question of discretion; it is a question of right; and in my view there is no question of arguable case and balance of convenience so far as concerns the trespass."

Failure to comply with party wall legislation

"So far as concerns the building in breach of the London Building Acts, it seems to me that similar propositions must apply ... As it seems to me, it is just as bad a case of invasion of a legal right, to tear down a party wall and put up your own wall in flagrant defiance of the most important provisions of the London Building Acts, enacted now for over a century to protect neighbouring owners in this crowded city, as to commit a straightforward trespass.

To my mind, although I agree it is an unusual order, and although it may be the defendants will suffer very substantially thereby, I should make a mandatory order for demolition. I bear in mind that [the defendants' counsel] asserts that it might be that one day his clients might get a party wall award permitting them to maintain the [new] wall within the boundaries of the former [party] wall.

As against that, there has been no attempt whatever to take advantage of any of the regular processes available, all of which were known expressly to the defendants and which they deliberately chose not to exercise and take ... In my judgment, they have disabled themselves from any claim to maintain that wall in that position unless and until an award has been made which permits such a wall to be built.

In my view, I should make orders for the removal of the infringing part of the structures. That that may, or may not, cause the whole of the structure to fall down is something which the defendants have wholly and entirely brought upon themselves. It is a case where they have rushed on beyond any question, hastening to get their building up, regardless of anybody else's ... rights."

The decision

The court therefore made interlocutory orders restraining the use of the crane in the plaintiff's airspace and requiring the removal of the offending building work.

LONDON, GLOUCESTER & NORTH HANTS DAIRY COMPANY v MORLEY & LANCELEY (1911)

King's Bench Division

In brief

- **Issues:** 1894 Act – building owner's entitlement to raise wall – approach to statutory interpretation – definition of type 'b' party wall – wall performing separating function for only part of its height – whether wall a party wall for its whole height.

- **Facts:** A freestanding wall stood on the Dairy Company's land. They had erected a two-storey building against it and Morley & Lanceley had enclosed on the other side of the wall. The wall continued upwards as a parapet above the level of the roof of the two-storey building. Morley & Lanceley served notice under the 1894 Act of their intention to raise the wall.

- **Decision:** Morley & Lanceley had no right to raise the wall. The wall was owned entirely by the Dairy Company. It was a party wall within the 1894 Act but only to the extent that it performed a separating function between buildings. It only did this up to the roof level of the two-storey building. There was no intention in the Act that a wall performing a separating function for part of its height should be a party wall for its entire height. Morley & Lanceley only had the right to undertake work to those parts of the wall which constituted a party wall. The parapet did not fall within the Act's definition so they had no right to raise it.

- **Notes:** See also: *Weston v Arnold* (1873); *Knight v Pursell* (1879); *Drury v Army & Navy Auxiliary Co-operative Supply Ltd* (1886).

The facts

The Dairy Company had a tenancy of some land in Lambeth. This was separated from adjoining land to the east by a freestanding wall which was included in their demise. The adjoining land was owned by Morley & Lanceley. This consisted of some shops, and a passageway which ran to the rear of the shops, along the whole length of the wall (Figure 13).

In 1890 Morley & Lanceley roofed over the passageway and extended the ground floors of the shops backwards to the wall. Then, in 1898, they constructed a walkway on the roof of this extension to provide access to the rear of the shops at first floor level. As part of this work they also raised the height of the wall by several courses to form a parapet for the walkway.

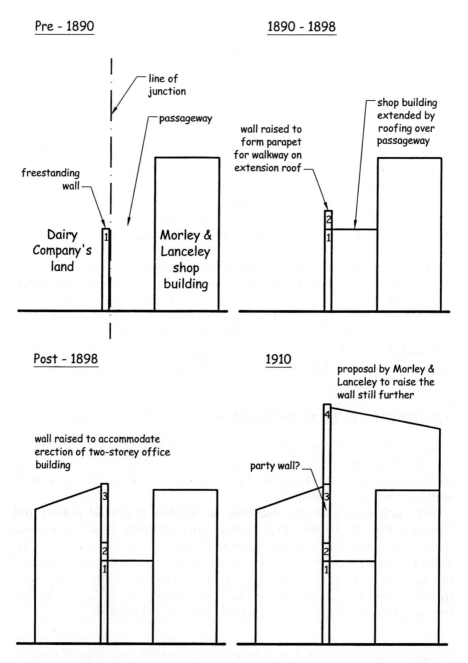

Figure 13: London, Gloucester & North Hants Dairy Company v Morley & Lanceley

Subsequently the Dairy Company erected a two-storey office building against the wall on their side. As part of this work they raised the wall to first floor roof level, and inserted airbricks into it to ventilate their new building.

Raising the height of a wall

The present case arose when Morley & Lanceley decided, once again, to raise the height of the wall. They served a party structure notice on the Dairy Company under the 1894 Act and, in due course, a surveyors' award sanctioned the work. The Dairy Company denied that the wall was a party wall. They objected to the work on the ground that it involved a trespass to their property and appealed to the county court against the award.

The county court had to adjudicate on the status of the wall. It found that the wall had been constructed entirely on the company's premises. Because Morley & Lanceley had enclosed on it and made use of it up to the level of their parapet, it was now a party wall up to this level under section 5(16)(a) of the Act[1]. However, above this level it was not a party wall and Morley & Lanceley therefore had no right to raise it under the Act. The court modified the surveyors' award accordingly.

Interpretation of party wall legislation

Morley & Lanceley appealed to the High Court against the county court's decision. They argued that the wall was a party wall in its entirety and that the legislation did not recognise the concept of a wall which was a party wall for part only of its height.

They acknowledged the decisions in *Weston v Arnold* (1873) and *Knight v Pursell* (1879) which contradicted this suggestion[2]. However, these cases arose under earlier legislation[3] and could not be relied on. The 1894 Act was a codifying statute which now contained a complete code of the law as between building owner and adjoining owner[4]. It should therefore be construed, not according to the law as it stood before the Act, but by giving the words of the statute their natural meaning[5].

---

[1] Because it was a wall "used or constructed to be used for separation of adjoining buildings belonging to different owners" within the meaning of that section. The concept is the same as the type 'b' party wall within section 20 of the 1996 Act.
[2] And also the dicta of Wright and Collins JJ in *Drury v Army & Navy Auxiliary Co-operative Supply Ltd* (1896) which had been decided under the 1894 Act.
[3] Bristol Improvement Acts (1840 and 1847) and the 1855 Act respectively.
[4] *Re Stone and Hastie* (1903).
[5] *Bank of England v Vagliano Brothers* (1891), applied in *R v Fulling* (1987).

> # London Building Act 1894
>
> ## PART I: INTRODUCTORY
>
> ### 5. Definitions
>
> In this Act unless the context otherwise requires:
>
> (16) The expression "party wall" means:
>
> > (a) A wall forming part of a building and used or constructed to be used for separation of adjoining buildings belonging to different owners or occupied or constructed or adapted to be occupied by different persons; or
> >
> > (b) A wall forming part of a building and standing to a greater extent than the projection of the footings on lands of different owners.
>
> ## PART VI: CONSTRUCTION OF BUILDINGS
>
> ### 58. Cases in which a wall to be deemed a party wall
>
> In either of the following cases:
>
> > (a) When a wall is after the commencement of this Act built as a party wall in any part; or
> >
> > (b) Where a wall built before or after the commencement of this Act becomes after the commencement of this Act a party wall in any part;
>
> the wall shall be deemed a party wall for such part of its length as is so used.

Can part of a wall be a party wall?

They argued that the natural meaning of these words clearly indicated that once a wall became a party wall it must be classified as such for its whole height. They cited numerous provisions within the legislation which purportedly supported this view. In particular they made reference to sections 5(16) and 58.

Section 5(16) defined a party wall as including a wall forming part of a building and used for the separation of adjoining buildings. Morley &

Lanceley maintained that a wall was so used if any part of it separated adjoining buildings. However, the definition referred to a party wall in its entirety and made no attempt to limit its extent only to those parts which performed a separating role. The natural meaning of section 5(16) was therefore that a whole wall would be treated as party if any part of it separated adjoining buildings from each other.

They argued that this interpretation was further supported by section 58 which did impose limitations on the extent of a party wall in circumstances where it was only used as such for parts of its length. In these circumstances the section limited the definition only to those parts of the wall which were so used. They argued that by limiting the extent of a party wall along its length, but by placing no such limitation throughout its height, the legislature clearly intended the whole height of a wall to be party if any part of it separated adjoining buildings.

The decision

The court accepted Morley & Lanceley's general approach to the interpretation of the statute but rejected their interpretation of the particular words.

> Bankes J: "The [case] raises a question as to the proper construction of the London Building Act 1894. I confess I have felt and still feel considerable doubt as to what the proper construction of that enactment should be. Counsel for the appellants invite us to regard it as a codifying Act, and, so treating it, to give the language its natural meaning. I assent to that. The difficulty is to discover what is the natural meaning of the language."

In particular, the court rejected the notion that section 58 was a limiting section which limited the broad definition of a party wall contained in section 5(16). The court held that section 5(16) itself limited the extent of a party wall only to those parts which were actually used for the separation of adjoining buildings.

The role of section 58 was simply to emphasise this in the case of walls which only performed a separating role for parts only of their length. The fact that no express provision appeared in respect of walls which were so used only for parts of their height did not prevent the same principle from applying.

> Phillimore J: "Obviously what is meant [by section 5(16)] is that the wall shall be a party wall so far as it separates, none the less because it does not separate in other parts and none the more because it is larger than is required for separation.

Then section 58 emphasises this and no more as the meaning of the Act ... I think that the best explanation of section 58(b) is that its object is to make it quite clear that section 5(16)(a) means all that it says and no more."

The court rejected the appeal. The top of the wall was not a party wall and Morley & Lanceley therefore had no right to raise it. The county court's modification of the surveyors' award was therefore upheld.

LOOST v KREMER (1997)

West London County Court

In brief

- **Issues:** 1939 Act – validity of party structure notice – definition of "owner" – definition of "party wall" – validity of surveyor's appointment – appointment of project architect – third surveyor – jurisdiction to determine matters of law – whether required to hold a hearing.

- **Facts:** Kremer proposed to raise the party wall which separated his maisonette from adjoining property owned by Loost. He served a party structure notice on Loost to this effect. Loost challenged the validity of the party wall procedures on various grounds. The third surveyor ruled that the procedures were valid. Loost appealed against the third surveyor's award.

- **Decision:** The party wall procedures, and the third surveyor's award, were valid. Kremer's landlord was not a "building owner" within the Act and she did not have to join in the service of the party structure notice. The landlord was not "desirous of building" merely by virtue of having consented to the works under the lease or having entered into a deed of variation to facilitate the works. Kremer was entitled to undertake work to the party wall irrespective of whether the fabric of the party wall was included in his demise. The party wall performed a separating function and this was sufficient to bring it within the Act's definition of party wall. The appointment of Kremer's surveyor was valid. In the absence of a conflict of interest there was no impediment to a building owner appointing the project architect as their surveyor. The third surveyor was entitled to determine matters of law although his decision would not be final and could be challenged in the courts. The third surveyor had conducted his adjudication correctly. He was not required to hold a formal hearing before making his award.

- **Notes:** See also: *Sims v The Estates Company* (1866); *Crofts v Haldane* (1867); *Gyle-Thompson v Wall Street (Properties) Ltd* (1974); *Lehmann v Herman* (1993).

The facts

The case concerned two terraced houses at 36 and 37 Upper Addison Gardens, London W14. The properties were separated by a party wall (Figure 14).

Mr and Mrs Loost were the owner-occupiers of number 36. Number 37

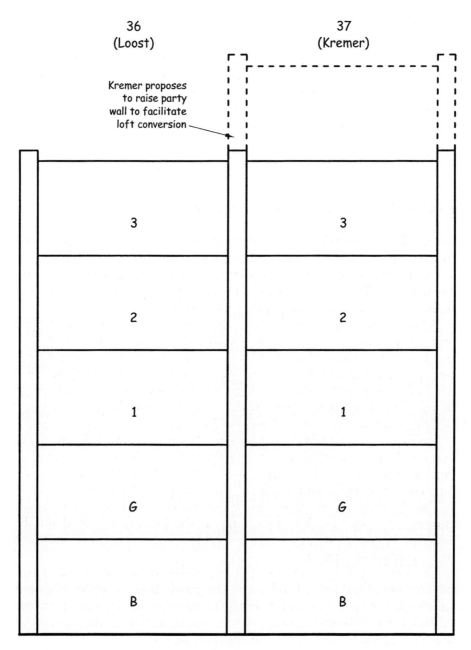

Figure 14: Loost v Kremer

had been converted into two flats. The freehold of this property was owned by Mrs Bartholomew, who also occupied the ground and first floors. The long leasehold interest in the top (second and third) floors was owned by Mr Kremer.

Mr Kremer wished to convert his roof space into living accommodation. The works would involve raising the parapet party walls and constructing a mansard roof in place of the existing pitched one.

He obtained Mrs Bartholmew's licence, as landlord, to carry out the alterations. He also obtained a variation of the lease to bring half the thickness of the party wall into his demise so that he could undertake the work. This had formerly been retained by the freehold title. He then served a party structure notice on Mr and Mrs Loost, under the 1939 Act, in respect of the proposed works.

Express dissent to party structure notice

Mr and Mrs Loost failed to respond to the notice so a dispute was deemed to arise. Each party appointed a surveyor and a third surveyor was selected. Mr and Mrs Loost then served an express "notice of dissent" on Mr Kremer. This, rather mischievously, listed three supposed procedural objections to the work being undertaken under the Act.

Firstly, they claimed that the party structure notice was invalid as it had only been served by Mr Kremer. According to Mr and Mrs Loost, Mrs Bartholomew should also have joined in the notice as she was also a building owner within the meaning of the Act. By granting the licence and deed of variation she had demonstrated that she too was "desirous of doing ... work affecting a party wall..." within the relevant definition of building owner in section 5 of the 1930 Act.

Secondly, they claimed that, in reality, Mr Kremer had no interest in the party wall and so had no right to undertake the work under the Act. The deed of variation was, in reality, a sham. It was also quite insufficient as it only conferred an interest in part of the wall on Mr Kremer.

Thirdly, they claimed that the appointment of Mr Kremer's surveyor (surveyor A) was invalid. According to the Loosts, he could not be appointed under the Act as he was also the project architect and therefore had a conflict of interest.

The parties' surveyors were unable to agree on these three issues so they were referred to the third surveyor. The third surveyor made an award which ruled in favour of Mr Kremer on all three points. Mr and Mrs Loost appealed to the county court against the award. In considering the appeal the court again addressed each of the three issues.

Definition of "building owner"

Firstly, the court held that Mrs Bartholomew was not a building owner within the meaning of the Act. She was not involved in undertaking the work herself and was not, in a legal sense, a joint owner with Mr Kremer.

Cowell J: "It seems to me perfectly clear that the person who is desirous of doing the work is Mr Kremer. It is quite true that Mrs Bartholomew appears to be perfectly happy to see the work done. She may even have encouraged Mr Kremer to do it. She certainly permitted it. But she herself will not be doing the work. It will be Mr Kremer by his architects and his contractors and agents and workmen, and so forth, who will be doing the work.

Whatever encouragement Mrs Bartholomew may be giving or may have given by virtue of the licence, the variation of the lease and her consent, does not, in my judgment turn her into a person desiring to do the work. It is simply a matter of ordinary English. That being so, although she is clearly an owner of number 37 because she has a sufficient estate, she is not a building owner.

The alternative way in which the matter is put is that she is a joint owner with Mr Kremer and therefore she must join in the notice. I have been referred to the case of *Lehmann v Herman* (1993) ... In that case the building owners were joint tenants of the leasehold estate. So in terms of ordinary property law, it necessarily meant that the owner was the two of them. Only the husband gave the notice and it was held to be bad. There are arguments that can be put forward to suggest that that decision was wrong but I do not need to deal with these.

It seems to me that that decision is entirely distinguishable from the situation here. It may be said that both Mrs Bartholomew and Mr Kremer are owners, and it might be said that in a sense they are therefore joint owners, but in law that is not the case because they do not jointly hold the same legal estate. Mrs Bartholomew is the freeholder, Mr Kremer is the long leaseholder. They cannot, in my judgment, possibly be joint owners in the sense in which the words were used in *Lehmann v Herman*."

Ownership of party wall

The court held, secondly, that the ownership of the party wall was irrelevant to the entitlement to undertake works to it under the Act. Providing it was a party wall within the meaning of the Act, a building owner, within the meaning of the Act, had the right to undertake the permitted works.

Cowell J: "It seems to me that what the legislature was doing in passing section 46 was saying that if we look at a wall such as the wall between 36 and 37, it is not possible to know whether the line [of junction] is definitely on the left side of the wall or on the right side of the wall or somewhere in the middle ... that that being, in my judgment, the reason why one finds the wide definition of 'party wall' in sections 44(i) and (ii).

... It follows that the party wall in a case of buildings which are already there, because of the definition in section 44(ii), may be a wall wholly on one side or the other of the line of junction. In that case, the building owner may be on either side and the adjoining owner may be on either side. It seems to me that ownership of the wall is immaterial to the definition.

Put shortly, the situation relevant to this case, which gives rise to the operation of section 46[1] is stated in this way in the Act: 'Where lands of different owners adjoin and at the line of junction the said lands are built on ... the building owner shall have the following rights...' [These rights] relate to doing various things to, in this case, a ... party wall, and by section 44(ii) it seems to me that the ownership of the party wall is immaterial.

...It is, in my judgment, the ownership, as defined, of the adjoining entity, not of the party wall or party structure, which is crucial."

Project architect as appointed surveyor

Thirdly, surveyor A was held to have been validly appointed. An appointed surveyor is a "quasi arbitrator" and owes certain duties in this capacity. However, in the absence of any conflict of interest, there is no impediment on a project architect being so appointed. He will simply owe different duties in his new role.

> Cowell J: "... it has been decided by Mr Justice Brightman in the case of *Gyle-Thompson v Wall Street (Properties) Ltd* (1974) that [the surveyor] is a quasi arbitrator. But as such he owes duties and if they are infringed there is no doubt a right of action.
>
> But the fact that he is an architect does not seem to me to invalidate his appointment. There is no evidence of a conflict of interest, nor did [the third surveyor] find that there was any. I really find it extremely difficult to see that this point can possibly be right because it must frequently happen that parties in this situation do appoint persons who have acted for them because they know very much more about the building than any new appointee would.
>
> The mere fact that he has acted as architect does not, in my judgment, mean that he must disqualify himself. What happens – and there have been a number of cases cited to me and I have read them – is that he changes in his capacity from being simply an agent to a quasi arbitrator and he has to bear in mind that those are his duties.
>
> That is really all there is to this point. He has those new duties as quasi arbitrator on his appointment, and by law he must act accordingly. But in no way does his previous capacity prevent him from acting in the new capacity."

---

[1] Broadly equivalent to section 2 of the 1996 Act.

London Building Acts (Amendment) Act 1939

PART I: INTRODUCTORY

1. Short title construction and citation

This Act may be cited as the London Building Acts (Amendment) Act 1939 and shall be read and construed as one with the London Building Acts 1930 and 1935 and may be cited with those Acts as the London Building Acts 1930 to 1939.

PART VI: RIGHTS ETC. OF BUILDING AND ADJOINING OWNERS

44. Interpretation of Part VI

In this Part of this Act unless the context otherwise requires the following expressions have the meanings hereby respectively assigned to them:

"party wall" means:

(i) a wall which forms part of a building and stands on lands of different owners to a greater extent than the projection of any artificially formed support on which the wall rests; and

(ii) so much of a wall not being a wall referred to in the foregoing paragraph (i) as separates buildings belonging to different owners.

London Building Act 1930

PART I: INTRODUCTORY

5. Definitions

In this Act, save as is otherwise expressly provided therein and unless the context otherwise requires, the following expressions have the meanings hereby respectively assigned to them (that is to say):

"building owner" means such one of the owners of adjoining land, as is desirous of building or such one of the owners of buildings, storeys, or rooms separated from one another by a party wall or party structure, as does or is desirous of doing a work affecting that party wall or party structure.

Third surveyor's jurisdiction

During the appeal, the Loosts also challenged the third surveyor's jurisdiction to determine the three issues addressed in his award. As matters of law, these could only be decided by the courts.

The court upheld the third surveyor's jurisdiction to decide these issues. Although matters of law, they were fundamental to his ability to make an award. He was therefore at liberty to decide them, although his decision should not be regarded as final. It could always be challenged in the courts by an aggrieved party.

> Cowell J: "It seems to me that those ... points were absolutely fundamental to the matter going ahead at all. It seems to me that an arbitrator, a third surveyor, does have jurisdiction to decide a matter, even if it is a matter of law, which is fundamental to the question of whether he makes an award or not.
>
> It is possible for an arbitrator to say: 'This is a matter of law, it ought to be decided by a court first, then bring it back to me', but I can see nothing wrong in the arbitrator saying: 'I must decide this point because it is fundamental. I will decide it and I will say what my award would be on one basis or another', and then leave it to the party aggrieved to appeal to determine the point of law.
>
> A number of cases have been cited to me which show what a third surveyor cannot do, and of course it is quite clear in all those cases that he could not decide on matters which were not in any way within the sections of the Building Act.
>
> But it seems to me that the matter really could not start without a decision on those ... fundamental matters and he was bound to decide it and, as I have indicated, it seems to me that he was simply right ... In my judgment, he had jurisdiction to decide those matters and he decided them correctly."

The words of Cowell J's judgment may conceal a distinction between surveyors making pragmatic decisions to enable them to proceed with an award and them actually having, in a legal sense, jurisdiction to determine matters of law. He was no doubt aware of these ambiguities, as he continues:

> "But in case I am wrong and he had no jurisdiction to decide them, I think it would be correct for me to make the declaration which is counterclaimed by the respondent to the effect that the party structure notice ... was valid and correct, and to declare that the respondent is and was a building owner and was entitled to serve the party structure notice, and that he was entitled to serve it without making Mrs Bartholomew a party to it and to make the other declarations sought."

Third surveyor's duties

Finally, the Loosts also suggested that the third surveyor had acted improperly in failing to hold a hearing before making his award. The court held that, as he was not an arbitrator, he was under no duty to do so. In any event, the only issues that could have been further aired by a hearing were matters of law and these were better addressed through the appeals procedure.

> Cowell J: "Finally, I should just add that at one stage of the argument it was suggested ... that [the third surveyor] should have held some kind of hearing so that [the Loost's surveyor] had a further opportunity of advancing his case, he having been away on holiday for a bit of July and most of August before the award was made.
>
> There is authority that the third surveyor is not an arbitrator and does not have to hold any kind of hearing but, even if that is wrong, these appeal points of law are of the kind which are simply corrected, if the third surveyor was wrong, on appeal.
>
> It seems to me that it is far from clear what further arguments, which would in any event be arguments of law, could have been put by [the Loost's surveyor] to [the third surveyor], even if he had not been on holiday. It is evident ... that the matter was fully ventilated in correspondence. It seems to me that [the third surveyor] had quite enough to go on and there was no need for anything more to be put before him."

The decision

The court upheld the third surveyor's award. All necessary procedural steps under the 1939 Act had been correctly followed. The court therefore made declarations, requested by Mr Kremer, to this effect.

LOUIS v SADIQ (1997)
Court of Appeal

In brief

- **Issues:** 1939 Act – failure to serve party structure notice – damage to adjoining owner's property – injunction – nuisance – consequential financial losses – causation of damage – remoteness of damage – consequences of non-compliance with 1939 Act – whether retrospective award has curative effect on earlier unlawful works.

- **Facts:** Louis and Sadiq owned adjoining properties separated by a party wall. Sadiq carried out work to the party wall without first serving a party structure notice. The work caused damage to Louis' property with the result that it became impossible to sell it. They incurred additional financial losses connected with their abortive purchase of a property in Guadeloupe. The work was restrained by an injunction. Sadiq subsequently complied with the 1939 Act. The injunction was then lifted and the work was completed.

- **Decision:** Sadiq was liable for Louis' losses, including all consequential financial losses. These related to the loss of a prospective purchaser for their property as this event had been caused by the unlawfulness of Sadiq's work. Furthermore, the losses were not too remote to be recoverable. Although the precise nature of the losses could not have been foreseen, it was sufficient that they were of a kind which was reasonably foreseeable. Compliance with the 1939 Act would have conferred immunity on Sadiq from liability in tort but he had chosen not to seek this protection. His subsequent compliance with the Act provided him with protection for the later works. However, this had no curative retrospective effect on ongoing liabilities arising from the earlier unlawful work. Damages for consequential financial losses were therefore calculated for the whole period until Louis were able to sell their property on the completion of Sadiq's works. The fact that an award was in force for part of this period was irrelevant.

- **Notes:** See also: *Standard Bank of British South America v Stokes* (1878); *White v Peto Brothers* (1888); *Leadbetter v Marylebone Corporation [No. 1]* (1904); *Adams v Marylebone Borough Council* (1907); *Emms v Polya* (1973); *London & Manchester Assurance Company Ltd v O & H Construction Ltd* (1989).

The facts

Mohammed Sadiq owned an end of terrace house at 50 Jenner Road, London N16. Paul and Paula Louis owned the adjoining terraced house at number 52. The two properties were separated by a party wall.

In August 1988 Mr Sadiq began substantial reconstruction works to his property which included work to the party wall. He did so without serving the requisite party structure notice on Mr and Mrs Louis under the 1939 Act. The works included the complete demolition of the front and rear walls of Mr Sadiq's property. This caused damage to Mr and Mrs Louis' house, including cracking to the front wall.

Mr and Mrs Louis issued proceedings against Mr Sadiq in nuisance. They sought damages (discussed below) and an injunction. In October 1988 the court granted an injunction which restrained Mr Sadiq from undertaking any further work until he had complied with the Act.

In December 1988 he served a party structure notice. A (third) surveyor's award was eventually made in June 1991 regulating the conduct of the work. During the whole of this period Mr Sadiq's property remained open to the elements with no front wall and with partly built flank and rear walls.

The injunction was lifted in December 1991 and Sadiq resumed the works. By October 1992 he had completed the work to his own property and, at his own expense, repaired all the damage which he had caused to number 52.

Damages for unauthorised works

The present case dealt with Mr and Mrs Louis' outstanding claim for damages in respect of the nuisance caused by the earlier unauthorised works. They claimed damages under the following three heads:

(1) General damages[1] for nuisance, including damages for distress and inconvenience.
(2) Special damages[2] in respect of mortgage interest paid on a further mortgage advance between November 1988 and 31 March 1993. (They had been in the process of moving to Guadeloupe. They had taken out a further advance on their property in April 1988 to finance the purchase of a building plot in that country. They had intended to sell their property to finance the building of a home on the plot. Due to the state of Mr Sadiq's house, their purchaser withdrew in November 1988. They were then unable to sell their property until March 1993. They therefore incurred additional interest charges for this period.)

---

[1] Damages which are not capable of precise calculation and which therefore have to be assessed by the trial judge.
[2] Damages awarded in respect of pecuniary losses which are capable of precise calculation.

(3) Special damages in respect of the increase in building costs in Guadeloupe for the same period.

The court awarded damages to Mr and Mrs Louis under each of these heads. Mr Sadiq then appealed to the Court of Appeal, solely against the award of the special damages. He based his appeal on issues of causation, remoteness and the nature of liability for breaches of the party wall legislation.

Causation

He argued that Mr and Mrs Louis' losses had not been caused by his actionable nuisance. They had failed to sell their property, not because the building works were unlawful, but because they were taking place at all. They would still have incurred them if the works had been undertaken lawfully and no nuisance had been committed. Their losses were therefore irrecoverable.

The court rejected this ground of appeal. The trial judge had found, and was entitled to find, that the sale to the prospective purchaser would not have fallen through if the works had been properly regulated by the Act. The fact that the works were unauthorised and unregulated was clearly the cause of Mr and Mrs Louis' inability to sell their house in 1988.

Remoteness

He also argued that the losses in respect of additional interest payments and increased building costs were too remote to be recoverable as these could not have been foreseen. The court rejected this argument too.

In the context of the additional interest payments, it was sufficient that Mr Sadiq should reasonably have foreseen that Mr and Mrs Louis might be unable to sell their house because of the unlawful works, and that they might thereby incur a continued liability to pay mortgage interest on a loan which would otherwise be repaid. It was not necessary for him to have foreseen the particular circumstances surrounding their further advance.

In the context of the increased building costs it was not necessary for Mr Sadiq to have foreseen Mr and Mrs Louis' intention to build in Guadeloupe. It was sufficient that he should reasonably have foreseen that they might have intended to build a property, in this country or elsewhere, and that increased building costs might have been the consequence of delay.

Liability for breaches of party wall legislation

Finally, he argued that a failure to comply with the 1939 Act did not per se amount to an actionable nuisance.

He also disputed that he could be liable for the special damages for the whole of the period in question because, for part of this period, the works were properly authorised by the Act. During this period he had fully complied with the surveyor's award and he was therefore protected from any liability in tort which might otherwise have arisen.

The court did not have to directly address the question of whether a failure to comply with the Act was actionable in itself. The trial judge had found, on the facts, that Mr Sadiq's conduct, between July and October 1988, amounted to a nuisance. The question of whether liability would have arisen in the absence of any such conduct was therefore irrelevant.

It could not escape the more difficult question of whether compliance with the Act conferred an immunity from liability in tort. In this context it examined the decisions in *Major v Park Lane Co* (1866), *Standard Bank of British South America v Stokes* (1878) and *Selby v Whitbread & Co* (1917). It concluded that Mr Sadiq would indeed have been immune from liability in nuisance if he had undertaken the offending works after obtaining proper authority for them under the Act.

> Evans LJ: "These authorities establish, in my judgment, that the appellant would not have been liable in nuisance if he had given notice, or obtained consent, in accordance with the Act and then done no more than was agreed or was approved by the surveyors. But then, no damage would have been caused to the respondents' house, save in the party wall itself, and in that respect no liability would have arisen...
>
> ... The adjoining owner's common law rights are supplanted when the statute is invoked, which can have the effect of safeguarding the building owner from common law liabilities when he complies with the statutory procedures, just as he may incur liabilities under the statute which did not exist at common law (the *Standard Bank* decision)."

Retrospective awards

However, the question to be decided was whether Mr Sadiq's continuing liability for a pre-existing nuisance would determine upon his subsequent compliance with the Act's procedures.

The court held that it would not. In undertaking the work without appropriate authority, Mr Sadiq had performed an unlawful act. He was liable for all the consequences of this particular act, irrespective of the fact that other acts were subsequently authorised by the statute.

> Evans LJ: "The issue raised in the present case is whether the appellant's liability at common law is either excluded or reduced by the provisions of the Act, which he invoked, eventually, after the nuisance had arisen. I would have no hesitation in rejecting this submission even without reference to authority,

because, in my judgment, there is nothing in the Act which can be said to have this effect...

... if [the building owner] commits an actionable nuisance without giving notice and without obtaining consent, he cannot rely upon a statutory defence under procedures with which *ex hypothesi* he has failed to comply. If he does then give notice he will in due course acquire statutory authority for whatever works are approved or agreed, but, in my judgment, this does not relieve him from liability for the continuing nuisance which he has unlawfully committed, until such time as and to the extent that such authority is obtained."

The court was satisfied that this conclusion was consistent with the decision in *Adams v Marylebone Borough Council* (1907) which could be distinguished from the present case. Mr Sadiq had cited this case in support of his contention that a subsequent award could have a curative effect on previous unlawful acts.

In *Adams*, works had been undertaken without service of appropriate notices. The works had caused loss to an adjoining owner for which damages were awarded by the court. An award was then made and the works continued under its authority.

When the adjoining owner made a claim (through the surveyors) for subsequent losses, the court held that no claim could lie "before any tribunal whatever" because the work was "in its nature rightful".

Evans LJ explained the basis for distinguishing the present case from the situation in *Adams*:

"In the present case ... [Mr Sadiq's counsel] submits that no claim lies for damage caused after the notice was given, in December 1988. If *Adams v Marylebone* was authority for that proposition then, of course, it would be binding on us, even if that result seemed not in accordance with principle, for the reasons stated above. But, in my judgment, there were two significant features of that case which are not present here.

First, there was an agreement between the parties when the notice was given 'that the work already done should thenceforth be treated as if it had been done under a notice duly given under the Act.' No such agreement is alleged or found to have been made in the present case. Therefore, the basis for the court's decision that what was done was lawful under the Act does not exist here.

Second, a related point is that there is no finding in the present case that the statutory approval when it was given covered all the works which created the nuisance in 1988. It would be surprising if that finding were even suggested, because the surveyors could hardly have agreed to approve works which caused such widespread damage to the adjoining house. For example, the appellant failed to provide shoring when the front wall of his house was removed.

So it cannot be said, in my judgment, that the works which created the nuisance were subsequently authorised, whether by agreement or by surveyors under the statutory procedures."

The decision

For the reasons discussed, Mr Sadiq was liable for the losses which Mr and Mrs Louis had incurred in respect of additional interest payments and increased building costs. Moreover, the award which was subsequently published related only to the works which were subsequently carried out. It had no retrospective curative effect on the earlier unlawful works. Mr Sadiq therefore incurred an ongoing liability for Mr and Mrs Louis' losses arising out of these works, even after the publication of the award. The appeal was therefore dismissed.

MAJOR v PARK LANE COMPANY (1866)
Court of Chancery

In brief

- **Issues:** 1855 Act – demolition of building owner's building – exposure of party wall – whether service of party structure notice required – consequential damage – whether a requirement to make good.
- **Facts:** Park Lane demolished their building. They left the party wall standing which separated their property from that owned by Major. The demolition works included the removal of a bressumer which was embedded in the party wall. No party structure notice was served prior to the work being undertaken.
- **Decision:** Park Lane were entitled to undertake the work without serving a party structure notice. On the facts there had not been an interference with the party wall requiring service of a notice. The bressumer was not part of the party wall and could also be removed without service of a notice. There was no obligation to make good the opening in the party wall left by the removal of the bressumer.
- **Notes:** (1) A party structure must now be served under section 2(2)(n) of the 1996 Act before demolishing a building where this will expose a party wall. See also: *Lewis & Solome v Charing Cross, Euston, & Hampstead Railway Company* (1906). (2) The published law report does not state whether the bressumer was only embedded in Park Lane's half of the wall or whether it also affected Major's half. It is unclear whether this would have had any influence on the court's decision. See also: *Upjohn v Seymour Estates Ltd* (1938). (3) See the following cases on the status of particular boundary features under the legislation: *Johnston v Mayfair Property Company* (1893); *Thornton v Hunter* (1898); *Moss v Smith* (1977).

The facts

Major owned 5 Park Lane, London W1. Park Lane Company owned the adjoining terraced property. The two properties were separated by a party wall (Figure 15).

Park Lane started demolishing their building in order to redevelop their site. They served no prior notice on Major under the 1855 Act. They considered that this was not required as the work left the whole thickness of the party wall standing.

Major became alarmed that damage would result to his property due to

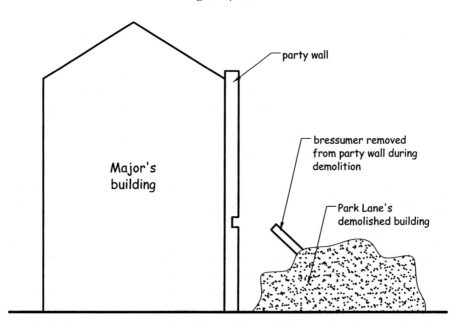

Figure 15:   Major v Park Lane Company

the careless manner in which the works were proceeding. He was particularly concerned that the end of a wooden bressumer had been removed from the party wall without the damage being made good. He therefore issued proceedings against Park Lane.

He argued that the work was unlawful because of their failure to serve a party structure notice on him before starting the work. Having regard to the interconnected nature of the two buildings, Park Lane's building could not be removed without interfering with the party wall. They therefore had no right to undertake the work without first serving notice on him under the 1855 Act. He requested an order that Park Lane should make good the damage caused by the removal of the bressumer.

Park Lane responded that service of a party structure notice was not required for a building owner to exercise his common law right of removing his house from the party wall, even if this involved some interference with the wall.

Notice required for removal of building from party wall?

The court considered whether service of a party structure notice was required for the removal of a building where the party wall was left standing. It considered that this would always depend on the degree of connection between the demolished building and the party wall. If any part of the demolished building could be said to form part of the party wall then a notice would be required.

> Sir W Page Wood VC: "The question raised upon [the] Act is one of considerable moment, and if I had been satisfied that any part of the defendants' building itself formed part of the party structure, I should have been disposed to agree with the plaintiff's contention that the removal of the defendant's building would amount to a dealing with the party structure so as to render notice under the Building Act necessary."

However, on the facts, the court was satisfied that the bressumer was not a sufficiently integral part of the wall for it to constitute part of the party wall:

> Sir W Page Wood VC: "This is a grave question which, however, it is not necessary to decide, because I am of opinion upon the whole, that this bressumer cannot be considered as part of the party structure. I cannot accept the argument that notice under this Act is required in every case of the mere removal of a building from the adjoining premises; but neither can I say that the building about to be removed may not form such a part of the party structure as that it cannot be taken away without notice being given under the Act."

The decision

The court held that Park Lane were not required to serve a party structure notice on Major before demolishing their building as, on the facts, this did not involve interfering with any part of the party wall. The bressumer which they had removed was not part of the party wall. They were free to remove it without making good the opening left by it. The demolition works were therefore lawful and Major's claim was dismissed.

MARCHANT v CAPITAL & COUNTIES plc (1983)
Court of Appeal

In brief

- **Issues:** 1939 Act – type 'b' party wall – interpretation of surveyors' award – power to impose continuing obligations – obligation to provide protection from the weather.

- **Facts:** Many years previously Marchant's mews house had been erected against, and had enclosed on, the rear wall of a warehouse owned by Capital & Counties. Capital & Counties now demolished the warehouse in compliance with the party wall procedures in the 1939 Act. The portion used by Marchant's house was left standing. The surveyors' award provided that they should be "at liberty" to carry out certain protection and stabilisation works to this portion. Years later, rain penetrated through the exposed face of the wall and damaged Marchant's property.

- **Decision:** Capital & Counties were in breach of a continuing obligation in the award to maintain the wall in a weatherproof condition. The award had to be read as a whole and it made no sense unless Capital & Counties were obliged to comply with the protection and stabilisation provisions. Furthermore, these imposed a continuing obligation to maintain the wall in a weatherproof condition. It will often be undesirable for surveyors to impose continuing obligations but it was within their power to do so.

- **Notes:** (1) See also: *Phipps v Pears* (1964); *Bradburn v Lindsay* (1983); *Rees v Skerrett* (2001). (2) Section 2(2)(m) of the 1996 Act now requires a building owner to provide "adequate weathering" as a condition of his statutory entitlement to expose a party wall.

The facts

Mrs Marchant owned a small mews house at 17B William Mews, London SW1. When the property had been erected it had enclosed upon a warehouse building to the rear, called the Pantechnicon (Figure 16).

The assumption in the case was that the rear wall of the house (which formed only a small part of the warehouse's wall) continued to form part of the Pantechnicon's title and that it was a type 'b' party wall. The Pantechnicon was owned by Capital & Counties.

In 1967 Capital & Counties decided to demolish the Pantechnicon, whilst leaving standing the portion of wall used by Mrs Marchant's

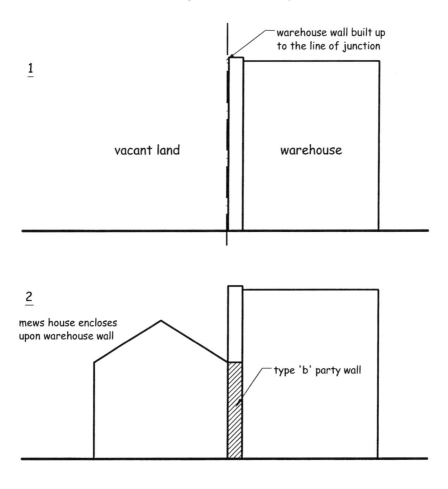
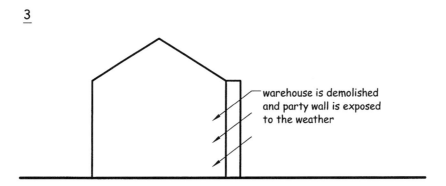

Figure 16: Marchant v Capital & Counties plc

property. They therefore served a party structure notice on her to this effect. Surveyors were then appointed and an award was published in 1969.

Clause 2 of the award provided that Capital & Counties were "at liberty" to undertake certain protection and stabilisation works to the wall. These included the following subclauses:

> "(d) Finish the top of the party wall and buttresses with a pre-cast concrete coping. Fill in the voids in face of party wall to form a reasonably flush face, form flashings to all ledges, and paint the whole wall on completion with two coats of white emulsion paint.
> (e) Maintain the exposed face of the wall in a weatherproof condition."

Once the award was published Capital & Counties demolished the Pantechnicon and carried out all the works to the party wall which had been referred to.

From about 1970 damp started to appear inside Mrs Marchant's house on the inside face of the party wall. This became progressively worse and in 1981 she eventually brought proceedings against Capital & Counties in connection with this. She argued that they were in breach of the obligation, contained in clause 2(e) of the award, to maintain the party wall in a weatherproof condition.

The trial judge dismissed Mrs Marchant's claim so she appealed against this decision to the Court of Appeal. The case raised two main issues. The first of these was the extent to which clause 2 of the award could be construed as imposing obligations on Capital & Counties.

Obligations in surveyors' award

The trial judge had held that the wording of clause 2 of the award indicated that it was not intended to impose liabilities. It referred to Capital & Counties simply being "at liberty to carry out the following". They therefore had an entitlement to carry out the works if they so wished but were subject to no obligation to do so.

After considering clause 2 in the context of the award as a whole, the Court of Appeal came to the contrary view. They held that it did impose obligations on Capital & Counties in respect of the party wall, including a continuing obligation to maintain it under subclause (e).

> Dillon LJ: "It is true that the wording of clause 2 begins: 'That upon the signing hereof the building owner shall be at liberty to carry out the following' whereas clause 3 provides that the building owner is to do various things. However, the award makes no shape or sense if all the essential provisions as to works in clause 2 are wholly optional on the building owner: he can construct brick

buttresses if he wants to; he can fill in voids in the face of the party wall if he wants to; he can rebed and make good the coping stone and finish the top of the party wall if he wants to; but Mrs Marchant ... must rest on his goodwill and have no enforceable rights under the award at all.

That to my mind does not make sense. I think that the true construction and effect of clause 2 is that the building owner is under an obligation to do all that is required by (a) to (e) of clause 2 if he elects to undertake the permitted works referred to in clause 12 at all; and as he did so, he became clearly bound to carry out the whole of the work prescribed by clause 2.

... As a further matter of the construction of the award [Capital & Counties have argued] that (e), 'Maintain the exposed face of the party wall in a weatherproof condition' is related only to the period while the works are being carried out and means that the exposed face of the party wall is to be maintained in a weatherproof condition until the earlier works are completed ending with the painting of the whole on completion with two coats of white emulsion paint.

That is not, in my judgment, a natural reading of the words. 'Maintain' is a continuous process. It means 'maintain the exposed face of the party wall in a weatherproof condition' without limit of time, and the natural placing of (e) as a separate obligation that is to happen after the voids have been filled in, the flashings have been formed, and the painting on completion of the whole has been done under (d).

As a matter of construction, therefore, I read the award as imposing a continuing liability on the defendants to maintain the face of the party wall in a weatherproof condition."

Surveyors' powers to impose continuing obligations

The second main issue concerned the question of whether the surveyors had the power to impose a continuing obligation on Capital & Counties to repair the party wall.

The trial judge had held that this was beyond the surveyors' powers (ultra vires). In his judgment surveyors had no power in their awards to do other than order the immediate carrying out of works. They therefore had no power to impose indefinite and unlimited obligations on the parties.

Once again, the Court of Appeal came to the opposite conclusion. Although it noted that there were limits to the surveyors' statutory authority, there was no restriction on them imposing obligations on the parties, simply because these were of an ongoing nature.

Dillon LJ: "It is clearly established, for instance in *Leadbetter v Marylebone Corporation* (1904) that the function of surveyors ... is to adjust the differences which have arisen between the adjoining owners and the building owners under the Act, the question for the arbitrators being how far the building owners are right in imposing their requirements and how far the adjoining owners are right in their cross demands made upon the building owners in relation to particular works proposed to be carried out to the party wall by the building owners.

It is beyond the powers of the surveyors to go any further and purport to decide future disputes that have not yet arisen or to lay down how the wall is to be dealt with in uncertain future circumstances. If surveyors go beyond their powers, then, notwithstanding that there is no appeal to the county court under the express appeal provisions of the 1939 Act, the courts have held that the award is *pro tanto* invalid and unenforceable."

However this was not the situation in the present case. The surveyors' task was to decide the differences that had arisen between the parties. One of these differences related to the newly exposed party wall's need for protection against the weather.

Dillon LJ: "The surveyors addressed their minds to practical ways of dealing with this, and, it seems to me, the need for a pre-cast concrete coping, for flashings, and for the voids to be filled in are all concerned with this problem. There would be no doubt, as it seems to me, that, had they thought fit, it would have been within their power to require that the wall be rendered weatherproof by a vertical damp-proof course being inserted and by plastering over so as to provide a surface impervious to rain and weather. If that had been an immediate work, it would have been beyond argument that it was something that could have been directed by the surveyors in their award.

The direction that building owners are to maintain the exposed face of the wall in a weatherproof condition is designed to achieve the same result; but one may infer that it would achieve that without the same initial expense to the building owners. I cannot see that it is in any way beyond the powers of the surveyors to impose any such requirement. It may be that it is in practice undesirable, save in somewhat unusual circumstances, that surveyors should seek to impose a continuing obligation by a party wall award rather than prescribing for particular works which will have a certain long-term result – or are expected to have a certain long-term result – to be carried out at once, but I cannot think that it is beyond the powers of the surveyors to deal with this matter as, in my judgment, they have on the proper construction of 2(e) of this award."

The decision

The Court of Appeal allowed Mrs Marchant's appeal. It also made a declaration that Capital & Counties were under a continuing obligation to maintain the exposed face of the party wall in a weatherproof condition.

MASON v FULHAM CORPORATION (1910)
King's Bench Division

In brief

- **Issues:** 1894 Act – party wall agreement – raising of party wall by predecessor in title – purchaser's entitlement to receive contribution towards expenses where adjoining owner subsequently makes use of raised portion – retention of title provision – contingent liabilities.

- **Facts:** Mason raised the party wall which separated his property from the Fulham library. In accordance with their statutory requirement the library agreed that they would contribute towards the cost of raising the wall if they ever made use of the raised portion in the future. By the time they eventually did so, Mason's property was owned by Galsworthy. The contribution was therefore paid to Galsworthy in accordance with a surveyors' award to this effect. Mason claimed that the contribution should have been paid to him as the building owner who had paid for the works.

- **Decision:** The contribution had correctly been paid to Galsworthy as the owner for the time being of the original building owner's property. The Act could not require payment to Mason as, at the relevant time, he had ceased to be a building owner within the terms of the Act. The Act's retention of title provision did not alter this situation. Its reference to "the building owner at whose expense the [wall] was built" referred to immediate liabilities arising out of repair work to a party wall. It was not intended to protect the original building owner's entitlement to receive payment of a contingent liability which might never arise. A building owner's contingent right to receive a contribution passed to a purchaser on a sale of the property. If this had a significant value then it was open to the building owner to reserve this right in the conveyance. This was not the situation in the present case so Mason was no longer entitled to receive payment of the contribution.

- **Notes:** (1) Contrast this case with *Re Stone and Hastie* (1903). In that case the right to receive the contribution did not pass to a tenant on the grant of a 21-year term. The freehold owner who had originally raised the wall retained his entitlement to payment. (2) See also the following cases which also address changes of ownership: *Carlish v Salt* (1906); *Selby v Whitbread & Co* (1917); *Observatory Hill v Camtel Investments SA* (1997).

The facts

Mason was the former owner of a house in Fulham Road, London SW6. The house adjoined the Fulham Library. In 1895, whilst still the owner of the property, he had raised the height of the party wall which separated his house from the library's garden.

He had served a party structure notice, under the 1894 Act, on the library's commissioners. Rather than appoint surveyors the parties had then entered into a written party wall agreement in May 1895. Under this agreement the commissioners agreed that Mason could pull down the wall and rebuild it to an increased height of 40 ft.

The agreement also provided that the commissioners should have the right to use the party wall at any time in the future upon paying one half of the cost of its erection. Mason then demolished and rebuilt the wall in accordance with the agreement. His surveyor certified the cost of erection at £111 9s.

The following year Mason sold the house to Slater and, in 1899, Slater sold it to Sir Edward Galsworthy. In the intervening years the powers of the commissioners had also been transferred to Fulham Corporation under local government legislation.

Adjoining owner making use of party wall

In 1908 Fulham Corporation decided to extend the library into the garden. This involved making use of the previously raised party wall. A party structure notice was therefore served on Sir Edward Galsworthy, as the current adjoining owner, in respect of the work. Surveyors were appointed and an award was made.

The award included a requirement that Fulham Corporation should pay half the original cost of raising the wall (£55 14s 6d) in accordance with the requirements in section 95(2) of the 1894 Act[1]. The Corporation then paid this to Sir Edward Galsworthy, as required by the award.

When Mason became aware of this he issued proceedings against the Corporation to recover the contribution from them under the terms of the agreement of May 1895. He submitted that he had never intended to assign his rights under the agreement to Slater, and that he had not in fact done so.

He also submitted that section 95(2) of the Act required the contribution towards the cost of erection to be paid to the building owner who had actually incurred the expenses, rather than to a subsequent purchaser of the property. This was clear from the words used in section 99. That

---

[1] Broadly equivalent to section 11(11) of the 1996 Act.

section provided for retention of title by the building owner in works under the Act, pending the payment of any contribution due from the adjoining owner[2]. In this context it referred specifically to retention of title by "the building owner at whose expense the same was built".

Finally, he cited the decision in *Re Stone and Hastie* (1903) and quoted the judgment of Collins MR in that case in support of his position: "I think it is obvious when this code is looked at that it provides for the recoupment of the owner at whose expense the wall was raised and not of anyone else".

The matter was dealt with, on appeal from the county court, by the King's Bench Division.

Retention of title in works executed under the Act

The court did not accept Mason's interpretation of section 99, upon which most of his argument was based. In particular, that section provided the necessary security for a building owner at a time the adjoining owner's liability to contribute had already arisen.

In the case of repair work, in situations where a party wall was defective or out of repair, that liability arose straight away. However, where a party wall was raised for the benefit of the building owner, no such liability arose until the adjoining owner subsequently made use of the raised portion. It was therefore only at that moment that the retention of title provision came into operation. The identity of the building owner must therefore be determined at that time.

As Mason had ceased to be the owner at the time that Fulham Corporation made use of the raised wall he could not be the "building owner at whose expense the same was built" within the meaning of the section.

> Phillimore J: "[The plaintiff] says that he is the only person answering the description of 'the building owner at whose expense the same was built'. The difficulty in the way of that contention is that the plaintiff is no longer the owner at all.
>
> The section contemplates a person who is the building owner at the time when the money has to be paid. It gives a security only upon the occasion of the adjoining owner becoming liable, the words being.....'where the adjoining owner *is* liable', whereas the contention of the plaintiff involves the proposition that the owner who rebuilt the wall is to stand possessed of it from the date of the rebuilding as security for the discharge of a contingent liability that may never arise.
>
> Until the adjoining owner wants to use the party wall there is no reason why the building owner should have any security. But where the adjoining owner

---

[2] A similar provision appears in section 14(2) of the 1996 Act.

> # London Building Act 1894
>
> Part VIII: RIGHTS OF BUILDING AND ADJOINING OWNERS
>
> ### 95. Rules as to expenses in respect of party structures
>
> (2) As to expenses to be borne by the building owner:
>
> If at any time the adjoining owner make use of any party structure or external wall (or any part thereof) raised or underpinned as aforesaid or of any party fence wall pulled down and built as a party wall (or any part thereof) beyond the use thereof made by him before the alteration there shall be borne by the adjoining owner from time to time a due proportion of the expenses (having regard to the use that the adjoining owner may make thereof):
>
> > (i) of raising or underpinning such party structure or external wall and of making good all such damage occasioned thereby to the adjoining owner and of carrying up to the requisite height all such flues and chimney stacks belonging to the adjoining owner on or against any such party structure or external wall as are by this part of this Act required to be made good and carried up;
> >
> > (ii) of pulling down and building such party fence wall as a party wall.
>
> ### 99. Structure to belong to building owner until contribution paid
>
> Where the adjoining owner is liable to contribute to the expenses of building any party structure then until such contribution is paid the building owner at whose expense the same was built shall stand possessed of the sole property in the structure.

does use the wall he becomes liable to pay his proportion of the cost, and from that time the building owner is entitled to hold the whole wall as security for the payment – if he is still the owner.

I say 'if he is still the owner', because when once it is conceded that the right of the building owner to stand possessed of the whole wall does not arise until the adjoining owner wants to build it is obvious that it is not a right which can be exercised by a person who having once been the owner has parted with his property.

The words 'shall stand possessed of the sole property' point to the person being already part owner. If he is not already part owner the section does not apply. The words 'the building owner at whose expense the same was built' must be taken to mean the person who at the material time happens to be the owner of the plot whether there has in the interval been a change of ownership or not.

It is only reasonable to assume that the rights of the building owner to contribution pass with the assignment of the property. There is no hardship in that view. The plaintiff, if he thought that the contingent right of contribution had any substantial value, might have made a provision in the conveyance reserving that right to himself."

The decision

A contribution towards the cost of works, where these are subsequently used by an adjoining owner, is payable to the owner for the time being of the original building owner's property. Sir Edward Galsworthy was therefore entitled to receive payment of the contribution and Mason's appeal was dismissed.

MATANIA v NATIONAL PROVINCIAL BANK LTD AND THE ELEVENIST SYNDICATE LTD (1936)

Court of Appeal

In brief

- **Issues:** Leases of separate floors in a single building – construction operations by one tenant – generation of excessive noise and dust by tenant's independent contractor – landlord's liability – breach of covenant for quiet enjoyment – contractor's liability – nuisance – tenant's liability – non-delegable duties.

- **Facts:** The Syndicate carried out alterations to their first floor premises without the Bank's consent as landlord. The works generated excessive dust and noise. As a result Matania had to move out of his premises on the second and third floors.

- **Decision:** The Bank were not liable for breach of their covenant for quiet enjoyment as they had neither carried out the work nor authorised its commission. The Syndicate's contractor had not exercised reasonable care and skill when undertaking the works. The dust could have been minimised by the use of sheeting. The effects of the noise could have been mitigated by staggered working hours to avoid times which were most intrusive to Matania. The contractor's conduct therefore amounted to an actionable nuisance. The Syndicate owed a non-delegable duty to Matania in respect of the works as they involved a serious risk of nuisance. They were therefore liable for their contractor's negligence.

- **Notes:** See also (on the effects of disruptive building operations): *Andreae v Selfridge & Co* (1937); *Emms v Polya* (1973); *Video London Sound Studios Ltd v Asticus (GMS) Ltd and Keltbray Demolition Ltd* (2001) – (on non-delegable duties): *Bower v Peate* (1876); *Dalton v Angus* (1881); *Hughes v Percival* (1883); *Alcock v Wraith and Swinhoe* (1991).

The facts

The case concerned a four-storey building at 128 and 130 Edgware Road, London (Figure 17). The freehold in the property was owned by the Portman Estate. A headlease of the whole building had been granted to National Provincial Bank who occupied the ground floor and cellars for their own use. The headlease was subject to a tenant's covenant against alterations without the landlord's consent.

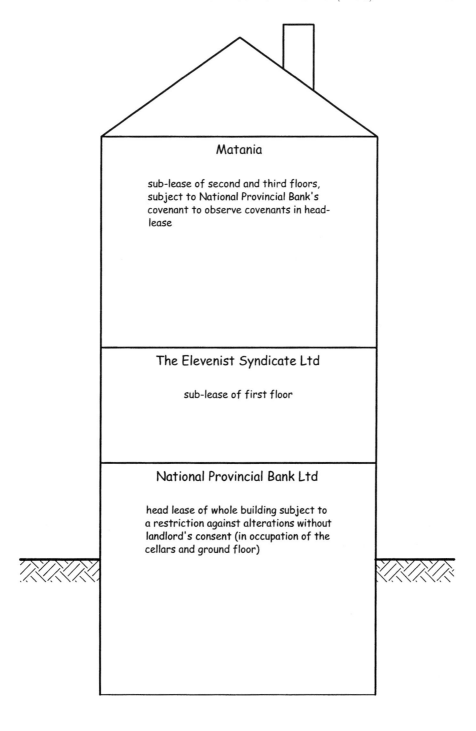

Figure 17: Matania v National Provincial Bank Ltd and The Elevenist Syndicate Ltd

In 1933 the Bank granted a sublease of the second and third floors to Matania. He was a singing teacher and professor of music. He used the premises as a studio for teaching and singing, for holding concerts, and for the occasional use by an operatic society. His sublease was subject to landlord's covenants to observe the landlord's covenants in the head lease, and for quiet enjoyment.

In 1934 the Bank also proposed to grant a sublease of the first floor to the Elevenist Syndicate, for use as a reading room. This was a religious organisation connected with the Church of Scientology.

Considerable alterations were necessary before the Syndicate could use the first floor premises for their intended purpose. These included the installation of a steel joist in the floor immediately below Matania's premises, and the temporary removal of the plaster ceiling immediately below his floor joists.

The Bank therefore obtained the Portman Estate's consent to the alterations, but this was expressed to be subject to consent being first obtained from Matania, as subtenant. The Bank then issued its own consent to the Syndicate but also made this subject to the same condition. Neither the Bank, nor the Syndicate made any attempt to obtain Matania's consent and the alterations therefore commenced without it, on 26 March 1934.

Noise and dust from construction operations

The works generated a considerable amount of noise and dust which made Matania's premises uninhabitable. The noise was said to be so great that it was impossible to hold a conversation there. For the 14 weeks that the works lasted he therefore had to vacate his premises which were incapable of use as a studio.

Matania had written to the Syndicate four days after the work first started. He had alerted them to the problem and requested a meeting "to see that some friendly arrangement can be reached in order to reduce my loss". The Syndicate's architect drew the matter to the contractor's attention but the contractor took no steps to reduce the noise or the dust.

Once the works had been completed Matania brought proceedings to recover damages for his losses. He sued the Syndicate in nuisance and the Bank for breach of their covenant for quiet enjoyment.

The court at first instance awarded judgment for Matania against the Bank, but dismissed his claim against the Syndicate. The Bank appealed against the judgment and Matania therefore brought a cross appeal against the Syndicate in respect of the nuisance claim. All the issues were therefore considered afresh by the Court of Appeal.

Covenant for quiet enjoyment

The Court of Appeal held that the Bank were not in breach of their covenant for quiet enjoyment and allowed their appeal.

The work which caused Matania's losses had not been undertaken by them, but by another tenant in the building. They could only be responsible for this work to the extent that they had authorised its commission. Their consent had been expressed to be conditional on the Syndicate obtaining Matania's prior consent. As this was never obtained the Bank could not be said to have authorised the work which caused the damage so they had not breached their covenant for quiet enjoyment.

Although their appeal was successful, the court refused to make an order for costs in their favour. By failing to prevent the alterations, they were in breach of their tenant's covenant in their headlease with the Portman Estate, not to allow alterations without their consent. This also placed them in breach of their landlord's covenant, in the sublease to Matania, to observe all covenants in the head lease.

The whole of the proceedings could have been avoided if the Bank had taken the trouble to obtain Matania's consent before allowing the alterations to proceed. They had failed to do so and the proceedings had resulted. In these circumstances they were not entitled to recovery of their costs.

Contractor's nuisance

The Court of Appeal had to consider two issues in connection with Matania's appeal against the dismissal of his nuisance claim. Firstly, there was the question of whether the contractor's conduct of the work was sufficiently culpable to constitute an actionable nuisance. Secondly, if this was found to be the case, the court would have to determine whether the Syndicate could be held liable for a nuisance committed by their independent contractor.

In the context of the first issue, the Syndicate argued that their contractor had done no more than he was entitled to. Although the works had undoubtedly caused an interference with Matania's premises, this did not, in itself, make them unlawful. The syndicate placed particular reliance on Vaughan Williams J's judgment in the case of *Harrison v Southwark & Vauxhall Water Co* (1891) to support this argument[1]:

> "It frequently happens that the owners or occupiers of land cause, in the execution of lawful works in the ordinary user of land, a considerable amount

---
[1] See also the later case of *Andreae v Selfridge & Co* (1937)

of temporary annoyance to their neighbours; but they are not necessarily on that account held to be guilty of causing an unlawful nuisance. I emphasise the word 'necessarily'.

The business of life could not be carried on if it were so. For instance, a man who pulls down his house for the purpose of building a new one no doubt causes considerable inconvenience to his next door neighbours during the process of demolition; but he is not responsible as for a nuisance if he uses all reasonable skill and care to avoid annoyance to his neighbour by the works of demolition.

Nor is he liable to an action, even though the noise and dust and the consequent annoyance be such as would constitute a nuisance if the same, instead of being created for the purpose of the demolition of the house, had been created in sheer wantonness, or in the execution of works for a purpose involving a permanent continuance of the noise and dust.

For the law, in judging what constitutes a nuisance, does take into consideration both the object and duration of that which is said to constitute the nuisance."

Whilst accepting that this represented the correct legal position, Slesser LJ emphasised that Vaughan Williams J had qualified his remarks by the use of the words: "if he uses all reasonable skill and care to avoid annoyance to his neighbour by the works of demolition".

Contrary to the finding at first instance, the Court of Appeal held that the contractor had failed to do this. This was explained by Slesser LJ:

"It is clear from the evidence, both from the experts called on behalf of the plaintiff and those on behalf of the defendants, that the dust, to a large extent, if not entirely, could have been avoided by means of sheeting.

Then as to the noise, it is said that there was no evidence that the noise could have been mitigated. I am not satisfied, speaking for myself, that it was shown that this work causing this noise need have been done during the hours when Mr Matania did his work.

The learned judge, having found the nuisance to be intolerable, has come to the conclusion that there was no actionable nuisance because he says that he was unable to see that anything in the carrying out of the operations by the contractors was carried out other than in a proper and reasonable way, avoiding unnecessary noise, unnecessary dirt and unnecessary dust.

The learned judge seems to have overlooked the fact I have already remarked upon more than once that so far as dust is concerned, the evidence was that it was not inevitable but evitable and that as regards the noise, for the reason which I have said, I think that that also could at least have been mitigated, if not avoided, by an arrangement about times when this building work took place."

Liability for contractor's torts caused by "extra hazardous acts"

The Court of Appeal noted the general rule that an employer will not be liable for torts committed by his independent contractor, and that there are a number of exceptions to this rule. It made particular reference to *Honeywill & Stein Ltd v Larkin Brothers Ltd* (1934) which had recently established the "extra-hazardous acts" exemption.

In that case the defendants, a firm of photographers had been employed by the plaintiffs to take some photographs inside a theatre. This necessarily involved the use of magnesium flares which, in the event, caused a fire on the premises. The plaintiffs were held liable for the photographers' negligence in causing the fire. Due to the extra-hazardous nature of the operation the plaintiffs were subject to a non-delegable duty to take reasonable precautions to see that no damage resulted from it. They had failed to discharge this duty and were therefore liable in negligence.

The Court of Appeal held that the principle in *Honeywill*, which related to negligence, could also apply to the tort of nuisance. In negligence cases the issue was whether the particular act was of an "extra-hazardous" nature for which precautions were necessary to prevent harm. However, in the context of nuisance, the court simply had to be satisfied that there was "a serious risk that nuisance would be caused" unless precautions were taken.

Liability where a "serious risk of nuisance"?

Finlay J expressed some reservations about this logic but was eventually convinced by it:

> "I thought – to some extent I still think – that there was force in the argument of [the Syndicate], which was to this effect, that this was simply the case of an ordinary contract to carry out repairs to buildings, things which in their nature are bound to cause some inconvenience and discomfort to persons of adjoining premises, or in premises over the ones where the operations are going on, but I have been convinced by what Slesser and Romer LJJ say that this case does fairly come within the exception.
>
> It is not at all, of course, a case of the same nature as *Honeywill & Stein Ltd v Larkin Bros Ltd*. Inevitably these cases differ in their facts. There the question was as to danger; here the question is as to nuisance, but I have been convinced that when the facts here are carefully examined it does appear that this is not a case of mere ordinary building; it is a case where unless precautions were taken there was a great and obvious danger that nuisance would be caused, as indeed it was caused."

Romer LJ also offers a less-than-convincing justification for applying the *Honeywill* exception to the present circumstances. A key ingredient in the thinking of all three of their Lordships' thinking was that the *Honeywill* case was decided on the same basis as *Dalton v Angus* (1881) and *Bower v Peate* (1876). However, whilst the *Honeywill* decision is clearly based on the existence of extra-hazardous acts, the decisions in these two cases were based on the existence of easements of support by neighbouring owners. Romer LJ seeks to reconcile these two principles in his judgment:

> "[In *Honeywill & Stein Ltd v Larkin Bros Ltd*, the] exception, as I understand it, is this, that where a man employs an independent contractor to do work which of its very nature involves a risk of damage being occasioned to a third party, that person is responsible to the third party if such damage be occasioned and cannot shelter himself under the general principle of non-liability for the negligence of an independent contractor.
>
> Now apply that principle to the present case ... The work that the independent contractor was employed to do was this. He was to remove a considerable part of the support of the floor of Mr Matania's premises on the second floor. He was further to remove the lath and plaster ceiling of the first floor room which would have the effect of laying bare the boards and joists constituting the floor of Mr Matania's room on the second floor.
>
> It is obvious that that work would involve the risk of Mr Matania being damaged. It would involve the risk of Mr Matania's floor being let down unless the work was done very, very carefully. It was not let down...
>
> The cases to which Slesser LJ has referred of *Dalton v Angus* and the earlier one of *Bower v Peate* were both cases of the independent contractor removing the support of adjoining premises, a nuisance for which the employer of the independent contractor was in both cases held liable.
>
> Just as, in my opinion, they would have been liable in this case if they had let the floor down, so, in my opinion, they are liable for the damage caused to Mr Matania by the dust rising up through the cracks in the floor, and the noise which would be very great in his rooms by reason of his only being protected by a floor constituting what has been called a grid or a grate, instead of a floor with a lath and plaster ceiling underneath."

Although Slesser LJ's judgment also identifies the principles in the three cases as being connected, his argument relies on the application of the principle laid down in *Honeywill* to the current circumstances:

> "Here, of course, we are not concerned with danger such as might found an action in negligence. We are here concerned with annoyance such as may found an action for nuisance, but the principles in my opinion are the same as regards the liability of a person who employs an independent contractor, that is to say, that if the act done is one which in its very nature involves a special danger of nuisance being complained of, then it is one which falls within the

exception for which the employer of the contractor will be responsible if there is a failure to take the necessary precautions that the nuisance shall not arise.

... It is really not in dispute that as regards the place where this work was to be done this noise and this dust were inevitable.....The only question which I see is whether in that state where the production of noise and dust is inevitable, sufficient precautions were taken to prevent that noise and dust affecting Mr Matania.

... In the case to which I referred, the case of *Honeywill & Stein Ltd v Larkin Bros Ltd* ... it was a hazardous operation to bring the fire into the theatre. So it was hazardous as regards the possible nuisance to Mr Matania to bring the noise and dust immediately below his apartment. What is said is with sufficient and proper precaution the result of that hazardous operation could have been avoided without detriment to him.

... I am of opinion that this was a hazardous operation within the meaning of the exceptions stated in *Honeywill & Stein Ltd v Larkin Bros Ltd*, that the principle which is there dealing with a case of negligence applies equally to the tort of nuisance, and that the Elevenist Syndicate are responsible for the fact that neither they nor their contractors, Messrs Adamson, took those reasonable precautions which could have been taken to prevent this injury to the plaintiff."

On this basis, the Court of Appeal was satisfied that the Syndicate were liable for the nuisance which had been caused by their contractor. Matania's appeal on the nuisance question was therefore allowed, with costs.

The decision

The Bank's appeal against the finding that they had breached their landlord's covenant for quiet enjoyment was allowed. However, they were not awarded costs as the proceedings could have been avoided if they had complied with their other obligations by obtaining Matania's consent to the works. Matania's appeal against the dismissal of his nuisance action against the Syndicate was also allowed. In failing to mitigate the effects of noise and dust the contractor had committed an actionable nuisance. The Syndicate were also liable for these acts in view of the nature of the work which involved a serious risk that nuisance would be caused by them.

MATTS v HAWKINS (1813)
Court of Common Pleas

In brief

- **Issues:** Party fence wall – at common law – under the 1774 Act – longitudinally divided party walls – rights of owners – raising half the width – removal of raised portion – trespass.

- **Facts:** Matts and Hawkins owned adjoining plots of land. These were separated by a party fence wall which they had jointly erected. Matts now started to raise his own half of the wall but Hawkins removed the raised portion. Matts alleged that this amounted to a trespass on his half of the wall.

- **Decision:** Had the wall been owned by the parties as tenants in common Hawkins would have been within his rights to remove the raised portion. However, the wall was known to stand equally on the land of each party. In these circumstances they retained ownership in their own half of the wall. Matts was therefore entitled to raise his own half of the wall and Hawkins' interference with the raised portion was indeed a trespass. The 1774 Act did not operate to make the wall the common property of the parties. It conferred certain rights in the wall for the benefit of each party but did not have the effect of transferring property in it.

- **Notes:** See also: *Cubitt v Porter* (1828); *Wiltshire v Sidford* (1827); *Standard Bank of British South America v Stokes* (1878); *Watson v Gray* (1880).

The facts

Matts and Hawkins owned adjoining plots of land which were separated by a party fence wall. The wall had been built about 25 years previously. At that time they had each contributed half of the construction costs and the wall had been built with its centre line running along the boundary between the two plots.

Matts decided to raise the height of his half of the wall, to facilitate the erection of buildings against it on his own plot. He therefore started constructing a number of additional courses on the top of the wall, on his side of the boundary line.

When Hawkins became aware of Matts' work, he entered his land and pulled down the raised portion. Matts therefore issued proceedings against him for trespass.

Entitlement to raise a party wall

Hawkins maintained that he had committed no trespass. He argued that the two parties owned the wall as tenants in common by virtue of the fact that they had jointly erected and paid for it. All parts of the wall were therefore in the common ownership of the two parties rather than being longitudinally divided according to ownership of the land on which each half stood.

Because the whole wall was owned in common, the additional courses were also owned on the same basis. As Hawkins shared the ownership of the whole wall with Matts he could not, unless he completely dispossessed him from it[1], be liable for trespass for interfering with what was his own property.

He claimed that he was therefore within his rights to remove the additional courses from the top of the wall. He also claimed to be justified in so doing as Matts had constructed these without first serving notice upon him as required by section 38 of the 1774 Act[2].

Legal status of party wall

The requirement for service of notice did not figure in the court's deliberations. The court regarded the status of the wall as being the dominant consideration. It agreed with Hawkins that, had the parties owned it as tenants in common, he would have committed no trespass.

However, this was not the case. Because it was known that the wall stood equally on the land of each party they must each own half of its thickness. Matts was entitled to raise his half of the wall and Hawkins had therefore committed a trespass in demolishing the additional courses.

> Mansfield CJ: "If these parties are tenants in common, no trespass lies; but I see not how they became tenants in common. Under the circumstances, each has a right to the use of this wall; but the wall stands, part on the ground of each, and therefore is not the property of them as tenants in common; and each party, for any injury done to the part which stands on his own land, must have the ordinary remedy."

The court rejected the notion that the 1774 Act had any bearing on the case and specifically held that it did not have the effect of making the wall

---

[1] A process referred to as 'ouster'.
[2] The section required an owner to serve three months' notice on his adjoining owner "to repair, pull down, or rebuild any party wall or party fence wall or any part of parts thereof". It is not known whether the provision would have properly applied to the circumstances of the present case where the works involved the raising of the wall.

the common property of the two parties. The Act simply conferred rights on parties to undertake certain works to a party wall without having any effect on title.

> Chambre J: "At common law, no action of waste lay by one tenant against another ... The same reasons that prevented a tenant in common from maintaining waste at common law, equally hinder him from maintaining trespass: the question then is, whether these persons are tenants in common.
>
> Now the statute which gives each party certain rights in a wall built in this way, does not make it a common property; it only confers on each a right to use it for certain purposes. There is no transfer of property here, and the parties are severally owners of their respective land as before."

The decision

Half the thickness of the wall was owned by each party because half of it had been built on each plot of land. The parties retained their rights in their own half and had not become tenants in common of the whole wall, either at common law or by virtue of the 1774 Act. Matts was therefore entitled to raise his half of the wall and Hawkins was liable in trespass for removing the additional courses which Matts had erected. Judgment was entered for Matts, and Hawkins was ordered to pay damages for the losses which Matts had suffered.

RE METROPOLITAN BUILDING ACT *EX PARTE* McBRIDE (1876)

Chancery Division

In brief

- **Issues:** 1855 Act – failure to select third surveyor – court's power to appoint arbitrators – whether court's power extends to appointment of third surveyor – whether frustrated by pending suit between the parties – interference with easement of light – surveyors' jurisdiction.

- **Facts:** McBride wished to undertake works affecting a party wall. Notices were served and surveyors appointed but the parties and their two surveyors were unable to agree on a number of issues. One of these related to McBride's entitlement to brick up an opening which his neighbour alleged would interfere with an easement of light. Proceedings on the easement of light issue were pending. The two surveyors had failed to select a third surveyor. McBride therefore requested the court to appoint one under powers in the arbitration legislation.

- **Decision:** The court appointed a third surveyor. The court's statutory power to appoint an arbitrator applied to all arbitrations. This was the case whether the reference to arbitration was by written document, word of mouth or (as in the present case) by statute. The pending action between the parties was no barrier to an appointment. The surveyors had no jurisdiction to adjudicate on the easement of light issue. However, there was no suggestion that the appointment of a third surveyor was required simply to resolve this issue. If he attempted to do so his award would be invalid and that would have to be dealt with as a separate matter. There were other issues to be resolved by surveyors under the Act and a third surveyor should therefore be appointed to resolve these.

- **Notes:** (1) It is not known whether the courts would be prepared to appoint a third surveyor under section 18 of the Arbitration Act 1996. In any event this would now be unnecessary in view of the appointing officer's power to make the appointment under section 10(8) of the 1996 (Party Wall) Act. (2) See also (on the question of the surveyors as arbitrators): *Re Stone and Hastie* (1903) – (on the surveyor's lack of jurisdiction to decide questions relating to easements): *Crofts v Haldane* (1867).

The facts

McBride owned 10 Jerwin Crescent in the City of London. The adjoining terraced property, at 9 Jerwin Crescent, was owned by Warner. McBride wished to demolish and rebuild his property and served the appropriate party structure notice on Warner under the 1855 Act. Surveyors were appointed by each of the parties but they neglected to select a third surveyor.

The parties were in disagreement over various aspects of the works and the appointed surveyors were unable to agree an award. One area of disagreement between the parties related to whether McBride was entitled to brick up openings in the party wall between the properties.

In April 1876, Warner issued proceedings against McBride over this single issue. He requested an injunction to prevent the openings being bricked up on the basis that this would infringe his easement of light. The court refused to grant an interim injunction but left the question of a final injunction to be considered at the main hearing of the suit.

A few days later McBride served notices on the two appointed surveyors, requesting them to select a third surveyor so that the outstanding matters arising out of the party structure notice could be resolved. McBride's surveyor was willing to comply but, on the instructions of his appointing owner, Warner's surveyor refused to accede to the request.

McBride therefore issued the current proceedings in which he requested the court to appoint the third surveyor under section 12 of the Common Law Procedure Act 1854. He argued that unless the third surveyor was appointed Warner's refusal to co-operate would frustrate the rights which had been given to him under the 1855 Act.

Section 12 of the Common Law Procedure Act 1854 related specifically to arbitrations. It provided the mechanism for the appointment of arbitrators where there had been some failure of the agreed appointment procedures[1]. It made provision for the service of a notice requesting the appointment, in default of which the court had power to appoint. It related to the appointment of umpires and third arbitrators, as well as to the more usual appointment of a sole arbitrator.

Court's power to appoint third surveyor

Interestingly, there was no suggestion, either by Warner's counsel or by the court, that this procedure could not apply because appointed surveyors were not strictly arbitrators. Warner's objections to the appointment were instead presented in different terms.

---

[1] Equivalent (although different) procedures are now contained in sections 17 and 18 of the Arbitration Act 1996.

He argued, firstly, that section 12 did not apply to arbitrations under the 1855 Act as it made reference to a "document" authorising the reference to arbitration. References under the 1855 Act arose by statute and the statute was not a "document" within the meaning of the section. This must be all the more so, as in the present case, where the statute authorising the reference was not in existence at the time that section 12 was enacted. The court made reference to the case of *Re Lord* (1854) as authority for section 12 applying to any type of arbitration and dismissed Warner's first objection on this basis.

> Malins VC: "... in my opinion, the clause applies in every case where an arbitration is to take place, whether the reference to arbitration is to be by a document in the strict sense of the word, or by parol[2], or (as in the present case) by an Act of Parliament, and the plain object of the Legislature was that, if the parties to an arbitration refused to appoint an umpire, then the court might appoint one for them.
>
> I am glad to find this view was entertained by Vice Chancellor Wood in *Re Lord*; that case meets with my entire approbation, and I think, therefore, that this case does come within section 12 of the Common Law Procedure Act."

Appointment during pending litigation

Warner's second objection to the appointment of a third surveyor was that it represented an attempt, by McBride, to prejudge the outcome of the other litigation that was still pending between the parties.

The real issue was the question of whether McBride was entitled to block up the openings in the party wall. Although the interim injunction had been refused, this did not finally settle the issue, which would be considered afresh at the full hearing of the matter. The surveyors had no power to adjudicate on this issue, which related to the law of easements rather than to matters under the 1855 Act[3]. It should therefore be left for the court to decide in due course.

The court also dismissed this objection. It accepted that the surveyors had no power to authorise an infringement with an easement of light. However, this should not prevent the appointment of a third surveyor to enable legitimate matters under the 1855 Act to be resolved by the surveyors' tribunal. If the tribunal exceeded its authority then this was a separate issue and the possibility of this should not frustrate the proper appointment of a third surveyor.

---

[2] (In this context) by word of mouth.
[3] The following authorities were cited in support of this: *Titterton v Conyers* (1813); *Wells v Ody* (1836); *Crofts v Haldane* (1867); *Weston v Arnold* (1873).

> **Common Law Procedure Act 1854**
>
> **12.** If in any case of arbitration the document authorising the reference provides that the reference shall be made to a single arbitrator, and all the parties do not, after a difference has arisen, concur in the appointment of an arbitrator, or if where the parties or two arbitrators are at liberty to appoint an umpire or third arbitrator, such parties or arbitrators do not appoint an umpire or third arbitrator, then in every such instance any party may serve the remaining parties or the arbitrators, as the case may be, with a written notice to appoint an arbitrator, umpire or third arbitrator respectively; and if within seven days after such notice shall have been served, no arbitrator, umpire, or third arbitrator be appointed, it shall be lawful for any judge of any of the superior courts of law or equity at Westminster, upon summons to be taken out by the party having served such notice as aforesaid, to appoint an arbitrator, umpire, or third arbitrator, as the case may be, who shall have the like power to act in the reference and make an award as if he had been appointed by the consent of all parties.

Malins VC: "... it is argued that on account of this pending suit there can be no appointment of arbitrators; if so, it would follow that whenever an adjoining owner wished to vex his neighbour, the building owner, and prevent him from proceeding with his building operations, he has only to file a bill in this court and refuse to appoint an umpire, and the mere fact of the pending litigation (however ridiculous the litigation may be) will have the effect of paralysing the arm of the court.

It is unnecessary to refer to the argument that this is an attempt by a side-wind to effect what is the object of the suit, because it is quite certain that a surveyor appointed under the Building Act can have no power to decide the question of ancient lights which is raised in the suit. All that surveyors are to decide is, in what manner and under what circumstances the party wall is to be built, and if they shall fail in their duty, and should do anything to affect the question as to ancient lights, it will be disregarded by the court as being beyond their powers, and entitled to no consideration. I cannot, therefore, come to the conclusion that the right of the building owner to have an umpire appointed is to be interfered with because of the pending suit...

... if the building owners, with full notice of the plaintiff's claim to this window, go on to erect their house and to infringe any right which the adjoining owner may have, I shall have no hesitation in ordering them to undo what they have improperly done. Under these circumstances I accede to the application, and decide that the opposing party has entirely failed."

The decision

The court had power to appoint a third surveyor under the Common Law Procedure Act 1854 and the fact that there was pending litigation between the parties was no impediment to this. The court therefore appointed a third surveyor in accordance with McBride's request[4].

---

[4] The application made in this case would be unnecessary in the event of a failure by the surveyors to select a third surveyor under the 1996 Act. In such circumstances, section 10(8) makes provision for appointment by the appointing officer.

MIDLAND BANK PLC v BARDGROVE PROPERTY SERVICES LTD AND JOHN WILLMOTT (WB) LTD (1992)

Court of Appeal

In brief

- **Issues:** Adjacent excavations – interference with natural right of support – probability of future damage – nuisance – whether actual damage must have occurred for cause of action to arise.

- **Facts:** Bardgrove carried out excavations on their land which interfered with the support for the Bank's adjacent land. The initial damage was made good but the probability of further damage remained. The Bank carried out remedial works and sought to recover the cost from Bardgrove and from Willmott (their contractor).

- **Decision:** Bardgrove and Willmott were not liable. At the time that the Bank undertook the work there was no actual damage but only a potential for future damage. This would have been sufficient to apply for a *quia timet* injunction but the Bank had not done so. It had no cause of action in nuisance for damage which had not yet occurred so it could not recover the monies expended to remove the risk of such damage.

- **Notes:** See also: *Bower v Peate* (1876); *Dalton v Angus* (1881); *Ray v Fairway Motors (Barnstable) Ltd* (1968); *Holbeck Hall Hotel Ltd v Scarborough Borough Council* (2000).

The facts

Midland Bank owned a property at Oxgate House, Oxgate Lane, London NW2. Bardgrove was a property development company. In 1985 they developed a site which immediately adjoined the Bank's property. John Willmott were employed as the main contractor.

In March 1985 Willmott excavated the site and this involved exposing a vertical earth face at the common boundary with the Oxgate House. They provided temporary support to the land in the form of shoring. This consisted of a series of vertical steel frames set in concrete which supported a number of horizontal wooden railway sleepers, placed against the exposed earth face. There were gaps between the sleepers which were filled with unconsolidated clay.

Despite the presence of the shoring, and as a direct result of the excavations, the land subsided. This caused damage to a roadway on the Bank's premises as well as to a plinth, fence and gas supply pipe. Willmott undertook the necessary repairs and modified the shoring by replacing the unconsolidated clay with mass concrete. Later on they replaced the shoring with a permanent retaining wall.

No further movement took place. However, there was evidence (agreed by all parties) that the retaining wall did not provide a permanent solution. There remained the probability that it would fail through rotational collapse within 10 to 20 years of its construction.

The Bank therefore, on their own initiative, took steps to prevent further movement by restoring the stability to their land. They did this by installing sheet piling on their own side of the retaining wall. The cost of this work was £230,000.

The Bank then commenced proceedings against Bardgrove and Willmot (as well as two consulting engineers who were involved in the design of the shoring) to recover their losses. The claim was brought in nuisance, on the basis of an interference with their natural right of support for the land.

Requirement for actual damage

The judge at first instance dismissed the Bank's claim on the basis that they had no cause of action. Although they had a natural right of support and the two defendants had undoubtedly removed support from the land, this had caused no damage at the time that the action was commenced. In the absence of physical damage there was no interference with a natural right of support and therefore no cause of action. The Bank appealed to the Court of Appeal against this decision.

The Court of Appeal considered a number of earlier authorities on the question of whether the potential for future damage was a sufficient basis for a claim for interference with a natural right of support to land. They noted that the entitlement to support had originally been independent of any damage occurring: *Nicklin v Williams* (1854). However, this position had been overruled by the House of Lords in *Backhouse v Bonomi* (1861).

The later decision in *Lamb v Walker* (1878), which was relied on by the Bank in the present case, had been wrongly decided and a dissenting judgment by Lord Cockburn CJ had been correct.

This dissenting judgment, and the earlier decision in *Backhouse v Bonomi*, were reaffirmed in two subsequent cases by the House of Lords, *Darley Main Colliery Co v Mitchell* (1886) and *West Leigh Colliery Co v Tunnicliffe & Hampson Ltd* (1908). The authorities therefore demonstrated that an adjacent owner's cause of action did not arise until actual physical damage was caused by the excavation.

The balance of convenience

The Bank's counsel therefore submitted that the "balance of convenience" in the present case demanded that his client should be entitled to recover for the cost of installing the sheet piling.

He submitted that the Bank had acted prudently to eliminate the potential risk of future collapse. If they had not done so they would have been able to recover damages at such future time as a collapse occurred. If the court dismissed the present claim for damages then, as the potential for any future tort had now been eliminated, the Bank would be prejudiced by their prudent actions.

The court rejected this submission on three grounds. These were explained in Sir Alistair Slade's judgment:

> "In [the Bank's] submission, it is offensive and against common sense that the Bank is not now to be entitled to compensation for the cost of preventative works. I have three observations to make in response to this point.
>
> First ... the Bank (at least so far as we know) took it upon themselves voluntarily to execute the works without giving Willmott the opportunity of further supporting the land.
>
> Second, it would appear from the decision in *Redland Bricks Ltd v Morris* (1969) that, notwithstanding the absence of a present cause of action in respect of the prospective damage, (a) it would have been open to the Bank to apply to the court by way of *quia timet* injunction for a mandatory injunction compelling Willmott to carry out the sheet-piling works; (b) the court would have had the jurisdiction to grant such injunction, provided only that the likelihood of future damage was sufficiently proved and the injunction was in terms clear enough to give Willmott adequate indication of the work that had to be done. No such injunction was sought by the bank...
>
> Third, if the Bank's contentions were correct and the claim now under discussion were allowed to proceed to trial, the trial judge, before allowing in full the Bank's claim for the cost of the sheet-piling works, would surely have to compare this cost with the amount of the future damage which the bank would have been likely to suffer if these works had not been carried out. The assessment of this possible future damage would present formidable difficulties; the court and the parties would be exposed to the inconvenience of a wholly speculative inquiry. Policy considerations of this kind ... appear to have carried considerable weight in determining the decision ... in *Backhouse v Bonomi* (1861) by which the decision in *Nicklin v Williams* (1854) was overruled, and from which the relevant present law developed, requiring that actual damage is an essential feature of the cause of action."

The decision

The Bank had no cause of action as they had suffered no damage. Their appeal was therefore dismissed.

MOSS v SMITH (1977)
Court of Appeal

In brief

- **Issues:** Bristol Corporation Act 1926 – freestanding garden wall – continuation on same line as type 'a' party wall – party fence wall – entitlement to raise whole thickness of the wall – trespass – statutory interpretation – permanent interference with property rights.

- **Facts:** Moss and Smith owned property separated by a party fence wall. Smith raised the whole thickness of the wall and incorporated it into a house extension. Moss brought proceedings in trespass.

- **Decision:** There is only a right to raise one half of the thickness of the wall at common law. The Bristol Corporation Act 1926 did not confer rights to raise the whole thickness of the wall. In the absence of clear wording in the Act to the contrary it was presumed that it did not authorise the permanent appropriation of an adjoining owner's property.

- **Notes:** See also (on the right to raise a party wall): *Matts v Hawkins* (1813); *Cubitt v Porter* (1828); *Alcock v Wraith and Swinhoe* (1991) – (on the status of a wall which continues along the same line as a party wall): *Johnston v Mayfair Property Company* (1893) – (on the presumption against interference with property rights); *Barry v Minturn* (1913); *Burlington Property Company Ltd v Odeon Theatres Ltd* (1938); *Gyle-Thompson v Wall Street (Properties) Ltd* (1974).

The facts

Numbers 91 and 93 Whiteway Road, St Georges, Bristol were two adjoining terraced houses. Each had a rear garden. The Smiths owned number 91 and Moss owned number 93 (Figure 18).

All properties in the road were constructed with a traditional terraced house layout. Each property therefore had a projection to the rear of the main structure to provide additional accommodation. Each projection was built in two sections. The section nearest the main structure was erected to a greater height than that at the rear. The side wall of each projection separated the house of which it was a part from the garden of the adjacent house. Brick built garden walls then ran from the rear of each extension for the full length of the boundary between adjoining properties. In each case the face of the garden wall was built flush with the face of the side wall of the relevant projection.

Before

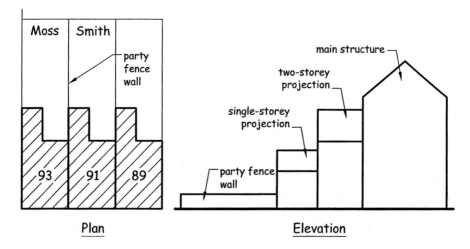

After

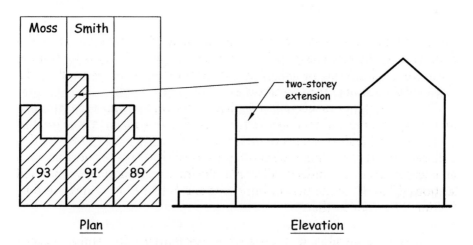

Figure 18: Moss v Smith

The Smiths wished to enlarge their property. They did this in 1974 by raising the height of the rear section of the projection, and also by extending the projection to the rear. This latter operation involved building up a portion of the garden wall to form the side wall of this extended portion of the projection.

Moss objected to the Smiths using the garden wall as he regarded this as an encroachment on his own property. He voiced his concerns to them at the time of the work. He eventually issued proceedings against them in

1976, after the work had been completed. He claimed damages for trespass and also an injunction, requiring them to remove that part of their extension which was built on "his half" of the garden wall.

Any right to raise a party wall under the Bristol Acts?

Both parties agreed that all the walls separating the properties were party walls within the meaning of section 38 of the Law of Property Act 1925. As this included the garden wall to the rear of the property Moss argued that the Smiths had no right to erect their extension on his half of the wall. The Smiths maintained that, as this wall was a party wall, they had a statutory right to raise its whole thickness under section 93(1) of the Bristol Corporation Act 1926. As they had authority to build on Moss' side of the wall they could not therefore be liable for trespass. Section 93(1) provided that:

> "... it shall be lawful for the owner or part owner of any external or party wall to raise the same provided that the wall when raised will be of the substance required by any ... byelaw ..."

On a preliminary issue the county court judge decided that this section did entitle the Smiths to raise the whole thickness of the wall, notwithstanding that half of this was owned by Moss. Moss therefore appealed to the Court of Appeal against this finding.

The Court of Appeal considered whether the section had been correctly interpreted by the county court judge as authorising an encroachment on to Moss' land and an interference with his property rights. The Smith's counsel argued, by analogy with other parts of the Bristol code, that such an encroachment did indeed fall within the intentions of the Act. He cited sections 27 to 31 of the Bristol Improvement Act 1847 in support of this. Section 31 provided that:

> "... it shall be lawful for the ... part owner intending to take down and rebuild such party wall ... to enter upon and into any ground house or premises for the purpose of such work..."

The Court of Appeal rejected this analogy. Sections 27 to 31 of the 1847 Act provided rights over adjoining property only for the purpose of enabling a building owner to protect his own property. The rights granted were temporary rather than authorising the permanent expropriation of property and were subject to control by the courts.

> Shaw LJ: "In my view the analogy is inapt. Sections 27 to 31 of the Act of 1847 were dealing with a mode of redress where an owner or part owner

> **Law of Property Act 1925**
>
> **38. Party structures**
>
> (1) Where under a disposition or other arrangement which, if a holding in undivided shares had been permissible, would have created a tenancy in common, a wall or other structure is or is expressed to be made a party wall or structure, that structure shall be and remain severed vertically as between the respective owners, and the owner of each part shall have such rights to support and user over the rest of the structure as may be requisite for conferring rights corresponding to those which would have subsisted if a valid tenancy in common had been created.
>
> (2) Any person interested may, in case of dispute, apply to the court for an order declaring the rights and interests under this section of the persons interested in any such party structure, and the court may make such order as it thinks fit.

apprehended that some defect in an adjacent structure might cause damage to his own property.

Even so he had to pursue the procedure prescribed under the relevant provisions, which included a hearing before justices of the peace, before it became lawful for him to enter upon the property of the adjacent owner in order to do what work of repair had been certified as necessary for the preservation of his own property.

What was made 'lawful for the ... owner or part owner' was the procedure of enforcement albeit that it involved a temporary intrusion on the property of another. The owner apprehensive of damage to his own property was not accorded a permanent licence to derogate from his neighbour's title; when the work of repair was completed he had to withdraw."

Presumption against interference with property rights

The court then turned to the words of section 93(1) themselves. In the absence of any clear wording to this effect, it was unwilling to construe them as depriving the plaintiff of his property rights. It concluded that the words must instead be read in the context of an owner simply exercising rights over his own property.

The effect of section 93(1) was therefore simply to provide owners with permission, for building control purposes, to build in conformity with bye-laws, without the need to apply for further consent. The section had no effect on existing property rights so the Smiths had no right to build on property which was owned by Moss.

> Shaw LJ: "Section 93(1) of the Bristol Corporation Act 1926, where the words used, as has already been indicated, are 'it shall be lawful for the owner or part owner of any external or party wall to raise the same', makes the position clear, as it seems to me, for an owner needs no statutory title to build on his own property provided that he conforms to relevant building regulations under byelaws or statute.
>
> Thus the phrase 'it shall be lawful' cannot be construed as conferring a title to do that which involves encroachment upon the proprietary rights of a part owner of a party wall. What the phrase does is to confer a licence upon those otherwise having title to a party wall or part of it to build upon their part of it provided that the substance of the raised part will be in conformity with any relevant byelaw.
>
> Deputy Judge Magnus, by giving to the phrase 'it shall be lawful' an amplified meaning, came to the conclusion, in regard to the work of extension of the wall done by the defendants (and I quote from his judgment), that 'if it is lawful it cannot be a trespass'. With all respect to that conclusion, it seems to me to be founded on a construction which is too simplistic and literal.
>
> The true result of section 93(1) is that the building up of the wall has been exonerated from offending against any statutory or other building restriction or regulation or byelaw; but it has nonetheless violated the plaintiff's right of ownership and possession of his half of the party wall, for which the section provides no exculpation."

The decision

On this basis the Court of Appeal allowed Moss' appeal against the decision on the preliminary issue. The case was then remitted back to the county court to decide the outstanding matters.

OBSERVATORY HILL LTD v CAMTEL INVESTMENTS SA (1997)

Chancery Division

In brief

- **Issues:** 1939 Act – obligation to make good damage – rights of entry on to adjoining land – contents of award – whether rights in award are an interest in land – whether rights in award are capable of registration as a caution – change of ownership.

- **Facts:** An award was made in respect of works undertaken by Observatory Hill under the 1939 Act. It required them to make good any damage to Camtel's property. In default of making good the award also authorised Camtel to undertake the repairs at Observatory Hill's expense. It also provided them with the necessary rights of entry on to Observatory Hill's land for this purpose. Camtel applied to the Land Registry to register their interests under the award as a caution against Observatory Hill's title.

- **Decision:** An adjoining owner's rights under an award could not be registered as a caution. They were not interests in land. They did not therefore fall within the definition of cautionable interests within the Land Registration Act 1925.

- **Notes:** The following cases also address the question of whether rights under the party wall legislation are rights in land or merely personal rights between the parties: *Re Stone and Hastie* (1903); *Carlish v Salt* (1906); *Mason v Fulham Corporation* (1910); *Selby v Whitbread & Co* (1917).

The facts

Observatory Hill and Camtel owned adjoining properties in Observatory Gardens, Kensington, London W8. These were separated by a party wall.

Observatory Hill planned to undertake some renovations to their property which would affect the party wall. They also involved adjacent excavations. Appropriate notices were therefore served under the 1939 Act. Surveyors were appointed and an award was published.

The award authorised the building works and also contained some provisions about the making good of damage. If the works caused any damage to Camtel's property, Observatory Hill were required to make this good at their own expense. In the event that they failed to do so the award authorised Camtel to carry out the necessary remedial work themselves, at Observatory Hill's expense. It also granted them access on to Observatory Hill's land to enable them to do this.

During the course of the work substantial damage was caused to Camtel's property by the excavations. Observatory Hill failed to make this good. To protect their position Camtel therefore registered a caution in the proprietorship register of Observatory Hill's title at HM Land Registry.

Registration of caution

Cautions can be registered against titles to land by persons having an interest in the land which is not already recorded on the register. The purpose is to make the interest binding on any subsequent purchaser of the land. Once the caution is registered, if the registered proprietor attempts to register any dealing with the land (for example a sale to a purchaser), the Land Registry will notify the cautioner who then has 14 days to substantiate his interest to the Land Registry.

Assuming that he is able to do so, the caution remains on the register and any purchaser will be bound by the cautioner's interest. However, as the registration of a caution is a hostile act, the registered proprietor has an opportunity to "warn it off" at the time that the cautioner first enters it on the register. Under this procedure either the Land Registry or the courts will decide whether the caution should remain on the register.

In the present case, Observatory Hill (the registered proprietors) applied to the court to remove the caution as soon as they received notification that it had been registered. They argued that Camtel's interest under the award was not sufficient to entitle them to register a caution under section 54 of the Land Registration Act 1925.

Can a surveyors' award be registered as a caution?

The court therefore had to consider whether interests under party wall awards were cautionable interests. There was no decided case law on the issue so the court made its decision after considering relevant extracts from *Emmet on Title*[1] and *Ruoff and Roper on Registered Conveyancing*[2].

The court noted that, under section 54(1), a person could register a caution where they were interested "under an unregistered instrument or interested as a judgment creditor or otherwise" in the land. As Camtel

---

[1] Farrand, J. & Clarke, A. (1986) *Emmet and Farrand on Title*, Sweet & Maxwell.
[2] Roper, R.B., Pryer, E.J. *et al.* (1991) *Ruoff & Roper: Registered Conveyancing*, Sweet & Maxwell.

> **Land Registration Act 1925**
>
> 54. Cautions against dealings
>
> (1) Any person interested under any unregistered instrument, or interested as a judgment creditor, or otherwise howsoever, in any land or charge registered in the name of any other person, may lodge a caution with the registrar to the effect that no dealing with such land or charge on the part of the proprietor is to be registered until notice has been served upon the cautioner.

were clearly not a judgment creditor, the court considered whether their interest under the award could amount to an interest under an unregistered instrument.

Their interest was a contingent one. At best it was "a potential interest to enter on to the plaintiffs' land if and so far as the award of the [surveyors] is not in due course observed and put into effect by the plaintiffs". The court referred to the following extract from *Emmet* which gave examples of interests which were and were not cautionable:

> "The interest of a contributory and creditor in land of a company in liquidation is *not* a minor interest which can be protected by caution. Nor is a provision for a sum additional to the purchase price to be payable if development occurred.
>
> Similarly, it is not thought that a *mareva* injunction[3], which does not strictly involve any interest in land, can properly be protected by a caution. However a receivership order obtained by a tenant against his landlord could be so protected."

The decision

After considering the various texts the court concluded that the interest of an adjoining owner under an award was not a cautionable interest.

> Levy J: "Having regard to the foregoing, in my judgment, there is no right in an adjoining owner who has the benefit of an award under the 1939 Act to become a person or a body who can enter a caution on the register as a person

---

[3] A form of injunction where the court freezes the English assets of a debtor who is currently abroad to prevent him from removing them from the jurisdiction. The name is derived from the case where the remedy was first used: *Mareva Compania Naviera SA v International Bulkcarriers SA* (1975).

having an interest in land. Such a person or body does not come within any of the categories which are now shown to exist as a person having an interest in land; I think it would be unwise to enlarge that category.

In the circumstances it seems to me that the submissions made [by Camtel] cannot be right. I recognise that in certain circumstances the defendant might in due course become entitled to enter on to the plaintiff's land but such a right is not sufficient as now to entitle the defendant to have such an interest in the plaintiffs' land as can be protected by a caution on the register. The fact that there is a party wall there, in some part of which the defendants may have some interest, does not seem to me to be a sufficient interest in land any more than does the fact that the implication of the 1939 procedures have been invoked."

The court therefore ordered that the caution be removed.

ORF v PAYTON (1905)
Chancery Division

In brief

- **Issues:** 1894 Act – line of junction notice – adjoining owner consenting to works – liability to contribute to cost of works – retention of title provision – definition of "owner" – tenant at will.

- **Facts:** Orf was a tenant at will. He served a line of junction notice and subsequently erected a party wall with Payton's consent as adjoining owner. Payton failed to make any contribution towards the cost of the work. Orf subsequently sought to prevent Payton from making use of the wall on the basis that he retained title in it until the contribution had been paid.

- **Decision:** Orf could not prevent Payton making use of the wall. As he was a tenant at will he was expressly excluded from the Act's definition of owner. The fact that he had certain additional rights (for example to remove sand and gravel) did not affect the basis on which he occupied the land. As he was not an owner he had no rights under the Act. He could not therefore have retained title in the wall and accordingly he had no right to prevent Payton from making use of it.

- **Notes:** See also (on line of junction notices): *Thornton v Hunter* (1898) – (on the definition of "owner"): *Cowen v Phillips* (1863); *Fillingham v Wood* (1891); *List v Tharp* (1897); *Crosby v Alhambra Co Ltd* (1907); *Spiers & Son v Troup* (1915); *Solomons v Gertzenstein* (1954); *Lehmann v Herman* (1993); *Frances Holland School v Wassef* (2001) – (on the 1894 Act's retention of title provision): *Mason v Fulham Corporation* (1910).

The facts

Orf and Payton were builders who had bought adjoining building plots in Mare Street, London E8, from the London County Council. The properties had been bought at auction.

Each auction contract required the builder to erect a house on the plot in accordance with approved plans. A long lease of the plot would then be granted to him. Pending the grant of the lease, the builder would occupy the plot as tenant at will. However, he was also granted certain additional rights, including the right to take away sand and gravel.

The proposed houses were to be terraced properties. Orf therefore served a line of junction notice on Payton, under section 87(1) of the 1894

London Building Act 1894

PART I: INTRODUCTORY

5. Definitions

In this Act unless the context otherwise requires:

(29) The expression "owner" shall apply to every person in possession or receipt either of the whole or of any part of the rents or profits of any land or tenement or in the occupation of any land or tenement otherwise than as a tenant from year to year or for any less term or as a tenant at will.

Part VIII: RIGHTS OF BUILDING AND ADJOINING OWNERS

87. Rights of owners of adjoining lands respecting erection of walls on line of junction

Where lands of different owners adjoin and are unbuilt on at the line of junction and either owner is about to build on any part of the line of junction the following provisions shall have effect:

(1) If the building owner desire to build a party wall on the line of junction he may serve notice thereof on the adjoining owner describing the intended wall.

(2) If the adjoining owner consent to the building of a party wall the wall shall be built half on the land of each of the two owners or in such other position as may be agreed between the two owners.

(3) The expense of the building of the party wall shall be from time to time defrayed by the two owners in due proportion regard being had to the use made and which may be made of the wall by the two owners respectively.

99. Structure to belong to building owner until contribution paid

Where the adjoining owner is liable to contribute to the expenses of building any party structure then until such contribution is paid the building owner at whose expense the same was built shall stand possessed of the sole property in the structure.

Act, of his intention to build a party wall on the boundary between the two plots. Payton consented to this in accordance with section 87(2) the Act. Orf then constructed the wall.

Liability for expense of building on line of junction

Payton refused to pay his contribution, anticipated by section 87(3) of the Act, towards the cost of building the wall. When he started erecting the house on his own plot against the wall Orf therefore applied for an injunction to restrain him.

Orf maintained that, under section 99 of the Act, the wall remained his property until payment of the contribution had been made. In making use of the wall, without payment having been made, Payton was therefore committing a trespass.

Payton denied that he was liable to pay the contribution under the Act. He maintained that Orf had no rights under the Act, as he was not an "owner" within the meaning of section 5(29). That section specifically excluded a tenant at will from the definition.

In response, Orf argued that, despite the wording of the contract, he was not strictly a tenant at will. In particular, he had the right to commit waste by removing sand and gravel which was inconsistent with the rights of a tenant at will.

The decision

The court held that the wording in the contract was conclusive. According to the contract Orf was a tenant at will and therefore had no rights under the Act.

> Swinfen Eady J: "It is quite clear, however, that he is tenant at will, because he has agreed with his landlords to be tenant at will. True it is that his landlords have given him certain rights to which he would not otherwise have been entitled, for example, to take away sand and gravel, but that does not affect the relationship between them.
>
> If, therefore, the relationship of tenancy at will exists between the plaintiff and his landlords, there is no reason why that relationship should be differently regarded as between the plaintiff and the defendant, the adjoining owner. The plaintiff, being a tenant at will, had no rights as 'owner' under the London Building Act 1894, and the motion must be dismissed with costs."

Orf had no right to recover a contribution towards the cost of building the wall under the 1894 Act. Consequently he did not retain title in Payton's half of the wall and Payton was free to build against it. His action was therefore dismissed.

PHIPPS v PEARS (1964)
Court of Appeal

In brief

- **Issues:** Demolition of adjacent building – resulting damage to exposed wall by rain and frost – whether an easement of protection from the weather.

- **Facts:** Many years previously Phipps' house had been built right up against Pears' house. Pears now demolished his house and exposed Phipps' flank wall to the weather. The wall suffered rain penetration and frost damage. Phipps sought to recover the cost of repairs from Pears on the basis that he had interfered with an easement to be protected from the weather.

- **Decision:** Phipps' claim failed. The courts are reluctant to recognise negative easements because they restrict an owner's use of his own property. For this reason there is no such easement known to law as an easement to be protected from the weather.

- **Notes:** (1) This case related to separate but adjacent buildings rather than to a party wall situation. (2) It seems unlikely that the decision has ever been applicable where a party wall is exposed by the demolition of one of its immediately adjoining buildings. There is no direct authority on the point but see *Upjohn v Seymour Estates Ltd* (1938) where an adjoining owner was held to be entitled to the protection previously afforded by his neighbour's half of the party wall. (3) The following recent cases now provide authority for a right of protection from the weather in party wall situations within the law of tort: *Bradburn v Lindsay* (1983); *Rees v Skerrett* (2001). (4) A right of protection from the weather under the 1939 Act was recognised in *Marchant v Capital & Counties* (1983). (5) Section 2(2)(n) of the 1996 Act now expressly requires a building owner to provide adequate weather protection for a party wall which is exposed by his works.

The facts

The case concerned adjacent properties at 14 and 16 Market Street, in Warwick. The history of the two plots is fully described in Figure 19 and is summarised here.

In 1930 a house was erected at number 16. It was built right up against an existing house at number 14. At that time both properties were in common ownership. The owner subdivided the plot in 1931 and by 1957 Phipps was the owner of number 16 and Pears owned number 14.

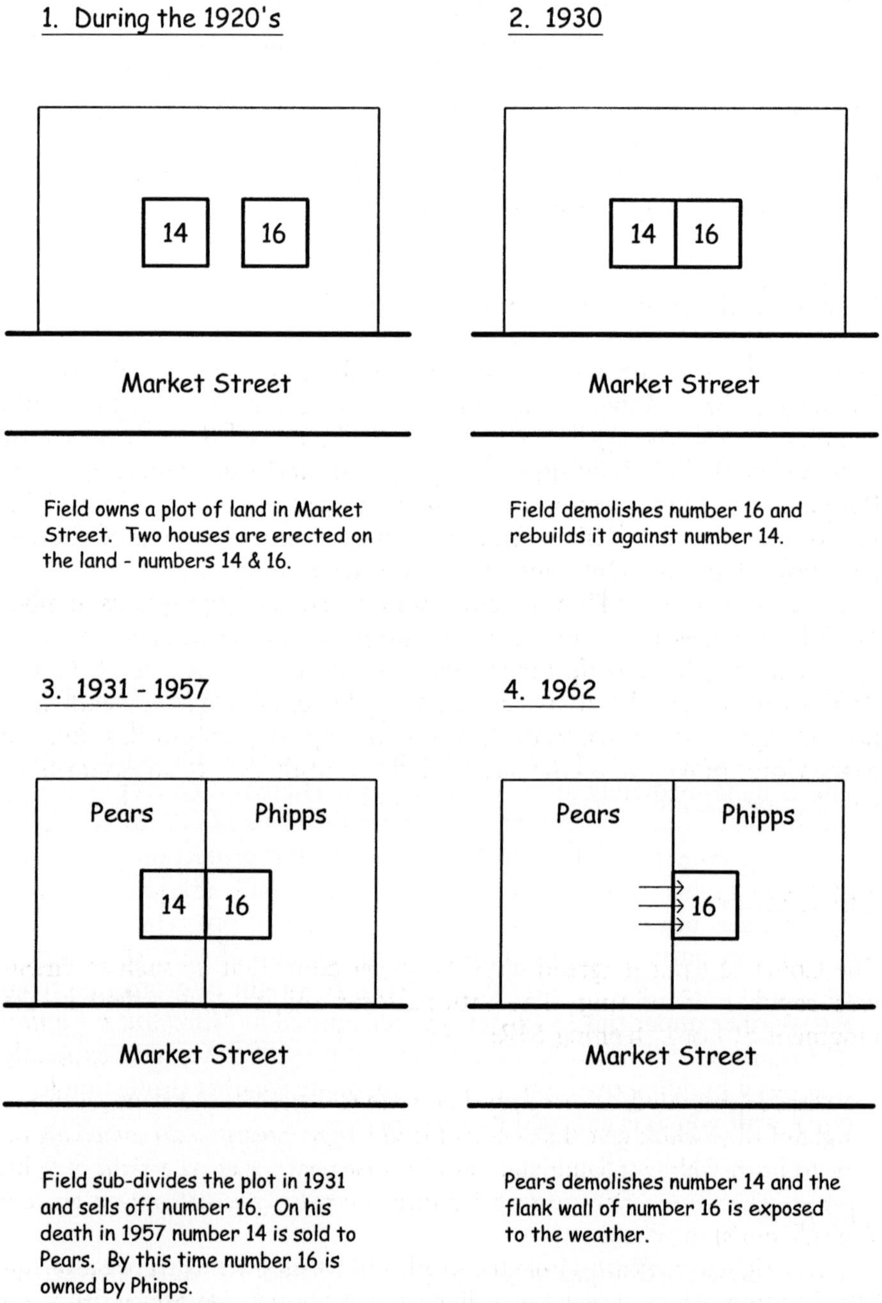
1. During the 1920's
14
16
Market Street
Field owns a plot of land in Market Street. Two houses are erected on the land - numbers 14 & 16.
2. 1930
14
16
Market Street
Field demolishes number 16 and rebuilds it against number 14.
3. 1931 - 1957
Pears
Phipps
14
16
Market Street
Field sub-divides the plot in 1931 and sells off number 16. On his death in 1957 number 14 is sold to Pears. By this time number 16 is owned by Phipps.
4. 1962
Pears
Phipps
16
Market Street
Pears demolishes number 14 and the flank wall of number 16 is exposed to the weather.

Figure 19: Phipps v Pears

In 1962 Pears demolished number 14 in compliance with a demolition order made by the local authority. This exposed the flank wall of number 16 to the weather. Because it had always been hidden by number 14, this wall had been roughly constructed and had never been properly pointed. The wall was therefore not weathertight. As a result of the number 14 being demolished, it became saturated with rainwater and subsequently suffered frost damage during the winter months.

Easement of protection from the weather?

Phipps issued proceedings against Pears for damages for the cost of repair. He alleged that number 16 had the benefit of easements of support and weather protection against number 14. These had been acquired by implication on the subdivision of the properties and also by prescription as Phipps and his predecessors in title had enjoyed these rights since 1931. Phipps therefore based his claim in nuisance, as the demolition of number 14 involved an unlawful interference with his easements.

The court dismissed Phipps' action at first instance. On the facts number 16 did not depend on number 14 for support and the damage had been caused entirely by the effect of the weather. However, the court held that Phipps was not entitled to an easement of protection from the weather, as no such right was known to the law. Phipps appealed against this decision to the Court of Appeal on the question of his entitlement to an easement of weather protection.

The decision

The Court of Appeal agreed with the lower court that no such easement was capable of existing. The rationale for this was described in the judgment of Lord Denning MR:

> "There are two kinds of easements known to law: positive easements, such as a right of way, which give the owner of land *a right himself to do something* on or to his neighbour's land: and negative easements, such as a right of light, which gives *him a right to stop his neighbour doing something* on his (the neighbour's) own land.
>
> ... a right to protection from the weather (if it exists) is entirely negative. It is a right to stop your neighbour pulling down his own house. Seeing that it is a negative easement, it must be looked at with caution, because the law has been very chary of creating any new negative easements.
>
> Take this simple instance: Suppose you have a fine view from your house. You have enjoyed the view for many years. It adds greatly to the value of your house. But if your neighbour chooses to despoil it, by building up and blocking

it, you have no redress. There is no such right known to the law as a right to a prospect or view: see *Bland v Moseley* (1587). The only way in which you can keep the view from your house is to get your neighbour to make a covenant with you that he will not build so as to block your view.

Take next this instance from the last century. A man built a windmill. The winds blew freely on the sails for 30 years working the mill. Then his neighbour built a schoolhouse only twenty-five yards away which cut off the winds. It was held that the miller had no remedy: for the right to wind and air, coming in an undefined channel, is not a right known to the law, see *Webb v Bird* (1861). The only way in which the miller could protect himself was by getting his neighbour to enter into a covenant.

The reason underlying these instances is that if such an easement were to be permitted, it would unduly restrict your neighbour in his enjoyment of his own land. It would hamper legitimate development. Likewise here, if we were to stop a man pulling down his house, we would put a brake on desirable improvement. Every man is entitled to pull down his house if he likes. If it exposes your house to the weather, that is your misfortune. It is no wrong on his part ... There is no such easement known to the law as an easement to be protected from the weather."

The court therefore dismissed the appeal.

PRUDENTIAL ASSURANCE CO LTD v WATERLOO REAL ESTATE INC (1999)

Court of Appeal

In brief

- **Issues:** Type 'a' party wall forming outside wall of building – building owner proposing to permanently demolish wall – adverse possession – whether building owner had acquired title to adjoining owner's half of the wall – *animus possidendi*.

- **Facts:** A 30 metre long party wall separated land owned by Waterloo from that owned by Prudential. A 7 metre length of this formed the outside wall of a building on Waterloo's land known as the Normandie Hotel. This wall had previously separated the Normandie from a pub on Prudential's land which had since been demolished. For many years since the demolition Waterloo had treated the whole wall as their own. They now claimed to be entitled to demolish it as they maintained that they had acquired title to Prudential's half of the wall by adverse possession.

- **Decision:** Waterloo had acquired title to Prudential's half of the Normandie's wall by adverse possession and were therefore at liberty to demolish it. A column formed a sufficient division between the 7-metre length and the rest of the wall to enable title to be acquired to part only of its length. Waterloo's conduct in respect of the wall clearly demonstrated the necessary intention to possess the wall and it could not all be explained away on other grounds. They were therefore entitled to demolish the wall.

- **Notes:** (1) In party wall procedures the possible effect of adverse possession should always be considered in cases where a type 'a' party wall does not also perform a separating function with an adjoining building. See the definition of "party wall" in section 20 of the 1996 Act. (2) In the case of registered land the law of adverse possession will change when the Land Registration Act 2002 comes into effect. It will then be necessary to apply for registration of title once the land has been adversely possessed for ten years. The current rules, whereby the paper owner's title is automatically extinguished after 12 years adverse possession, will then only apply to unregistered land.

The facts

The parties owned two adjacent sites in the Knightsbridge area of London. Waterloo owned the Normandie Hotel at the corner of Knightsbridge,

Knightsbridge Green and Raphael Street. Prudential owned a much larger site to the south and west of the Normandie site, known as the Caltrex site. Each of the parties proposed to develop their respective sites.

Unfortunately the two proposals were mutually incompatible. Waterloo planned to erect a luxury hotel on the Normandie site. This required vehicular access to the rear of the site from Raphael Street over land presently occupied by the Normandie's south wall, which they planned to demolish. Prudential's plans required the closure of Raphael Street, the subsoil of which formed part of their site.

Disputed ownership of wall

Prudential issued proceedings against Waterloo for an injunction to prevent them demolishing the south wall of the Normandie Hotel. They maintained that they owned the entire length of the wall and that any attempt to demolish it would be a trespass.

This wall was 5 metres high and 7 metres in length. It formed part of a much longer wall (30 metres in length) which marked the boundary between the two sites. The 7-metre length abutted directly on to the pavement of Raphael Street. The remaining 23 metres was a party wall separating buildings on the two sites from each other.

The trial judge found that the disputed wall had formerly separated the Normandie Hotel from the Pakenham Tavern and that the two properties had then been in common ownership. Each property had been conveyed to separate owners in 1929. At that time, the wall was declared to be a party wall, with ownership being severed vertically along its entire length. In 1957 the Pakenham had been demolished and Raphael Street had been moved northwards to its present position immediately adjoining the south side of the wall.

As a matter of paper title the wall therefore continued to be a party wall with the north half being owned by Waterloo and the south half by Prudential. However, the trial judge held that Waterloo had acquired title to the south half of the wall through adverse possession.

Adverse possession

Since 1957 the owners of the Normandie had, without any protest from the owners of the Caltex site, publicly and consistently treated the wall as their own property.

They had regularly cleaned, repaired and maintained it and had, on two occasions, granted leases of their site which included tenants' repairing covenants in respect of the wall. They had also cut into and attached items to the wall on various occasions. These included the erection of scaffolding against the wall, the installation of a security camera, security lighting and an entryphone, and the insertion of a night safe and overflow pipe through it. In addition to this, a mansard roof had been fitted to the Normandie in the 1970s which involved substantial alterations to the structure of the wall itself.

The trial judge considered the requirements for adverse possession set out in the judgment of Slade J in *Powell v McFarlane* (1977) and held that these were satisfied by the history of the site. The three requirements were that the person seeking to establish title must establish (1) possession (2) an intention to possess and (3) that the possession was adverse.

The facts, as outlined above, were not "trivial acts of trespass nor equivocal in the sense of being equally explicable on some basis other than that the person was exercising the possessory rights of an owner". On the contrary, they clearly demonstrated that Waterloo had exclusive possession of the south side of the wall. Such possession was also clearly adverse, as it was in defiance of Prudential's paper rights.

Evidence of intention to possess

In considering the requirement for an intention to possess (*animus possidendi*) the trial judge made reference to the following words of Slade J in the *Powell* case:

> "The courts will, in my judgment, require clear and affirmative evidence that the trespasser, claiming that he has acquired possession, not only had the requisite intention to possess, but made such intention clear to the world.
>
> If his acts are open to more than one interpretation and he has not made it perfectly plain to the world at large by his actions or words that he has intended to exclude the owner as best he can, the courts will treat him as not having had the requisite *animus possidendi* and consequently as not having dispossessed the true owner."

The trial judge held that Waterloo had the necessary intention to possess and explained his rationale for this in the following terms:

> "The Normandie did not, of course, make some ostentatious pronouncement to the world at large that it intended to exclude everybody ... from the wall ... but I do not believe that it needed to, and I do not believe that Slade J meant to indicate that it did.
>
> I think the learned judge had in mind the conscious trespasser – the dispossessing trespasser who knows what he is doing – not someone like the Normandie who does not realise that he is trespassing at all, especially when scarcely anyone else realises it either.
>
> In any case, it must be right that an intention to exclude everyone else from

possession can be taken to have existed when the common sense assumption which anyone would make is that, if someone else had tried to interfere with the claimant's possession, the claimant would have done something about it.

In this case if other persons ... had started to interfere with the Normandie's possession of the south face of the wall ... (for example by affixing posters to it, or by purporting to license others to affix posters) I do not need specific evidence to enable me to assume that the Normandie would have tried to stop them. In my judgement that is enough to satisfy me that the Normandie had the necessary intention to possess the crucial stretch of wall."

Prudential appealed to the Court of Appeal against the finding of adverse possession and this raised a number of issues.

Conscious and unconscious trespassers

Prudential argued that the trial judge had erroneously applied a less stringent test to Waterloo in determining whether the necessary *animus possidendi* existed. The words of his judgment suggested that different standards should be applied to conscious and unconscious trespassers.

The court rejected this submission. The requirement was that the claimant's conduct must make it plain to the world that he intended to possess the land and exclude the paper owner from it. The requirement was the same for both categories of trespassers. The trial judge's words were consistent with this and, on the facts, the necessary intention was demonstrated by Waterloo's conduct which was visible to anyone who looked at the wall.

Nature of paper owner's rights

Prudential argued that, for a successful claim to be made, the true nature of the paper owner's title must be known to him as well as to any given member of the world at large. If this was not the case then it was not possible to determine whether a claimant's conduct demonstrated the necessary intention to possess. As the status of the title to the wall was unknown to both parties before the present case was brought, it was not possible to determine whether Waterloo had *animus possidendi*. The court rejected this submission also.

> Gibson LJ: "There may be real uncertainty, incapable of being resolved save by litigation, as to the true nature of the paper owner's rights. But it cannot be correct that that could affect a claimant, who with the intention of taking exclusive possession overtly treats the disputed land as his own. What is important is whether the claimant's conduct is unequivocally that of a person

asserting possession to the exclusion of any other or whether it is referable to the rights that the claimant already has."

Easements of support and ancillary repairing rights

Prudential argued that no significance could be attached to the repair and maintenance work undertaken by Waterloo in demonstrating *animus possidendi*. The owner of each half of the party wall had an easement of support in respect of the half which was not in their ownership. They therefore had ancillary rights to effect repairs to that half to ensure that such support continued. The maintenance work undertaken by Waterloo could equally be viewed as an exercise of these rights and did not therefore provide unequivocal evidence of an intention to possess.

The court accepted the existence of such ancillary rights. However, some of the work which was undertaken (for example, cleaning and removal of graffiti) was not essential to the continuance of support and could not, therefore, be regarded as equivocal.

Significance of construction works

Prudential sought to minimise the significance of the various items of construction work affecting the wall. Whilst acknowledging that they constituted trespasses they maintained that they were too trivial to provide unequivocal evidence of *animus possidendi*.

The court held that some of the work (particularly the work associated with the mansard roof) had the effect of excluding Prudential from parts of their property. This work, as well as the installation of the overflow pipe and the night safe clearly demonstrated that Waterloo intended to take possession of the wall and could not be described as minor trespasses.

> Gibson LJ: "It seems to us quite unnatural to categorise, as [Prudential's counsel] would have us do, those permanent alterations as minor trespasses rather than manifesting a clear appropriation of the wall."

Significance of statements in legal documents

Prudential argued that the grant, by Waterloo, of leases which included tenants' repairing obligations in respect of the wall was of no significance. These were private documents. Like the evidence of private conversations, which was excluded in *Powell v McFarlane*, these had no evidential value in demonstrating *animus possidendi*.

The court disagreed. The private conversations in *Powell* had been between the claimant and his half-brother. The leases in the present case were formal documents made with independent third parties. As such, they were evidence that Waterloo considered that it owned and possessed the wall, which corroborated the other evidence.

Requirement for discrete feature to adversely possess

Prudential submitted that it was not possible for only 7 metres of a 30-metre wall to be adversely possessed. The whole wall had been described as "one geographical feature" by the trial judge and there was no defined boundary between the portion claimed by Waterloo and the rest of the wall.

The court accepted that there had to be some physical division between the two parts but, on the facts, found that this was marked by a column.

> Gibson LJ: "We cannot accept that submission which seems to us to defy common sense. It is of course important, as the judge recognised, that there should be a clear and definite physical division between that part of the wall possession of which the claimant claims and that part of the wall possession of which it does not claim.
>
> A glance at the photographs shows that there is a clear distinguishing point between that part of the wall that to any objective observer has been incorporated into the premises of the Normandie, that is to say the wall from point A up to the westernmost column at point C, and the rest of the wall to the east. In our judgement, that is enough."

The decision

Having decided each of these six issues in Waterloo's favour the court dismissed the appeal and upheld the trial judge's finding that they had adversely possessed the wall. Waterloo were therefore free to demolish the wall and erect the luxury hotel on the Normandie's site.

RAY v FAIRWAY MOTORS (BARNSTABLE) LTD (1968)
Court of Appeal

In brief

- **Issues:** Adjacent excavations – damage to adjoining owner's boundary wall and building – natural right of support – easement of support – negligence.

- **Facts:** Ray and Fairway owned adjacent yards. Fairway excavated their yard to a depth of 4 ft in the immediate vicinity of Ray's boundary wall. This caused damage to Ray's wall and to a shed which he had subsequently built against it. Ray sought to recover his losses from Fairway in negligence and for breach of his natural rights and easements of support.

- **Decision:** Fairway were not liable in negligence. In the absence of some further omission, a duty of care was not breached merely by the removal of support from the structures on a neighbour's land. There was also no liability for interference with Ray's natural right of support to his land. There was no evidence that there would have been any appreciable damage to the land if it had been in its natural state. The court did not accept that the excavations would have produced an inherently unstable four-feet-high precipice of soil along the boundary line. This ignored the fact that Ray's land had already been excavated at the time that the boundary wall was built. The actual height difference was therefore only 1 ft – the height of the exposed soil below the wall's foundations. However, Ray was able to recover his losses on the basis of an interference with his easement of support. The wall had acquired an easement by prescription. Fairway had failed to demonstrate that the subsequent erection of the shed had extinguished this by substantially increasing the burden on their land.

- **Notes:** See also (on natural rights of support): *Midland Bank plc v Bardgrove Property Services Ltd* (1992); *Holbeck Hall Hotel Ltd v Scarborough Borough Council* (2000) – (on easements of support): *Bower v Peate* (1876); *Dalton v Angus* (1881); *Bond v Nottingham Corporation* (1940) – (on negligence): *Dodd v Holme* (1834); *Southwark & Vauxhall Water Company v Wandsworth District Board of Works* (1898); *Leakey v National Trust for Places of Historic Interest or Natural Beauty* (1980).

The facts

Ray owned business premises at 3 Trafalgar Lawn in Barnstable. These included a yard at the rear of the premises (Yard C in Figure 20). Fairway owned premises at 1 Trafalgar Lawn. These also included a rear yard (Yard A). Fairway owned a third yard (Yard B) which lay to the rear of number 2 Trafalgar Lawn. This yard was situated between the other two and therefore immediately adjoined Yard C, which Ray owned.

A freestanding stone wall (the "east wall") ran along the back of all three yards. The owner of each yard owned the portion of this wall which immediately adjoined their property. A similar wall (the "north wall") separated Yard B from Yard C. This wall was owned by Ray. Both walls were very old and would certainly have existed for long enough to have acquired prescriptive rights over adjacent properties.

In 1962 Ray erected a workshop in the rear corner of his yard which he subsequently let out to a firm in the motor trade. The workshop was erected against both the east and north walls. These therefore formed two of the walls to the new building and now bore an increased load as they supported the workshop's roof structure.

In 1964 Fairway decided to level the surface of Yard B to bring it on to the same level as its other land in Yard A. At that time Yard B sloped down from the level of Yard C to that of Yard A, which was approximately four feet lower.

Damage caused by adjacent excavations

They appointed a contractor to undertake this work. The work involved excavating to a depth of four feet in the vicinity of the east and north walls. The contractor advised that support should be provided for each wall in the form of a bank of earth. In each case, his proposal was that a bank of earth should be left against each wall, two feet wide at the top and splaying out to about four feet at the bottom.

In fact, Fairway instructed him to undertake the work without the provision of any support to the walls and he proceeded on this basis. The excavations left the land of Yard B four feet below the level of Yard C and 1 ft below the level of the bottom of the foundations of the north wall. A month after the works, in early 1965, the north wall cracked and began to bulge, and that in turn caused damage to the workshop.

Ray therefore issued proceedings for damages against Fairway, and also against the contractor, as second defendant. He based his claim (a) on an interference with the natural right of support for his land, (b) on an interference with an easement of support for the north wall, and (c) on both defendants' negligence in excavating without providing adequate support for the wall. At the trial, the judge gave judgment for Ray on the

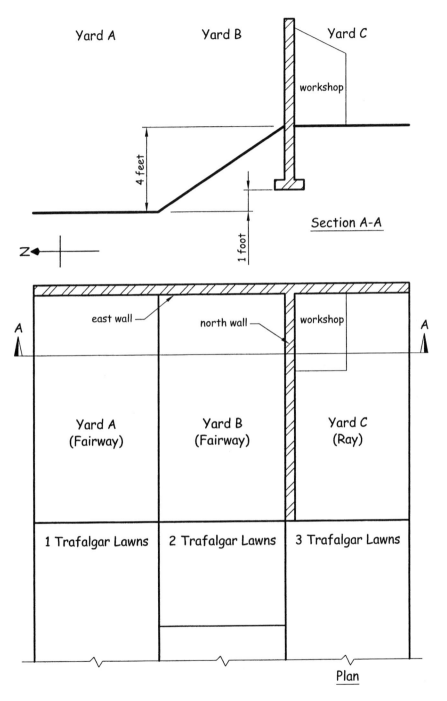

Figure 20: Ray v Fairway Motors (Barnstable) Ltd

basis of his claims under (a) and (b). He awarded him damages for the cost of repair and for loss of rent from the workshop.

In view of the fact that the method of excavations had been adopted on Fairway's express instructions, the judge held, in the third party proceedings, that the contractor was entitled to a complete indemnity from them for any damages which he might have to pay to Ray. Fairway appealed against the judgment (but not against the decision in the third party proceedings).

The Court of Appeal accepted the trial judge's findings of fact about the cause of the damage. Based on the evidence at the trial, he had found that the damage to the wall and the consequent damage to the workshop had been caused by the movement of soil underneath the foundations of the north wall. This, in turn, had been caused by the removal of support for the wall by Fairway's excavations.

Specifically, the movement of soil had not been caused (as alleged by Fairway) by the additional weight imposed on the wall when Ray erected the workshop.

The Court of Appeal therefore confined itself to the issue of whether Fairway's removal of support constituted a breach of a legal obligation which they owed to Ray. They considered each of the issues (interference with a natural right, interference with an easement, and negligence) in turn.

Interference with natural right of support

In considering the first of these issues, the court noted that Ray had an entitlement (a natural right) to support for his land in its natural state from Fairway's adjoining land. It also noted that he had no such right in respect of the buildings on his land.

However, where buildings were situated on the land, he would still have a claim for interference with his natural right if he could prove that the withdrawal of support complained of would still have produced appreciable damage to the land, even if no buildings had been erected on it. If he was able to so prove, he would be able to recover for the damage to the land and also, for the consequential damage to the buildings erected on the land.

According to Ray, it was a matter of common sense that the excavations would inevitably have resulted in a collapse of an appreciable amount of material from his land, even in the absence of any buildings upon it. If one envisaged the land in its natural state, Fairway's excavations had produced a 4 ft vertical precipice of clay and loam soil which was inherently unstable. Ray therefore maintained that he was entitled to recover for all the damage and consequential damage which he had suffered on this basis. The trial judge had found in his favour in this respect.

The Court of Appeal did not accept this. They were unconvinced, on the evidence presented to them, that the fall of soil would be anything more than *de minimis* whereas proof of "appreciable damage" was required. Ray had therefore failed to discharge the burden of proof which was upon him.

More significantly, the court did not accept the basis of the plaintiff's argument. The 4 ft high precipice was a fiction which ignored the fact that the plaintiff's predecessor in title had already excavated the soil on his side of the boundary when he erected the north wall. The actual height of the difference in height between the two yards was therefore only the exposed soil beneath the level of the north wall's foundations. This was 1 ft rather than the 4 ft which Ray was suggesting.

In these circumstances the fall of an appreciable amount of soil was rendered even less likely. The Court of Appeal therefore disagreed with the trial judge that Ray was entitled to succeed on this ground.

Interference with easement of support

The second issue to be considered by the court was whether Fairway's excavations had interfered with an easement of support in respect of the north wall.

There was no dispute that Ray and his predecessors in title had acquired this right by prescription, in the years prior to 1962. However, Fairway argued that this had been extinguished when Ray increased the burden on the servient tenement at the time that he erected the workshop.

The trial judge had held that the easement had not been extinguished and had therefore found in Ray's favour, on the basis of an interference with his easement of support. The Court of Appeal came to the same conclusion, although for slightly different reasons.

The Court of Appeal accepted the principle that an easement of support could be extinguished in this way and this was explained by Willmer LJ:

> "As I understand it, the principle, dating back at least to *Luttrel's Case* (1601) is that an easement is extinguished when the mode of user is so altered as to cause prejudice to the servient tenement.
>
> Thus, an easement of support in relation to a building may be extinguished if the building is so altered or reconstructed as to throw a substantially increased burden on the servient tenement to the prejudice of the owner thereof ... it is not easy in any given case to decide whether the additional burden thrown on the servient tenement is sufficiently substantial to cause prejudice to the owner thereof."

However, in the present case, their Lordships were unanimously of the view that, if there had been any increased burden on the servient land, there was no evidence that it was sufficient to have prejudiced the use of the land.

Fairway placed great reliance on expert evidence that the load on the wall doubled as a result of the construction of the workshop. They argued that it was self-evident that the lateral thrust on their own land must have increased by the same proportion. The Court of Appeal noted that Fairway, as the servient owner, had the burden of proving the extinguishment of Ray's easement. On the basis of the evidence presented to the court, they had failed to discharge this burden.

On the contrary, the evidence which was available to the court suggested that their land had suffered no prejudice as a result of the increased load on the wall. In particular, the evidence demonstrated that identical support measures had been proposed for both the north and east walls. Contrary to Fairway's suggestion, this tended to suggest that the lateral thrust from the north wall was no greater than that from the east wall, despite the fact that its vertical load had increased. The Court of Appeal therefore dismissed Fairway's appeal on this ground.

Negligence

Although the court's findings on the easement of support issue were sufficient to dispose of the case, they also delivered judgments on Ray's claim in the law of negligence.

The court made reference to numerous authorities on the question of whether a land owner owes a duty of care to the owners of adjacent buildings when excavating his land[1]. These demonstrated that, in the absence of a right of support, there was no duty of care requiring an excavating owner to provide shoring or other support. Therefore, no wrong is committed if the adjacent building collapses as a consequence of the excavations.

Although the authorities were all rather old, the court did not consider that the impact of *Donoghue v Stevenson* (1932) was sufficient to have reversed the law in this area. Fenton Atkinson LJ summarised the legal position:

> "So far as this court is concerned, the authorities are clear that no legal duty rested on the defendants to leave any part of the existing support of the wall, or to provide alternative forms of support.
>
> Although it may be thought unreasonable in such circumstances for a man to exercise his legal right in such a manner as he knows may well cause collapse

---

[1] Including *Dalton v Angus* (1881) and *Southwark & Vauxhall Water Co v Wandsworth Board of Works* (1898).

of his neighbour's building, equally it may be thought unreasonable for a man to erect a building so close to his boundary as to restrict his neighbour in the use which he may wish to make of his land."

However, although there may be no duty to provide support within the law of tort, it was acknowledged by Fenton Atkinson LJ that there is likely to be a more general duty of care in respect of the conduct of adjacent excavations:

"It will be observed that all the pleaded allegations of negligence in this case were allegations, either of removal of support, or of failure to provide alternative support.

If the plaintiff could have shown something beyond the mere removal or omission to take active steps to prevent its effect, for example, the adoption of an unnecessarily dangerous method of removal causing collapse, the position could well have been different."

The decision

In conclusion therefore, the Court of Appeal found that Fairway were not liable in negligence or for any interference with a natural right of support. However, they were liable for their interference with Ray's easement of support for the north wall. As this was sufficient for Ray's claim to succeed, Fairway's appeal was dismissed.

READING v BARNARD (1827)

Court of Common Pleas

In brief

- **Issues:** 1774 Act – rebuilding of defective party wall – adjoining owner's liability to contribute – service of account on adjoining owner – contents of account – validity of account – encroachment of new wall on adjoining owner's land.

- **Facts:** Reading demolished and rebuilt the defective party wall between his property and that owned by Barnard. He then served an account for half the cost on Barnard. Barnard refused to pay as the account was not in the form required by the Act. He also claimed that the new (thicker) wall encroached disproportionately on to his land.

- **Decision:** Barnard was required to contribute half the costs of rebuilding the wall. The account contained sufficient information about the quantities to enable the total amount due to be calculated according to the rates set out in the Act. It was not invalidated simply because it also claimed a total sum which was in excess of the amount recoverable under the Act. Reading had no authority, under the Act, to encroach disproportionately on to Barnard's land. However if this was of a minor nature due to inevitable inaccuracies in measurement then this had to be tolerated. The court was satisfied that any disproportionate encroachment in this case was only of a very minor nature and was sanctioned by the Act.

- **Notes:** See also (on the service of accounts): *Spiers & Son Ltd v Troup* (1915); *J. Jarvis & Sons Ltd v Baker* (1956) – (on encroachments on to adjoining land): *Barry v Minturn* (1913).

The facts

Reading and Barnard owned adjoining terraced houses. The party wall between them was only 8 in thick. This was of insufficient thickness according to the building control requirements in the 1774 Act.

Reading therefore demolished and rebuilt the wall to the 18 in required thickness, as was his right under the Act. He then served an account on Barnard, under the Act, claiming one half of the costs of the work from him.

Barnard refused to pay. He maintained that Reading had not complied with the Act's requirements in two respects. Firstly, the form of the account which he had served on Barnard was not in the form required by the Act. Secondly, he had erected the new wall in the wrong place. This encroached disproportionately on Barnard's land rather than being built half on the land of each owner, as required by the Act.

Reading brought the present action to recover the monies claimed in the account. The court addressed each of Barnard's two objections to payment.

Account for work carried out

Firstly, it considered the form of the account. Section 41 of the 1774 Act imposed limits on the amounts that could be claimed from adjoining owners. It specified allowable rates for labour and materials according to the dimensions of the wall and the materials actually used.

The account which Reading had served on Barnard was for a proportion of the amount which he had actually expended. This exceeded the statutory limits. Despite this, the court held that the account was valid as it also contained sufficient information to enable Barnard to understand how much he was required to contribute:

> Lord Tenterden C J: "All that the Act requires is a true account of the number of rods in the party wall, and of ... the materials. The Act itself fixes the price to be paid ... in respect of them.
>
> If, therefore, the quantities are correctly stated, the party charged has the means of knowing the amount of the claim against him. There is, then, no necessity for, and the Act does not require, any specification of the price; and the account will not be vitiated by the unnecessary insertion of a price which the party cannot be charged with."

Accuracy in positioning of rebuilt party wall

The court then considered the significance of the position of the rebuilt party wall (Figure 21). The parties gave conflicting evidence about this. Reading claimed that it was equally astride the boundary line. Barnard claimed that the entire 10 in of increased width had been built entirely on his land. Whilst the wall continued to project 4 in on to Reading's land he therefore alleged that 14 in now projected on to his land.

As the issue was a matter of evidence, rather than of law, Lord Tenterden CJ referred it to the jury to decide. However, in his directions to the jury he took a practical approach to the realities of precise measurement on site. He drew a distinction between major encroachments on to an adjoining owner's land which would not be sanctioned by the Act, and lesser encroachments resulting from minor inaccuracies in measurement, which must be tolerated.

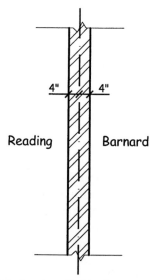

Original party wall of inadequate thickness. Remedial work required under the Act.

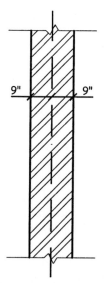

Rebuilt party wall sanctioned by the Act. Increased thickness constructed equally on the land of each owner.

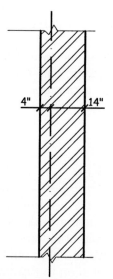

Major encroachment. Increased thickness constructed entirely on adjoining owner's land. Not sanctioned by the Act.

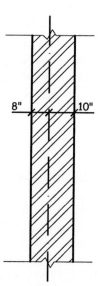

Lesser encroachment. Increased thickness placed disproportionately on adjoining owner's land due to inaccuracies in measurement. Sanctioned by the Act?

Figure 21: Reading v Barnard

Lord Tenterden C J: "I should ... have great difficulty in holding that a party, who having had due notice, as he must, of the intended alteration, lay by and made no objection while the building was going on, could afterwards, if on minute measurement it turned out that the centre of the wall was not exactly on the boundary, defeat such an action as the present.

On the other hand, if the alteration was really effected by a complete encroachment, as, for instance, if all the additional thickness were taken out of the defendant's premises, that would be so complete a violation of the Act that the plaintiff could have no remedy under its provisions.

The question for you will be, whether upon the evidence, you think the wall was fairly built half-and-half, or unfairly built, and intentionally encroaching on the defendant's premises? In the former case, the verdict should be for the plaintiff, notwithstanding any minute inaccuracy in the measurement; in the latter, it should be for the defendant."

The decision

The court rejected the notion that Reading's entitlement to recover his expenses should be frustrated by minor irregularities where these had had no significant effect on the parties' rights.

The judge had therefore ruled that the account was valid and that minor inaccuracies in the positioning of the rebuilt party wall were of no consequence. The jury then decided that, notwithstanding minor inaccuracies, the wall had not disproportionately encroached on Barnard's land. Judgment was therefore entered for Reading.

REES v SKERRETT (2001)
Court of Appeal

In brief

- **Issues:** Demolition of building owner's building – damage to exposed party wall – easement of support – interference with lateral restraint – failure to provide weather protection – negligence.

- **Facts:** Skerrett demolished his building leaving the party wall with Rees' building standing. He failed to provide adequate support or weather protection to the exposed party wall. The failure to provide weather protection resulted in damage to Rees' property. Further weather-related damage was caused because of the presence of cracks which had appeared in the wall. These had developed because of lateral movement in the now-unrestrained party wall following the demolition of Skerrett's building.

- **Decision:** Skerrett was liable for the damage to Rees' property. The law did not recognise an easement of protection from the weather. However it was not disputed that Rees had an easement of support. A removal of lateral restraint amounted to an interference with such an easement. The lack of restraint had led to cracking and this had resulted in much of the weather damage. Skerrett was therefore liable for this damage due to his interference with Rees' easement of support. He was also liable for the remaining weather-related damage which had been caused by his failure to provide adequate weather protection. Recent authorities demonstrated that an occupier of land owes a general duty of care to neighbours in respect of hazards occurring on his own land which may damage his neighbour's land. This duty of care included a duty to provide weather protection to the exposed wall in the present case. On the facts Skerrett was in breach of this duty and therefore liable in negligence.

- **Notes:** See also: *Jones v Pritchard* (1908); *Bond v Nottingham Corporation* (1940); *Phipps v Pears* (1964); *Bradburn v Lindsay* (1983); *Leakey v National Trust for Places of Historic Interest or Natural Beauty* (1980); *Marchant v Capital & Counties plc* (1983); *Holbeck Hall Hotel Ltd v Scarborough Borough Council* (2000)

The facts

Rees owned a terraced house at 14 Hastings Street, Plymouth. The neighbouring house, number 14A, was owned by Skerrett. The two properties were separated by a party wall which straddled the boundary between them. Both properties were over 100 years old.

In January 1990 Plymouth City Council served Skerrett with a demolition order, under the Housing Act 1957, in respect of number 14A. At the same time they also served a notice on him under the Building Act 1984 in respect of the conduct of the demolition works. This required him to provide shoring and weather protection to the exposed wall of number 14.

In February 1990 Skerrett demolished number 14A but made only token gestures in respect of shoring and weather protection. Despite the existence of planning consent for the erection of flats, he also took no immediate steps to redevelop the site. As a result, the party wall, which remained standing following the demolition, was left exposed and unsupported for a considerable period.

Damage caused by demolition of adjoining building

Although number 14A had not previously provided lateral *support* for the weight of number 14, it had provided lateral *restraint* for the party wall. Following the demolition the party wall was therefore entirely unrestrained for its whole height and width and was at risk of collapse due to wind suction. In addition, some minor lateral movement occurred due to the effects of wind suction, and this resulted in cracking.

The wall also suffered considerable damp penetration after the demolition. Part of this could be attributed to Skerrett's failure to provide adequate weather protection, and part to the presence of the cracking which now provided access for moisture penetration.

Rees therefore issued proceedings against Skerrett. He sought a declaration that his property was entitled to an easement of support. He also sought damages for the cost of repairs on the basis of Skerrett's interference with the easement (nuisance), and for his negligence in failing to provide adequate weather protection.

The trial judge granted the declaration that an easement of support existed but refused to grant any further relief.

He acknowledged that the demolition amounted to an interference with Rees' easement of support. However, he held that the damage which had occurred to his property was not attributable to this. The damage occurred due to the effects of the weather (wind and rain) and not because the building had been deprived of support.

He also rejected Rees' claim for damages in respect of the failure to provide weather protection. Relying on *Phipps v Pears* (1964) he noted that the law did not recognise an easement of protection from the weather. In the absence of such an easement he also refused to accept that Skerrett owed any duty of care to provide weather protection when demolishing his building.

Rees appealed against the decision to the Court of Appeal. Their Lordships considered his claims in nuisance and negligence separately.

Nuisance due to interference with easement of support

The trial judge had made a distinction between support for the weight of a building and support provided by way of wind restraint. Based on previous judicial pronouncements about the nature of the right of support, he had concluded that only the former could be included within the ambit of an easement of support. The latter related to the weather rather than to support and, on the basis of *Phipps v Pears*, was not recognised within the law of easements.

The Court of Appeal accepted that Rees could not succeed in a claim based on an easement of protection from the weather but considered whether the damage due to wind suction could legitimately be regarded as a consequence of the interference with the easement of support. The court was able to distinguish *Phipps* from the present case.

The damage in *Phipps* had been due to rain penetration alone in circumstances where the plaintiff had no easement of support. In the present case the plaintiff had an easement of support and the damage, although weather related, was due to movement having occurred as a result of removal of lateral restraint. In these circumstances the court considered that the plaintiff was able to recover for the damage which he had suffered.

> Lloyd J: "I have come to the conclusion that ... wind support is properly to be regarded as an aspect of the support which one of two adjoining buildings provides to the other.
>
> It is true that the cracking damage experienced in this case is not, on the judge's findings, part of the consequences of the pressure exerted on the flank wall, vertically or laterally, by the weight of the other parts of the structure, and so it is not within the terms in which judges have previously spoken of the right of support, as quoted ... above[1].
>
> But it is a respect in which the building is unstable, it is so because of the demolition of the structure which did provide support, and because the flank

---

[1] Lloyd's J's judgment had previously made reference to Lord Selborne's description of the right of support in *Dalton v Angus* (1881): "What is support? The force of gravity causes the superincumbent land, or building, to press downward upon what is below it, whether artificial or natural; and it also has a tendency to thrust outwards ... Using the language of the law of easements, I say that, in the case alike of vertical and of lateral support, both to land and to buildings, the dominant tenement imposes upon the servient a positive and a constant burden, the sustenance of which, by the servient tenement, is necessary for the safety and stability of the dominant."

wall itself is subject to outward lateral pressures to which it would not have been subject while 14A was in place.

It is also true that, whereas weight pressures operate on the structure continuously, being kept in equilibrium by the adjoining structure, the pressure from the effects of the weather is something to which, while the adjoining structure is in place, the dominant tenement is just not exposed. From that point of view, it occurred to me that the right which [Rees] claims might more properly be called a right of protection from these particular effects of the weather than a right of support for anything.

But on reflection it seems to me that it is correctly to be regarded as an aspect of support, even if what one is considering is the effect of the unsupported weight, largely if not exclusively, of the wall itself. As [the plaintiff's engineer] said in his first report, the flank wall is too high and too long to be stable as a free standing structure. Before the demolition it was supported and was therefore stable. Afterwards it is not supported and therefore it is not stable. Why is it not stable? Because it is too big, and therefore too heavy, to stand without additional support.

It is the effect of the wind suction that has made the instability manifest by cracking in the wall, but it does not seem to me that it matters what particular circumstances produce the actual cracking which is the symptom and result of the instability caused by the withdrawal of support.

It could be objected that for Mr Rees to be able to recover for this type of damage, caused by the withdrawal of support, would be anomalous when Mr Phipps might have suffered exactly the same damage, but could not recover. However, the distinction is that Mr Phipps' property had not been supported by the adjoining house, and it must therefore be presumed to have been stable as a free standing structure, and not subject to this particular risk.

For these reasons I have come to the conclusion that the judge was wrong to hold that the particular type of damage suffered in this case as a result of wind suction was not within the scope of the right of support. It may be that it was somewhat misleading to categorise it as something other than part of the destabilising effect of the pressure of the weight of the structure itself. As I see it, it is an effect of weight pressure, though of the weight of the wall itself, and is accordingly within the protection afforded by the right of support."

Negligence in failing to provide weather protection

The court's finding on nuisance liability did not dispose of the matter entirely. Some of the moisture damage to the wall had been caused by rain penetration which was unconnected to the movements in the structure. Rees therefore also had to establish that Skerrett was liable for a failure to provide weather protection if he was to succeed in this part of his claim.

The decision in *Phipps v Pears* prevented him from pursuing this claim on the basis of the law of easements so he framed his claim on the basis of a breach of a duty of care. He argued that, due to recent developments in the law of tort, there was now a duty to the occupier of an adjoining property to take reasonable steps to weatherproof a party wall, when demolishing an adjoining terraced property. He further argued that Skerrett was in breach of this duty in the present case.

In this context the court considered the principle established in *Leakey v National Trust for Places of Historic Interest or Natural Beauty* (1980). Lloyd J described this principle in the following terms:

> "It established the proposition that an occupier of land owes a general duty of care to a neighbouring occupier, in relation to a hazard occurring on his land, whether natural or man-made, that duty being to take such steps as were reasonable in all the circumstances to prevent or minimise the risk of any injury or damage to the neighbour or his property of which the occupier knew or ought to have known, and that what is reasonable in all the circumstances will depend, among other things, on the cost of the steps which might usefully be taken and, viewed broadly, the resources available to the occupier and the neighbour."

The court noted that the *Leakey* principle had been applied in *Bradburn v Lindsay* (1983). The defendant's failure to prevent the spread of dry rot had been a crucial feature of that case. However, there were also similarities with the present case as inadequate support and weather protection had been provided when the defendant's property was demolished. In that case the court had awarded damages for failure to provide support which had included an element for the rendering (i.e. weather protection) to the buttresses.

More recently the courts had also applied the principle in other cases which suggested that there might now be a positive duty to provide shoring in certain circumstances, even in the absence of an easement of support. Two cases were mentioned in this context, *Bar Gur v Bruton* (1993) and *Holbeck Hall Hotel Ltd v Scarborough Borough Council* (2000). Lloyd J explained the significance of these cases:

> "The latter case concerned support by land of buildings, and arose from the fall of a cliff which led to the hotel being damaged and having to be demolished. The complaint against the defendant was not that it had done something negligently, but that it had failed to take steps to maintain the support, or to prevent it from being withdrawn by a fall of the cliff.
>
> The Court of Appeal rejected the submission for the defendants that, if they were to be made liable, there had to be something that they had done, rather than mere omissions. They therefore held that the statement in *Bond v Nottingham Corporation* (1940) that the servient owner can with impunity stand by and let his property fall into decay, and thereby withdraw support ... is no longer good law in the light of the *Leakey* case.

In fact both in the *Holbeck Hall Hotel* case and in *Bar Gur v Bruton*, there by a majority, the court held that the defendant was not liable on the facts. But the decisions show that the duty of the owner of the supporting land or building has now to be seen as extending, in some circumstances, to a duty to take positive steps to maintain and continue support."

Although these cases dealt with a failure to provide support, the court considered whether the same principles should be applied to the failure to provide weather protection in the present case. It decided in the affirmative.

> Lloyd J: "The liability which [Rees] seeks to establish is dependent on showing, first, that the ... defendant knew or ought to have known of the risk of damage likely to result from his demolition works, if not accompanied by weatherproofing and, secondly, that the damage suffered would have been prevented by work which it would have been reasonable in all the circumstances for him to carry out.
>
> Unlike the *Leakey* case itself and the *Holbeck Hall Hotel* case, but like *Bradburn v Lindsay*, liability is sought to be established here for the consequences of an action, rather than omissions, by the ... defendant; this may be easier to establish on the facts.
>
> ... Having considered the matter in the light of the *Leakey* case and the cases since, especially the *Holbeck Hall Hotel* case, it seems to me that it would be right to hold that Mr Skerrett was under a duty in February 1990 to take reasonable steps to provide weatherproofing for the dividing wall once it was exposed to the elements as a result of the demolition of number 14A ...
>
> ... On that basis, I am satisfied that the judge was wrong to reject Mr Rees' claim against Mr Skerrett based on a duty of care requiring him to weatherproof the dividing wall after the demolition of 14A."

Rationale for extending the law

The court was conscious of the fact that its decision represented a significant extension of the law. A number of statements in the judgments seek to provide reassurance about the reasonableness of the decision in practice. In this context Waller LJ placed the decision in the context of the broader legal principle which was being applied as well as the circumstances of the two parties:

> "As regards the liability for damage on the basis of breach of duty I ... have some anxiety ... that to give judgment on this aspect in favour of the claimant involves an extension of the law.
>
> ... Such anxiety as I have is, however, dispelled by the following. If someone can owe a duty to take reasonable steps to prevent damage to his neighbour's

property resulting from something naturally on his land: *Leakey v National Trust for Places of Historic Interest or Natural Beauty* or placed on his land by trespassers: *Sedleigh-Denfield v O'Callaghan*, it would seem strange that he should not owe some duty when he pulls down the house which he appreciates protects the wall of the neighbouring house.

The only suggestion is that a duty be imposed which is reasonable having regard to the position and resources of the two house owners. That does not seem extreme. In this case, furthermore, a notice had been served by the local authority requiring weatherproofing to be done. It could thus scarcely be said either that it was not obvious that weatherproofing would be necessary if the house was demolished, or that it was unreasonable as between the claimant and the ... defendant to require the ... defendant to carry out that weatherproofing."

Lloyd J addressed the policy considerations against the creation of new negative easements, which had been articulated by Lord Denning in *Phipps v Pears*:

"Lord Denning MR in *Phipps v Pears* ... identifies a policy as underlying the denial of a right of protection, namely that to recognise the right which Mr Phipps asserted would stop a man pulling down his house and would put a brake on desirable development.

Such a policy is no doubt relevant, especially as regards the establishment (or not) of property rights, the infringement of which carries absolute liability. As regards the law of negligence and nuisance developments since 1965, especially the *Leakey* case, show that the balance between the position of those who are neighbours (both in fact and law) is now to be drawn differently in these areas of law.

Moreover, it is not a question of preventing a man from pulling down his house altogether, any more than a right of support prevents demolition. Here Mr Skerrett was obliged to pull down his house. Rather, it is a question of requiring him, if and when he does demolish the house, to provide substitute protection, to the extent that this can be done by works which in all the circumstances it is reasonable to expect him to undertake."

Effect of Party Wall etc. Act 1996

Finally, Lloyd J emphasised that the enactment of the 1996 Act had, in practice, achieved much the same result as the decision in the present case:

"... the carrying out of works such as those undertaken by Mr Skerrett would now be governed by the provisions of the Party Wall etc. Act 1996. By section 2(2)(n) of that Act a building owner has the right to expose a party wall hitherto enclosed subject to providing adequate weathering. Though an owner who is subject to a demolition notice does not have to serve a party structure notice under section 3 (see section 3(3)(b)) he is still liable to compensate his neighbour for loss or damage resulting from the work done under section 7.

Accordingly, circumstances of the kind at issue in this action which arise now would be likely to be dealt with under that statutory regime. The application of common law liability to a new, and far from uncommon, set of circumstances which [the plaintiff] invites us to make may, therefore, be of less far-reaching significance than it would be if the point had not in the meantime been affected in practical terms by this statutory provision."

The decision

The loss of restraint which Rees had suffered to the party wall constituted an infringement with his easement of support. Because of the principle established in *Leakey*, Skerrett was also in breach of a duty of care in failing to provide weather protection to the exposed face of the party wall.

The Court of Appeal therefore allowed Rees' appeal and entered judgment for him for damages for the cost of repairs and of weather protecting the party wall.

RILEY GOWLER LTD v NATIONAL HEART HOSPITAL BOARD OF GOVERNORS (1969)

Court of Appeal

In brief

- **Issues:** 1939 Act – appeal against third surveyors' award – time limit for appeal – date of "delivery" of the award.

- **Facts:** The third surveyor made an award and indicated that it could be collected on payment of his fee. Riley Gowler paid the fee and collected the award. A copy of the award was then posted to the Hospital and they received it on the following day. Riley Gowler appealed against the award 15 days after collecting it but only 14 days after the Hospital received it.

- **Decision:** The appeal was out of time. As against Riley Gowler, "delivery" of the award was effective at the moment that they received it. They had therefore missed the 14-day time limit for making an appeal by one day. It was irrelevant, so far as they were concerned, that the Hospital had not received their copy until the following day.

- **Notes:** This case should be read in the context of changes introduced by the 1996 Act. This now makes provision for awards to be "served" on the parties. Appeals must now be filed within 14 days of "service" upon a particular party rather than within 14 days of "delivery". See sub-sections 10(14) to (17) of the 1996 Act.

The facts

Riley Gowler and the National Heart Hospital were the owners of adjoining properties in London. A difference had arisen between them, under the 1939 Act, following the service of a party structure notice. The party-appointed surveyors had been unable to agree on a particular matter so this had been referred to the third surveyor.

The third surveyor made his award on 27 May 1968. He then wrote to both parties, informing them that it was available for collection on payment of his fee. Both parties received his letter on the following day (28 May). Riley Gowler's surveyor immediately paid the fee on his appointing owner's behalf and collected the award. The third surveyor then posted a copy to the Hospital and they received this on 29 May.

On 12 June Riley Gowler filed an appeal against the award with the county court. At the hearing of the appeal the Hospital submitted that this was out of time as it had not been filed within 14 days of delivery of the award, as required by section 55(n) of the Act. The award had been delivered to Riley Gowler's representative on 28 May and the appeal had been filed on 12 June. The appeal was therefore out of time by one day.

Riley Gowler argued that delivery of the award had not taken place until it was also received by the Hospital on 29 May (14 days before the filing of the appeal). The appeal was therefore on time. The county court dismissed the appeal on the basis that it had been filed out of time so Riley Gowler appealed to the Court of Appeal.

Receipt of award by appealing party

Their Lordships were careful to limit the scope of their decision to the facts of the case before them but unanimously held that this particular appeal had indeed been out of time.

Having received the award on 28 May, it was not for Riley Gowler to argue that they should have an additional day in which to appeal on the basis that the Hospital did not receive it until the following day.

> Sachs LJ: "In those circumstances, there arises the narrow issue before this court, namely whether, as against the appellants, the delivery of the award took place on 28 May. It is really sufficient to say that for my part I am fully in agreement with the view taken in the court below that, at any rate as against the appellants, it is impossible to say that the award was not delivered to them on 28 May, when thus the 14 days would begin to run. It does not seem to me that it lies in their mouth to quarrel with that
>
> Any other view would result in considerable confusion, for the person who took up the award would be quite unable to say when time began to run without making enquiries, which might turn out to be difficult and extensive, as to when exactly the copy of the award, if any, had reached the other side. All sorts of complications could ensue according to where the other party happened to be at the relevant time and whether delivery had been effected in accordance with such time-table as the post office may deem normal."

Is delivery of award a single event?

The decision relates only to the date on which delivery of the award was effective on Riley Gowler. It does not address the wider question of whether delivery was also effective on the Hospital at the same moment, or whether this took place the following day, when they actually received the award. The general question of whether "delivery" is a single event or whether it can take place on different occasions for different parties also remained unanswered.

Because of this Edmund Davies LJ stressed that his judgment only

> **London Building Acts (Amendment) Act 1939**
>
> PART VI: RIGHTS ETC. OF BUILDING AND ADJOINING OWNERS
>
> DIFFERENCES BETWEEN OWNERS
>
> **55. Settlement of differences**
>
> (m) The award shall be conclusive and shall not except as provided by this section be questioned in any court;
>
> (n) Either of the parties to the difference may, within 14 days after the delivery of an award made under this section appeal to the county court against this award...

addressed the narrow issue of delivery on Riley Gowler. He emphasised the complications that could arise if delivery was always taken as being effective at the moment that the first party received their copy of the award:

> "I agree, and like Sachs LJ, I desire my observations to be confined to the narrow question of when there was a delivery of an award under section 55 of the London Building Acts (Amendment) Act 1939, to the appellants. Immediately one goes outside that narrow question all kinds of matters of difficulty arise which are irrelevant to the determination of this appeal...
>
> ... What, for example, about the hypothetical case ... of a building owner being out of the country and having no knowledge of the existence of an award until, perhaps, two or three months after the adjoining owner had taken it up? Is it to be said in those circumstances that he has no right of appeal because 14 days have gone by? I raise the question and I avoid answering it. All I say is that it does involve consideration of matters which, happily, do not arise for consideration for the purposes of this appeal."

Sachs LJ acknowledged that the question of whether there can be more than one delivery of an award was still to be decided. However, he suggested that in the circumstances of the present case, delivery was probably effective on both parties at the moment that the first party collected the award:

> "That concludes the case, but I would like to add that it seems to me, as at present advised, that at any rate where a third surveyor notifies both parties that an award is available to be taken up and they are given what on the facts

amounts to equal opportunities to be there at the relevant time, there cannot be more than one delivery of an award, any more than there can be more than one delivery of a reserved judgment whether or not only one of the parties attends the court.

It must be, incidentally, kept in mind that section 55 is designed to provide a specific and speedy time-table having regard to the fact that building operations may be in progress. That, however, is a matter which can be left to be decided when the point arises. For my part I would dismiss the appeal."

The decision

Delivery of the award was effective on Riley Gowler when they actually received it. This was 15 days before they filed their appeal with the county court. The appeal was therefore dismissed as it was out of time.

SACK v JONES (1925)
Chancery Division

In brief

- **Issues:** Easement of support – disrepair of adjoining owner's building – subsidence – adjoining owner's building dragging building owner's building over – whether an interference with easement of support – whether an actionable nuisance.

- **Facts:** Jones owned an end-of-terrace house and Sack owned the adjoining house in the terrace. Damage occurred to Sack's house. She attributed this to disrepairs in Jones' house. She alleged that the flank wall of that house had subsided and that the house was rotating so as to drag her house with it.

- **Decision:** Her claim was dismissed. She had failed to produce sufficient evidence to prove the cause of the damage. Her claim would still have failed even had she done so as she had no cause of action. There was no interference with an easement of support where a servient tenement subsided and dragged the dominant tenement with it. Furthermore, it was well established that an owner was not subject to any obligation to his neighbour to keep his own property in repair. This was even the case where his property was subject to an easement of support in favour of his neighbour. Therefore, in the absence of some further act or omission, Jones could not be held liable in nuisance merely on account of the poor condition of his own property. Jones had simply allowed a tenant to occupy his premises and, in these circumstances, had committed no act which could make him liable in nuisance.

- **Notes:** This case restates the traditional rules relating to easements of support and the lack of any ancillary repairing obligations. See also *Jones v Pritchard* (1908); *Bond v Nottingham Corporation* (1940). These rules must now be read in the context of recent developments in the law of tort. See, in particular, *Bradburn v Lindsay* (1983); *Rees v Skerrett* (2001).

The facts

Jones owned an end of terrace house at 91 Torbay Road, London NW6 which he had let to a tenant. Sack was the owner-occupier of the adjoining house at 89 Torbay Road. Her property was separated from number 91 by a party wall. Both parties acknowledged that mutual party wall rights and mutual rights of support existed in respect of the two properties.

Between 1919 and 1924 cracks appeared in Sack's house and these became progressively worse. She attributed the damage to subsidence in the flank wall of Jones' house. She alleged that this had caused Jones' house to rotate, and that this was pulling the party wall over in his direction.

Jones denied this. Although the flank wall did have a slight tilt this was not unusual for houses built on clay soils and his house was actually in good repair. The damage had instead occurred due to the slight settlement of both houses and of the party wall between them.

Sack issued proceedings against Jones. She based these on his failure to provide the support to which she was entitled, and on nuisance due to the dangerous condition of his house. She claimed damages, an injunction and a declaration that Jones was bound to maintain adequate support.

The court held that Sack had failed to prove her case. As the plaintiff, the burden of proving the case fell on her. The expert evidence presented by each party, as to the cause of the damage, was in conflict. The court found that it was impossible to come to a definite conclusion on the evidence and therefore rejected the claim.

Failure to provide support?

However, although not material to its decision, the court also considered whether Sack would have had a cause of action if the facts relied on had been proved.

It first considered her argument that the dragging of the party wall over by the neighbouring property amounted to a failure to provide support, as required by the easement of support. The court rejected this notion due to any lack of authority for it.

> Astbury J: "But even assuming that the facts were as the plaintiff alleges – namely, that her party wall is being pulled over by the subsidence of the defendant's house and flank wall – she would have no justification in law for maintaining this action. Both parties have joint easements of support in respect of this party wall. But having regard to the authorities the plaintiff is unable to rely on her easement of support as giving her any remedy…
>
> … The plaintiff then contends that the owner is … liable if he withdraws support by allowing his house to fall away from and pull over his neighbour's house, but no authority is produced in support of that contention."

It then considered her alternative argument – that Jones was liable in nuisance because of the dangerous condition of his house. In particular, due to his neglect, he had allowed his house to subside, and her house had been dragged with it as a result.

Liability in nuisance for condition of adjoining property?

Sack cited two cases in support of this argument, where owners had been held liable in nuisance for damage caused to adjoining property. In *Attorney General v Roe* (1915) a quarry owner had been held liable where his excavations caused the adjoining highway to collapse. In *Broder v Saillard* (1876) damp had penetrated the plaintiff's house from an artificial mound of earth placed against it by the defendant. The defendant had been held liable for nuisance for causing the disrepair to the plaintiff's land.

The court distinguished the present circumstances from each of these decisions. The defendants, in each of these cases, had caused damage to the plaintiff through their own act or omission. Jones had done neither of these things. He had done nothing more than allow a tenant to reside in his house.

In drawing this distinction the court first had to satisfy itself that an owner was subject to no obligation, to neighbouring owners, to keep his own property in repair. The court noted that there was no such obligation, even where, as in the present case, the owner was required to provide support to his neighbour's property.

> Astbury J: "Now although the defendant's house is subject to an easement of support in favour of the plaintiff's house, the defendant is under no obligation to the plaintiff to keep her own house in repair for that purpose."

The court made reference to *Colebeck v Girdlers Co* (1876), *Dalton v Angus* (1881) and *Jones v Pritchard* (1908) in support of this position. Astbury J also quoted the judgment of Mellor J in the *Colebeck* case as demonstrating the established nature of the rule:

> "... it is well established that there is no obligation to repair on the part of the owner of the servient tenement, but the owner of the dominant tenement must repair, and that he may enter on the land of the owner of the servient tenement for that purpose: *Pomfret v Ricroft* (1669)."

In the absence of an obligation to repair, Jones could only have been liable in nuisance if there was some other wrongdoing on his part. In fact, he had simply allowed his tenant to occupy the house and had committed no wrongful act at all.

In its deliberations the court placed particular emphasis on the judgment of Parker J in *Jones v Pritchard* and cited the following words approvingly:

> "... if a grantor is doing, on land retained by him, only what ... was at the time of the grant in the contemplation of the parties ... and is guilty of no negligence

or want of reasonable care or precaution, he cannot be liable for nuisance entailed upon the grantee."

In allowing his tenant to occupy his house Jones had done no more than was contemplated by the parties and could not therefore have been liable in nuisance.

The decision

Sack's claim was dismissed. She had failed to prove that the damage to her house had been caused by movement in Jones' house. Even if she had done so, her claim would still have failed due to the absence of a cause of action.

SAUNDERS v WILLIAMS (2002)

Court of Appeal

In brief

- **Issues:** Damage to party wall during building operations – builders' liability in negligence – calculation of damages – delays by adjoining owner in undertaking repairs – foreseeability of damage – duty to mitigate losses – Civil Procedure Rules – proportionality.

- **Facts:** Williams was a builder. He caused serious damage to the party wall with Saunders' property whilst working on the adjoining building. Saunders could not afford to undertake the necessary repairs for eight years. For the whole of this time she was unable to use two rooms adjoining the party wall. She claimed damages for the cost of repairs and for consequential losses. The latter related to inconvenience and the loss of rent for the two rooms. The trial judge awarded damages for the cost of repairs but only £1000 for the consequential damages. This represented the rent for one year – the length of time for which it was reasonably foreseeable that the wall would remain damaged. Saunders appealed against the award in respect of consequential damages.

- **Decision:** The Court of Appeal substituted an award of £8000 for the consequential damages. As the kind of damage suffered had been foreseeable it was irrelevant that its full extent was not. Damages were therefore payable for each of the eight years. Damages could only have been reduced on the basis that Saunders had failed to mitigate her loss by delaying the repairs for so long. In view of her poor financial position it had not been open to her to undertake the repairs at an earlier date. The court was required, by the Civil Procedure Rules (CPR), to deal with the case in a way which was proportionate to the issues involved. It therefore substituted its own award on the basis of the available evidence rather than remitting the case back to the Construction and Technology Court to make a decision.

- **Notes:** See also *Louis v Sadiq* (1997) on the question of damages for consequential losses for damage to a party wall.

The facts

Mrs Saunders was the owner-occupier of a semi-detached house at 127 Tillery Street, Abertillery, Gwent. Mr Williams was a builder. In 1992, he was employed to carry out work to the adjoining semi-detached house at number 126.

During the course of his work Williams caused extensive damage to the party wall between the two properties. The necessary remedial work required the wall to be completely rebuilt. Unfortunately this was never undertaken, either by Saunders or by Williams, or by the owner of number 126.

Eight years later, Saunders issued proceedings against Williams in negligence. She claimed for the cost of the repairs and for distress and inconvenience due to the loss of two rooms which had become unusable. In particular, under this head, she sought damages representing the loss of rent for the two rooms which she had intended to let to tenants.

Williams admitted liability and the court at first instance awarded damages of £24,572 for the cost of repairs and a further £1000 for the consequential losses.

The court based this second sum on the fact that Williams could not reasonably have foreseen that the wall would remain damaged for eight years. It therefore awarded damages based on the wall remaining in this condition for only one year. After hearing testimony from a single joint expert it concluded that the value of the loss of the rooms was £1000 per annum.

Saunders appealed to the Court of Appeal against the court's award in respect of the consequential loss.

Consequential losses for damage to party wall

The Court of Appeal found that the trial judge's approach to the question of consequential losses was incorrect. Williams' foreseeability of the scale of the loss was irrelevant. Foreseeable damage caused to an unforeseeable degree is recoverable.

The correct test, in deciding whether the normal measure of damages should be reduced, was whether Saunders had failed to take reasonable steps to mitigate her loss. The burden of proving that she had failed to do so lay on Williams.

Unfortunately, the trial judge had failed to undertake the necessary findings of fact to enable him to determine whether Saunders had failed to mitigate her loss. If he had done so he would have considered her poor financial position and her consequent inability to afford the necessary repairs. He would not therefore have concluded that she had failed to mitigate her loss by failing to undertake the repairs for eight years.

The Court of Appeal did not consider it appropriate to remit the case back to the court at first instance in view of its duty, under the Civil Procedure Rules, to deal with cases in a way which is proportionate to the issues and costs involved. Therefore, although the court was not equipped to deal with detailed factual investigations which were a matter for the trial court, it nevertheless decided to substitute its own sum in respect of the consequential losses. Pill LJ explained the court's thinking in this regard:

> "... it does seem to me extremely unlikely that a rate of loss of significantly more than £1000 a year is likely to be established by Mrs Saunders under this head. Upon the evidence which was available, and upon the judge's findings, as far as they go, it seems to me highly unlikely that there was a real possibility that the rooms would have been let during the relevant period and, if they were let, would be let at a figure which would provide a regular net income significantly higher than the sum awarded.
>
> If the matter is regarded simply as one of distress and inconvenience, then, conscious though we are of the undoubted distress which Mrs Saunders has suffered ... a figure significantly in excess of £1000 a year seems to be very unlikely to be achieved.
>
> I mention that, which I acknowledge is contrary to the general proposition that it is not for this court to conduct its own factual investigation. Trials are for the trial judge. But this court has had to consider very carefully whether, having found that the judge's approach was inappropriate, this case should be the subject of remission to the technology and construction court.
>
> We must bear in mind the question of proportionality. Under the CPR it is the duty of the court to make decisions proportionate to the issues involved ... I have come to the conclusion that we should make our own estimate of damage on the material available to us, rather than remit the question ... We do not consider that there is any real prospect of Mrs Saunders obtaining significantly more damages than the amount we have in mind."

The decision

The court therefore allowed the appeal and substituted the sum of £8000 for the original award in respect of consequential losses, representing an annual loss of £1000 for each of the eight years.

SELBY v WHITBREAD & CO (1917)
King's Bench Division

In brief

- **Issues:** 1894 Act – change of ownership of building owner's land – whether original building owner's statutory liabilities pass to new owner – whether new owner takes subject to adjoining owner's existing statutory rights – surveyors' authority to make addendum awards – meaning of "difference" (dispute) – challenges to surveyors' awards on technical grounds – entitlement to damages at common law.

- **Facts:** Whitbread demolished their building and permanently exposed a part of the party wall shared with Selby's adjoining building. The work was undertaken in accordance with a surveyors' award and the relevant part of Whitbread's land was then transferred to the Council for dedication to the public. The surveyors subsequently made an addendum award requiring Whitbread to provide support to the exposed portion of Selby's building.

- **Decision:** The addendum award was valid and was binding on Whitbread. The required "difference" had arisen within the meaning of the Act as the term should be interpreted liberally. Surveyors had an ongoing entitlement to make addendum awards to address issues arising out of the service of notices. This was unaffected by a sale of the building owner's land as the new owner took subject to the adjoining owner's statutory *rights*. However, the original building owner continued to be responsible for performance of the statutory *liabilities* under the Act and these did not pass to the new owner. It was too late for Whitbread to challenge the surveyors' awards on technical grounds as the statutory 14-day appeal period had expired. Damages (but not specific performance) were awarded to Selby for breach of the addendum award. He was not entitled to damages for interference with an easement of support as his common law rights had been entirely replaced by rights under the Act.

- **Notes:** See also (on change of ownership): *Mason v Fulham Corporation* (1910) – (on challenges to surveyors' awards): *Re Stone and Hastie* (1903) – (on the replacement of common law rights): *Standard Bank of British South America v Stokes* (1878); *Lewis & Solome v Charing Cross Railway Co* (1906) – (on situations where an entitlement to common law damages still exists): *Bower v Peate* (1876); *Dodd v Holme* (1834).

The facts

Selby owned premises at 11 Royal Mint Street, London E1. Whitbread & Co owned the 'Rising Sun' public house at 12 Royal Mint Street. The properties had been built together about 200 years previously and were described as forming a continuous structure. They were separated by a party wall and were mutually dependent on each other for support (Figure 22).

Whitbread wished to demolish and rebuild the Rising Sun as this was in a dangerous state of repair. They had therefore obtained the necessary consent for the works from the licensing justices. This was conditional on the new building being set back from the street so as to provide a wider footpath at the front of the premises.

In August 1914 they had entered into an agreement with the London County Council in respect of the enlargement of the footpath. This provided that the vacant area of land to the front of the rebuilt premises would be conveyed to the Council who would then dedicate it to the public as a highway.

They subsequently served a party structure notice on Selby under the 1894 Act and the two party-appointed surveyors published an award in February 1915. The award included provision for the support and protection of Selby's property and also contained the usual provision reserving the power of the surveyors to make further awards. The relevant clauses (clauses 2 and 7) were in the following form:

> "2. That the building owner shall at their own expense take every precaution for the support of the building of the adjoining owner by providing all necessary temporary or permanent shoring or strutting or tie rods or other means as may be ordered by the district surveyor or mutually agreed, and hoard off or otherwise protect all parts of the adjoining owner's premises during the execution of the works with all necessary fans screens and other temporary protection to the reasonable satisfaction of the surveyor for the adjoining owner.
>
> 7. That power is reserved to the above named surveyors to make further awards as may be agreed between them in respect of any further matters in connection with the party structure."

Demolition of building owner's property

The Rising Sun was demolished in March 1915. The following month, in accordance with the earlier agreement, Whitbread conveyed the strip of land at the front of the plot to the London County Council. The Council also agreed to indemnify Whitbread for any future claims from Selby in respect of the conveyed land.

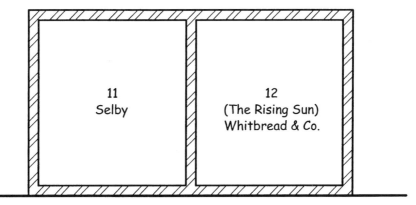
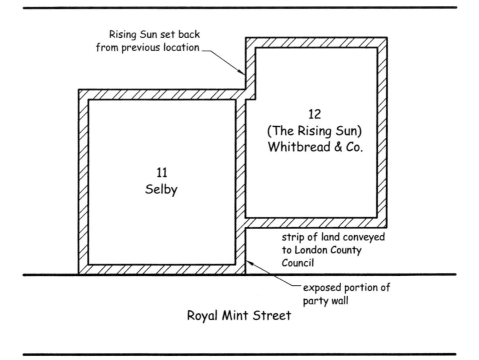

Figure 22: Selby v Whitbread & Co

By August 1915 the works had been completed and the rebuilt Rising Sun was again open to the public. As required by the licensing justices, the new building had been set back about 13 ft from the previous location. A large proportion of the party wall, which had previously separated the two buildings, was therefore now exposed.

This inevitable consequence of the work seems to have been overlooked by Selby's surveyor when he was first appointed, as the award made no specific provision for it. Once the works had been completed he therefore wrote to Whitbread's surveyor asking how the front portion of his appointing owner's building was now going to be protected.

Whitbread's surveyor responded that his appointing owner's responsibilities as building owner had been transferred to the London County Council. He therefore refused to take part in any further proceedings under the 1894 Act.

Selby's surveyor then approached the third surveyor and the two of them published an addendum award in January 1916. This required Whitbread to provide the necessary support for Selby's building by erecting a substantial pier on the newly cleared piece of land. Whitbread failed to comply with the award.

Selby issued proceedings against them to enforce the award and to recover damages at common law for the removal of support from the party wall. The case raised a number of important issues.

Building owner's liabilities following sale of land

Whitbread submitted that they had no liabilities under the 1894 Act in respect of the newly cleared piece of land. They had ceased to be a building owner, within the meaning of the legislation, from the moment that they transferred the land to the Council in April 1915. At that time their liabilities had passed to the Council, as the new owners of the land. Any requirement, under the Act, to erect the pier must therefore now be discharged by the Council.

They cited *Mason v Fulham Corporation* (1910) as authority for liabilities under the Act being transferred to a purchaser on sale and argued that the court was bound by this decision. In that case an adjoining owner (Fulham Corporation) had an obligation, under the Act, to contribute towards the cost of building a party wall if it ever made use of it in the future.

When, at a later date, it did make use of it the court held that the original building owner's successor in title was entitled to receive payment of the contribution rather than the original building owner himself. The court ruled that the original building owner had automatically assigned his contingent right to the contribution when assigning his interest in the property.

The court distinguished *Mason v Fulham Corporation* from the present case. The decision in *Mason* had addressed the right to receive payment of a debt (a benefit under the Act). The present case was concerned with the responsibility for discharge of liabilities under the Act (a burden).

The law treated the two situations entirely differently. There could be no objection to an owner transferring his benefits to a third party if he so wished. However, his burdens under the Act were for the protection of others. They were personal to him and he should not be able to defeat an adjoining owner's rights by transferring these to a man of straw. The court did not, therefore, regard itself as bound by *Mason*.

> McCardie J: "In my opinion *Mason v Fulham Corporation* turned upon the specific wording of sections 95 and 99 of the Act of 1894[1] as applied to certain specific facts. It has no application to the present case. The assignment of a chose in action such as a debt differs fundamentally from the assignment of an obligation.
>
> In *Mason v Fulham Corporation* there was a transfer of a mere debt. No question arose as to the transfer of a liability. The particular words of the Act upon which the decision turned differ in striking fashion from the provisions of sections 87 to 94[2] so far as they impose obligations which are relevant to the present dispute."

The court considered the party wall provisions in sections 87 to 94 of the Act. It could find none that, either expressly or by implication, supported Whitbread's argument that liabilities under the Act were transferred on a sale of the land. On the contrary, two of them suggested that, once a notice had been served, the obligations thereby created between the two individual parties would be ongoing until such time as the works had been satisfactorily completed.

Firstly, the adjoining owner's entitlement, under section 94[3], to require security for expenses from the building owner pending the satisfactory completion of the work suggested an ongoing, and personal, liability.

Secondly, section 91[4] gave the surveyors an ongoing jurisdiction to determine the differences between the original parties, pending the satisfactory completion of the works. Specifically, there was no provision for the original parties to withdraw from the process, and for new owners to become subject to the surveyors' jurisdiction.

---

[1] Sections 95 and 99 of the 1894 Act addressed the liability for expenses and the retention of title in works by a building owner pending payment of contributions due to him. They are broadly equivalent to sections 11 and 14(2) of the 1996 Act.
[2] Sections 87 to 94 contained the bulk of the Act's party wall provisions with the exception of those relating to liability for expenses. They are broadly equivalent to sections 1 to 10 of the 1996 Act.
[3] Broadly equivalent to section 12 of the 1996 Act.
[4] Broadly equivalent to section 10 of the 1996 Act.

McCardie J: "It seems to me that [section 94] contemplates that the person who serves the notice under section 90 shall be and remain liable for all the results which follow from such notice.

During the arguments I asked the learned counsel for the defendants to state the procedure by which the plaintiffs could have enforced against the London County Council the obligations under the award of February 22 1915. No satisfactory answer was given.

The conveyance of April 1915 took place after the demolition of the defendants' premises and before the erection of the new buildings, and any resultant transfer of obligations would be attended with unceasing confusion as to legal rights and duties and the possibility of conflicting or overlapping jurisdictions between several sets of arbitrators.

Section 91 of the Act provides that the arbitrators appointed after service of the notice shall settle from time to time all matters in dispute. Their jurisdiction is, I think, continuous and exclusive, subject to the rights of appeal given by section 91. It remains unimpaired until the final adjustment of all questions in difference between the building owners who gave the notice and the adjoining owner who received the notice, and until the operations involved in the notice are concluded."

The court therefore concluded that no transfer of obligations had taken place whereby Whitbread were freed from their liabilities to Selby under the Act.

Are adjoining owner's rights binding on a purchaser (and the public)?

It follows from the above discussion, not only that the original building owner remains liable to the adjoining owner following a transfer of his land, but also that a purchaser of the land does not become so liable. A purchaser will not, therefore, owe any positive obligations to the adjoining owner by virtue of the transfer.

However, a related issue was also raised by a further submission by Whitbread. They submitted that the addendum award of January 1916 must be invalid. In providing for the erection of a pier on the front portion of the plot the surveyors purported to authorise the performance of an illegal act.

Both the agreement to dedicate the land as a public highway (August 1914) and the conveyance of this land to the London County Council (April 1915) predated this award. In Whitbread's submission the surveyors therefore no longer had any authority to authorise the erection of the pier. In purporting to do so they would be requiring Whitbread to commit a criminal offence by obstructing the highway. They would also be requiring them to commit a trespass on land which was now owned by the Council.

The submission raised the question of whether existing rights under the Act (which arise when a notice is served) are binding on a purchaser, or whether the purchaser takes free from them. If the former, then the addendum award would be binding on the Council (and the public). If the latter, then the surveyors would have had no authority to make the addendum award for the reasons stated in Whitbread's submission. It also raised the related question of whether the public takes subject to rights under the Act on a dedication of the highway.

With regard to the dedication of the highway the court noted that this could be subject to various categories of pre-existing third party rights. These included market rights, and the right of an owner to deposit goods on the highway in front of his property. In the present case, Selby's rights sprang into existence as soon as the notice had been served upon him. The court held that, as statutory rights which contemplated an interference with the highway, these must be binding on the public.

> McCardie J: "It is unnecessary to analyse in detail the Act of 1894. It is obvious that the various building and other operations falling within the Act may involve a substantial interference with highways. The public are subject to the inconvenience thereby caused.
>
> The public are mere volunteers, and in my opinion the dedication of the space of land in question must be deemed to be subject to such interference therewith as was properly necessitated by the exercise by [Selby] of [his] statutory rights under the Act of 1894."

With regard to the position of a purchaser of the land, the court felt that it should not be open to Whitbread to defeat Selby's rights simply by conveying their property to a third party. After noting that the London County Council was fully aware of his rights at the time of the purchase, it concluded that it took subject to them (and also, apparently, that it would still have done so, even without notice of them).

> McCardie J: "The deed of April 9 1915 [conveying the land] indicates the full knowledge possessed by the London County Council. I hold that they took the land in question subject to the statutory rights of [Selby] under the Act of 1894. I should have come to such conclusion apart from the fact that the London County Council possessed the amplest notice at all material times of such rights.
>
> The London County Council were, I think, subject to such rights, both actual and potential, as sprang from the building notice given by their predecessors in title. The extent of such rights had not been defined in April 1915; but postponement of definition does not involve either the destruction or the impairment of the plaintiff's statutory privileges."

Hence, the court rejected Whitbread's submission that the surveyors had no authority to make the addendum award. An adjoining owner's rights under the Act arise as soon as notice is served. A purchaser of the building owner's land (or the public, where this land is dedicated to them) takes subject to these rights. Where the rights are "potential" rather than "actual" a subsequent surveyors' award may be required to give them actual effect. The surveyors' award did precisely this and did not therefore purport to authorise the performance of an illegal act.

Surveyors' authority to make addendum awards

Whitbread also argued that the surveyors had no authority to make an addendum award in respect of the support of the party wall as the first award dealt with this exhaustively. The award was therefore invalid for this reason also.

Recall that clause 2 of the first award provided for such support to be provided "as may be ordered by the district surveyor or mutually agreed". Whitbread argued that, in the absence of mutual agreement, any further provision of support could therefore only be required by the district surveyor, rather than by the appointed surveyors. Hence, the power reserved in clause 7 of the award to make further awards in respect of "further matters" could only be used to deal with issues that had not been exhaustively dealt with by the first award.

The court did not accept this interpretation. The protection of the adjoining owner could not properly be left to the district surveyor whose primary function was to safeguard the interests of the public. The appointed surveyors (the 'arbitrators') had a different function.

> McCardie J: "But the primary function of the arbitrators is to safeguard the interests of the adjoining owner; although they must, of course, consider the rights and interests of the building owner and follow the provisions of the Act of 1894. I cannot think that clause 2 left it to the district surveyor to determine the whole of the works required for the protection of the [plaintiff]...
>
> ... It is the duty of the district surveyor to guard against immediate danger, but it is the duty of the arbitrators to direct all such measures as shall protect the adjoining owners, not only from immediate danger, but from future risks which may reasonably result from the operations of the building owner."

Despite this interpretation the court did not suggest that the surveyors had wrongfully delegated one of their own functions to the district surveyor. Whilst accepting that clause 2 was rather vague, it felt that surveyors' awards should generally be interpreted liberally and that this clause preserved the surveyors' right to make further awards. Indeed, the words of the statute obviously anticipated the making of more than one award.

> McCardie J: "The words 'from time to time' in section 91 of the Act clearly contemplate that several awards may be made by the arbitrators...

"... I agree that clause 2 is somewhat vague. But it must be remembered that the award was made in February 1915, and the demolition did not begin until March 1915. It is clear from [Whitbread's surveyor's] letter of January 7 1915 that he realised fully the impossibility of ascertaining the full extent of future preventive measures required until demolition was complete.

The measure of precision desirable in an award must vary with the circumstances of the case. It is important to maintain the rights of arbitrators under the Act of 1894 to make interim awards, and I think, to construe awards without pedantic strictness or meticulous severity."

The surveyors therefore had authority to make the addendum award.

The nature of a difference (dispute) under the Act

Whitbread argued that the addendum award was also invalid because no difference (dispute) had previously arisen between the parties as required by section 91 of the Act. There had simply been a disagreement between the appointed surveyors as to the correct course of action in respect of a matter raised subsequent to the making of their (first) award.

The court noted that there was no authority as to the meaning of "difference". However, after considering the conduct and communications between the two surveyors leading up to the making of the award, it had no doubt that a difference had certainly arisen in the present case.

> McCardie J: "Under these circumstances I am clearly of the opinion that a difference within section 91 had arisen between the plaintiffs and defendants prior to the date of the second award.
> I may add the observation that I do not think it would be in conformity with the scheme of the Act of 1894 to give too rigid or confined a meaning to the word 'difference' as used in such Act. Moreover, a difference is none the less a 'difference' because the divergence of view as to law or fact has been indicated by phrases of courtesy rather than the language of vehemence."

Appeals against surveyors' findings of fact

Whitbread also challenged the validity of the addendum award on the basis that the erection of a pier against the flank wall was not, as a matter of fact, technically necessary.

As this was a question of fact and it involved expert knowledge, the court felt that it was a matter that fell most appropriately within the surveyors' jurisdiction. Such decision could be challenged by an appeal to the county court under section 91. As no such appeal had been made, however, the award was now conclusive as to the necessity for the pier.

McCardie J summarised the legal principles governing challenges to the validity of an award as follows:

> "I agree that an award is in no way conclusive if it is in excess of the jurisdiction of the arbitrators: see *Re Stone and Hastie*. But if the matters dealt with are not beyond the powers of the arbitrators, then the award cannot be challenged on such a point as that now arising save by appeal as indicated in section 91."

Party wall legislation and the common law

Recall that, as well as seeking to enforce the addendum award, Selby had claimed damages at common law for the removal of support to the flank wall of his property.

The court held that he had acquired a prescriptive easement of support. However, this had been entirely replaced by the statutory rights which he was entitled to under the 1894 Act. It therefore rejected his claim for damages at common law.

The relevant part of McCardie J's judgment explains the rationale for this decision:

> "An examination of the code at once shows that common law rights are dealt with in a revolutionary manner. The two sets of rights, namely, the rights at common law and the rights under the Act of 1894, are quite inconsistent with one another. The plaintiff's common law rights are subject to the defendants' statutory rights. A new set of respective obligations has been introduced. The common law was seen to be insufficient for the adjustment of modern complex conditions.
>
> Hence I think that the Act of 1894 is not in addition to but in substitution for the common law with respect to matters which fall within the Act. It is a governing and exhaustive code, and the common law is by implication repealed. I follow the views on this point of Jessel MR in *Standard Bank of British South America v Stokes* (1878), Warrington J in *Lewis v Charing Cross Railway Co* (1906), and also the dictum of Collins MR in *Leadbetter v Marylebone Corporation* (1904).
>
> I therefore hold that the [plaintiff] cannot succeed upon [his] claim at common law inasmuch as the defendants' party wall notice had been duly given under the provisions of the Act of 1894.
>
> In so holding I in no way negative the proposition that a plaintiff may bring his action for damages if he can establish that the defendant has exerted his statutory privileges so as to inflict injury on the plaintiff by negligence, improper obstructiveness, avoidable nuisance, or unreasonable delay: see *Pratt v Hillman* (1825), *Dodd v Holme* (1834), *Bower v Peate* (1876), *Percival v Hughes* (1882). Many authorities on this point are collected in *Browne and Allan on Compensation*, 2nd edn, pp. 117–119."

London Building Act 1894

Part VIII: RIGHTS OF BUILDING AND ADJOINING OWNERS

91. Settlement of differences between building and adjoining owners

(1) In all cases (not specifically provided for by this Act) where a difference arises between a building owner and adjoining owner in respect of any matter arising with reference to any work to which any notice given under this Part of this Act relates unless both parties concur in the appointment of one surveyor they shall each appoint a surveyor and the two surveyors so appointed shall select a third surveyor and such one surveyor or three surveyors or any two of them shall settle any matter from time to time during the continuance of any work to which the notice relates in dispute between such building and adjoining owner with power by his or their award to determine the right to do and the time and manner of doing any work and generally any other matter arising out of or incidental to such difference but any time so appointed for doing any work shall not unless otherwise agreed commence until after the expiration of the period by this Part of this Act prescribed for the notice in the particular case.

94. Security to be given by building owner and adjoining owner

An adjoining owner may if he think fit by notice in writing require the building owner (before commencing any work which he may be authorised by this Part of this Act to execute) to give such security as may be agreed upon or in case of difference may be settled by the Judge of the County Court for the payment of all such expenses costs and compensation in respect of the work as may be payable by the building owner.

The building owner may if he thinks fit at any time after service on him of a party wall or party structure requisition by the adjoining owner and before beginning a work to which the requisition relates but not afterwards serve a counter requisition on the adjoining owner requiring him to give such security for payment of the expenses costs and compensation for which he is or will be liable as may be agreed upon or in case of difference may be settled as aforesaid.

If the adjoining owner do not within one month after service of that counter requisition give security accordingly he shall at the end of that month be deemed to have ceased to be entitled to compliance with his party wall or party structure requisition and the building owner may proceed as if no party wall or party structure requisition had been served on him by the adjoining owner.

Specific performance of award?

Having decided that the addendum award was valid, the court considered the nature of the remedy which should be awarded. It noted that in some cases an order for specific performance could appropriately be granted to carry out the terms of an award. Typically one might be made where an award directs, for example, the conveyance of land: *Hall v Hardy* (1733), *Wood v Griffith* (1818).

Where the relevant obligation is to execute repairs or erect houses, *prima facie* the court does not grant specific performance. It does however have a discretion to do so and may do so where the defendant cannot be adequately compensated in damages where the work to be done is clearly and particularly specified.

Nevertheless, in the present case, the court did not feel that an order for specific performance should be granted. It considered that Selby could be adequately compensated by damages and that the circumstances rendered such an order undesirable.

The decision

The court held that Whitbread remained liable as building owners, notwithstanding the transfer of the site to the London County Council. It rejected all Whitbread's submissions in respect of the addendum award and issued a declaration that the award was valid and binding. As Whitbread were in breach of the terms of the award, in failing to erect the pier, the court awarded damages to Selby in respect of this. Selby's claim for damages at common law was dismissed.

SIMS v THE ESTATES COMPANY (1866)
Court of Chancery

In brief

- **Issues:** 1855 Act – damage to dividing wall – dispute whether a party wall – dispute as to validity of party structure notice – building owner's refusal to withdraw notice – surveyors' jurisdiction to determine whether wall is a party wall – statutory time limit for continued effectiveness of party structure notice – whether ten-day notice can be included in party structure notice.

- **Facts:** The Company were redeveloping their site which adjoined premises leased to Sims. They wished to demolish and rebuild the dividing wall between the two properties and served a party structure notice on Sims to this effect. Sims disputed that the wall was a party wall and requested that the notice be withdrawn. The Company refused to withdraw it but made various promises about their intentions, which failed to reassure Sims. Sims issued proceedings in trespass, seeking an injunction to restrain the proposed work. The matter was settled before the hearing but the court had to decide whether the proceedings were justified before making an order for costs.

- **Decision:** Sims had been justified in bringing the proceedings and costs were awarded in their favour. The wall was not a party wall. Furthermore the court was the correct tribunal to decide this. Specifically, the appointed surveyors had no jurisdiction to decide such matters. The Company's statement that the notice would expire after three months was untrue. No time limit appeared in the 1855 Act. Their statement that they could not proceed without first serving a ten-day notice was also untrue. Their request in the party structure notice that Sims should appoint a surveyor operated as the ten-day notice. The Company were therefore now free to appoint a surveyor on Sims' behalf and to proceed with the works straight away. In the circumstances the Company's statement that they no longer had any intention of undertaking the works did not provide sufficient reassurance for Sims. They faced the threat of the work in the notice and were entitled to insist that this was removed.

- **Notes:** See also (on ten-day notices): *Frances Holland School v Wassef* (2001) – (on the expiry of party structure notices): *Leadbetter v Marylebone Corporation [No.2]* (1905) – (on the surveyors' jurisdiction to determine matters of law): *Crofts v Haldane* (1867); *Loost v Kremer* (1997).

The facts

The Estates Company owned a long leasehold interest in two adjoining buildings in Bartholomew Lane, London EC2. They were in the process of redeveloping one of them, which was known as the Auction Mart. The other was at 3 Bartholomew Lane and they had sublet the ground floor of this building to Sims, a firm of stockbrokers, for a 12-year term.

At the time the case began, the Company had already demolished the Auction Mart and were nearing the completion of a new building on its site. During the course of the works some damage had been caused to the dividing wall between the two properties.

The Company now wished to install a staircase at the boundary between the two properties. This would have required the complete rebuilding of the dividing wall. Sims were unwilling to consent to the work so the Company served a party structure notice on them, under the 1855 Act, to demolish and rebuild the wall. The notice advised Sims that the Company had already appointed a surveyor and requested them to do so as well.

Sims disputed that the wall was a party wall and maintained that the Company had no right to carry out the work. They therefore wrote to them requesting that they withdraw the notice.

They refused to do so and justified this on three grounds. They claimed that the notice would expire in three months in any event and there was therefore no need to withdraw it. They also maintained that they no longer had any intention of proceeding on the basis of the notice. Finally, they claimed that, even if they did wish to proceed, they would have to serve another notice before doing so[1].

Proceedings for trespass

Sims were not reassured by this response and issued proceedings against the Company in trespass. They claimed damages for the damage which had already been caused to their wall and an injunction to restrain the work which was referred to in the party structure notice.

The published law report does not address the court's consideration of the substantive issues. It deals instead with the liability for costs of the suit which depended on the extent to which Sims had good reason for bringing the action in the first place.

---

[1] This appears to be a reference to the requirement to serve a ten-day request on a recalcitrant adjoining owner to appoint a surveyor before the building owner is free to proceed by appointing a surveyor on his behalf. The relevant provision was contained in section 85(9) of the 1855 Act and now appears as section 10(4)(b) of the 1996 Act.

Sims argued that it was necessary for them to bring the case because of the Company's refusal to withdraw the notice and the consequent risk that they would proceed with the work.

The Company argued that the proceedings were unnecessary. Firstly, Sims should have raised their objections to the notice with the appointed surveyors. They were the proper forum for deciding such matters rather than the courts. Secondly, the Company's response to Sims' request to withdraw the notice should have provided them with adequate reassurance even though the notice had not been expressly withdrawn.

Jurisdiction to determine status of party wall

The court rejected the Company's first submission. The courts were the proper forum to determine whether a wall was a party wall. The surveyors had no jurisdiction to decide such matters.

> The Vice Chancellor: "There was nothing in the Metropolitan Building Act to oust the jurisdiction of the Court of Chancery with regard to a structure which did not come within the Act.
>
> The Act did not say that the arbitrators were to decide whether the structure was a party wall or not. It did not prevent the court from determining whether a person's property was or was not *dehors* the Act."

Adequacy of protection from invalid notice

It then considered whether the Company had provided adequate reassurance to Sims to render the action unnecessary. It considered in turn each of their three justifications for refusing to withdraw the notice.

It described the statement that the notice would expire in three months as "most extraordinary" as this was simply not the case. Although subsequent legislation has included provisions that notices expire if work is not commenced within a certain time limit[2], no such provision was included within the 1855 Act. The three-month period referred to in section 85(1) simply related to the minimum notice period that must be given to an adjoining owner before notifiable works could be commenced.

It also found that the Company's statement that they had no intention of proceeding on the notice provided insufficient reassurance.

---

[2] Section 3(2)(b)(i) of the 1996 Act provides that a party structure notice will cease to have effect unless work has been commenced within 12 months of the date of service. See however, *Leadbetter v Marylebone Corporation [No. 2]* (1905).

> **Metropolitan Building Act 1855**
>
> PART III: PARTY STRUCTURES
>
> RIGHTS OF BUILDING AND ADJOINING OWNERS
>
> **85. Rules as to exercise of rights by building and adjoining owners**
>
> The following rules shall be observed with respect to the exercise by building owners and adjoining owners of their respective rights:
>
> (1) No building owner shall, except with the consent of the adjoining owner, or in cases where any party structure is dangerous, in which cases the provisions hereby made as to dangerous structures shall apply, exercise any right hereby given in respect of any party structure, unless he has given at the least three months previous notice to the adjoining owner by delivering the same to him personally, or by sending it by post in a registered letter addressed to such owner at his last known place of abode.
>
> (9) If either party to the difference makes default in appointing a surveyor for ten days after notice has been given to him by the other party, in manner aforesaid, to make such appointment, the party giving notice may make the appointment in place of the party so making default.

The Vice Chancellor: "It was true that they said they would not exercise the power which they would possess at the expiration of the three months without notice to the defendants.

But if the parties said, 'We will arm ourselves with the power given by this Act, and you must trust us to give you notice, but we will not withdraw our notice' (and at the end of the three months they would be able to pull down the wall at one month's notice), that was a promise which a party was not bound to rely on."

Finally, the court also rejected the Company's argument that they were not free to proceed on the notice without first serving a further notice. It appears to have taken the request to appoint a surveyor in the party structure notice as a ten-day request[3] which now entitled the Company to proceed without further notice to Sims[4]. The presence of this request

---

[3] Under section 85(9) of the Act. See also section 10(4)(b) of the 1996 Act.

[4] Most party wall surveyors today to not consider that a request under what is now section 10(4)(b) can properly be made until a dispute has actually arisen.

in the notice appears to have been particularly damaging to the Company's case:

> The Vice Chancellor: "If a further notice had been necessary before another surveyor could have been appointed by the defendants I would have been inclined to think that the application to the court had been somewhat too hasty.
>
> But that was not so, as appeared by the words at the back of the notice, and the defendants would be able to take immediate proceedings at the expiration of the three months."

The decision

The court decided, on the facts, that the dividing wall between the two properties was not a party wall. The Company therefore had no right to undertake the work under the Act and any attempt to do so would have constituted a trespass. The existence of the notice under the Act, threatening the works was a sufficient basis for Sims to bring proceedings. As the Company had, without adequate grounds, refused to withdraw the notice, Sims were justified in bringing the present proceedings against them. An order for costs was therefore made against the Company.

SOLOMONS v R. GERTZENSTEIN LTD (1954)
Court of Appeal

In brief

- **Issues:** 1939 Act – statutory interpretation – definition of "owner" – whether a receiver is an owner – breach of statutory duty.

- **Facts:** Gertzenstein owned the reversionary interest in a four-storey building which was let to tenants. Their duties as landlord were being undertaken by a receiver who had been appointed by a mortgagee following a default on their mortgage. The building's means of escape in case of fire had not been adequately maintained in accordance with provisions in the 1939 Act. As a result Solomons was injured in a fire. At first instance both Gertzenstein and the receiver were found liable for his injuries in the tort of breach of statutory duty. The receiver appealed.

- **Decision:** The receiver was not liable. A breach of the Act's obligations was capable of leading to civil, as well as criminal, liability. However, a receiver was not an "owner" within the meaning of the Act. The 1939 Act contained two definitions of "owner". A specific definition applied to Part V which addressed the *installation* of means of escape provision. This definition included a reference to agents and trustees and this was wide enough to include a receiver. However, the Act's general definition applied to the *maintenance* of those facilities which was the subject of the present case. That definition was based on interests in the land and could not therefore include a receiver.

- **Notes:** (1) The decision that a receiver is not an "owner" is unlikely to have been changed by the new definition in section 20 of the 1996 Act. The separation of the requirement for "possession" from a person in receipt of rent might suggest an intention to include receivers and others who act in an agency capacity. However, the specific exclusion of persons in possession as mortgagee contradicts this, at least in the context of mortgages. If a mortgagee in possession is excluded from the definition it seems unlikely that Parliament intended to bring receivers acting for mortgagees within it. (2) See also: *Cowen v Phillips* (1863); *Fillingham v Wood* (1891); *List v Tharp* (1897); *Orf v Payton* (1904); *Crosby v Alhambra Co Ltd* (1907); *Spiers & Son v Troup* (1915); *Lehmann v Herman* (1993); *Frances Holland School v Wassef* (2001).

The facts

The case concerned a four-storey building at 36 Gerrard Street, London W1. H&S Developments Ltd owned a long leasehold interest in the building. They had sublet the various floors to a number of business tenants. Gertzenstein Ltd were the tenants of the first floor. Solomons was an employee of the second floor tenant.

H&S retained control of the entrance, staircases and passages in the property. In 1925 the District Surveyor had required them to install adequate means of escape in case of fire. They undertook the work and this was approved by the District Surveyor. The work had included the installation of a trap door leading to the roof of the building, and the provision of a ladder with hooks which could be used to provide access to it.

H&S subsequently defaulted on their mortgage and the mortgagee therefore appointed a receiver. The receiver collected the rents from the tenants and also undertook certain management duties in respect of the common parts. These included the cleaning and lighting of the common parts in compliance with H&S's covenants with its tenants.

Failure to maintain means of escape in case of fire

One evening a fire started in Gertzenstein's first floor premises and this rapidly spread upwards. Solomons was working late on the second floor of the building and became trapped by the fire. He was unable to locate the ladder to the trap door as someone had placed this, out of view, in a passage behind a door which led to the ladies' lavatory.

He therefore escaped through a window and tried to climb down a rainwater pipe. He fell and broke his arm as it became caught in a junction of the pipe. He remained trapped and dangling from his broken arm for a considerable period until he was eventually rescued by the fire brigade.

Solomons therefore issued proceedings against Gertzenstein, H&S and the receiver for damages for his personal injuries. He claimed that Gertzenstein were liable in negligence for allowing the fire to start in the first place. His claim against H&S and the receiver was based on their breach of statutory duty. He maintained that, as the owners of the premises, they were liable to him for their failure to maintain the fire escape, in breach of section 133(2) of the 1939 Act.

The court at first instance held that the fire had started accidentally so Gertzenstein were not liable in negligence. However, both H&S and the receiver were indeed "owners" within the meaning of section 133(2). They were therefore in breach of the obligation to maintain the means of escape. Although the 1939 Act imposed criminal penalties for breach of its obligations Solomons was also entitled to recover civil damages for the tort of breach of statutory duty.

Is a receiver an "owner"?

The receiver appealed against this decision to the Court of Appeal. He argued that a receiver was not generally an "owner" for the purposes of the 1939 Act. Even if this was the case, the obligation to maintain the fire escape in section 133(2) imposed a criminal sanction. He argued that there was therefore no intention that it should be used as a basis for civil liability.

The court considered the meaning of "owner" for the purposes of the obligation contained in section 133(2). Two definitions were potentially relevant.

Firstly, a general definition in section 5 of the 1930 Act applied to the whole of the 1939 Act, unless the context required some other meaning[1]. This section defined "owner" in terms of persons in possession or in receipt of rents or profits from land otherwise than as a tenant from year to year or for any lesser term, or as a tenant at will.

However, section 33 of the 1939 Act contained a further definition. This related specifically to the "Means of Escape in Case of Fire" provisions in Part V of the statute. This section defined "owner" in terms of the person receiving the rack rent from premises whether on their own account or as agent or trustee for some other person.

The Court of Appeal was satisfied that the general definition in section 5 of the 1930 Act did not include a receiver or other agent. Section 33 made specific reference to an agent or trustee but no such reference appeared in section 5. This led the court to conclude that Parliament must have intended to exclude these roles from the definition in section 5.

> Romer LJ: "This language, if taken alone, might be wide enough to include an agent, for the expressions possession, receipt and occupation are not qualified by the word 'beneficial'.
>
> Nevertheless, it is clear to my mind that this definition, in its application to section 133 of the Act of 1939, does not comprehend a person who receives rent merely as an agent; for when there are two definitions of the same word applicable exclusively to different parts of the Act it would be against all canons of construction to introduce into one a characteristic which is only to be found in the other."

In contrast, the court had no doubt that the definition in section 33, complete with its reference to agent and trustee, applied to a receiver appointed by a mortgagee.

---

[1] 1939 Act, sections 1 and 33(1).

Which definition applies?

The section 33 definition applied to Part V of the Act which dealt with the provision of means of escape in case of fire. The question for the court was whether this definition could also be used in relation to the ongoing obligation to keep and maintain the means of escape in case of fire contained in section 133(2).

On the one hand, the subject matter of Part V and section 133(2) were substantially the same so it was logical that the same definition of "owner" should be applied to each. On the other hand, section 133(2) was contained in Part XII of the Act rather than Part V. The section 33 definition was expressed to relate only to Part V and made no reference to Part XII at all.

The court held that the latter, more literal, interpretation was correct. It rejected the suggestion that the section 33 definition could automatically be applied to Part XII after finding evidence of Parliament's contrary intention in section 141 of the statute.

That section, which was also contained within Part XII of the Act, made provision for an owner to gain access to property to discharge his obligations under Part V and section 133. Significantly, "owner" was specifically defined in the section by reference to the section 33 definition. As no comparable provision appeared in section 133 the court was satisfied that Parliament had not intended the section 33 definition to apply to it.

> Birkett LJ: "Much therefore turns on this question: Does the definition of 'owner' in section 33 of Part V apply to the 'owner' in section 133(2) of Part XII?
>
> It is true that section 133 begins with the words which are used for the heading of Part V, 'Means of escape in case of fire', and it is very natural to think that when the section goes on to impose on 'the owner of the building' the duty to maintain the means of escape in case of fire 'in good condition and repair and in efficient working order' that the 'owner' on whom the duty is laid to provide the means of escape in case of fire should be the same 'owner' upon whom the duty is laid of maintaining the means of escape.
>
> But a consideration of section 141 of the Act shows that the 'owner' for the purposes of Part V is not the same 'owner' for the purposes of section 133(2) in Part XII. Section 141(1)(a), which gives power to the 'owner' to enter the building for all the purposes of Part V, including, of course, the purpose of providing the means of escape in case of fire, also gives power to enter for the purpose of maintaining the means of escape as laid down in section 133. But by section 141(3) it is provided: 'In this section the expression "owner" in relation to any requirement in virtue of any of the provisions of Part V (Means of Escape in Case of Fire) of this Act has the same meaning as in that Part of this Act.'

This would seem to indicate that 'owner' in section 133(2) was not intended to have the same meaning as in section 33, Part V of the Act, and the definition of 'owner' for the purposes of section 133 must be found elsewhere."

The relevant definition of "owner" for the purposes of section 133 was therefore the general definition employed in the London Building Acts which was to be found in section 5 of the 1930 Act. As already noted, the definition in section 5 did not include a receiver. The receiver's appeal therefore succeeded as he was not an "owner" who was required to maintain the fire escape within section 133(2).

Civil liability for failure to maintain means of escape in case of fire?

In view of the court's decision on the definition of "owner" the receiver's second submission that there could be no civil liability for breach of the Act's provisions was not relevant to the outcome of the case.

Nevertheless, the court also considered this issue. It decided (Somerville LJ dissenting) that a breach of section 133 would give rise to civil, as well as criminal, liability.

Solomons was a member of the class of persons who the provision was intended to safeguard. As such, notwithstanding the criminal sanctions in the Act, a civil action would have been available to him, against persons properly falling within the definition of owner, for breach of statutory duty.

The decision

The court allowed the appeal. It held that the receiver was not an owner within the context of section 133. He was not, therefore, subject to the obligation to maintain the fire escape contained in that section.

London Building Act 1930

PART I: INTRODUCTORY

5. Definitions

In this Act, save as is otherwise expressly provided therein and unless the context otherwise requires, the following expressions have the meanings hereby respectively assigned to them (that is to say):

"owner" includes every person in possession or receipt of the whole or of any part of the rents or profits of any land or tenement, or in the occupation of any land or tenement, otherwise than as a tenant from year to year, or for any less term, or as a tenant at will.

London Building Acts (Amendment) Act 1939

PART I: INTRODUCTORY

1. Short title construction and citation

This Act may be cited as the London Building Acts (Amendment) Act 1939 and shall be read and construed as one with the London Building Acts 1930 and 1935 and may be cited with those Acts as the London Building Acts 1930 to 1939.

PART V: MEANS OF ESCAPE IN CASE OF FIRE

33. Interpretation of Part v

(1) In this Part of this Act, unless the context otherwise requires the following expressions have the meanings hereby respectively assigned to them:

"owner" in relation to any premises means the person for the time being receiving the rackrent of the premises whether on his own account or as agent or trustee for any other person or who would so receive it if the premises were let at a rackrent.

PART XII: MISCELLANEOUS

133. Maintenance of means of escape etc.

(2) All means of escape in case of fire and all safeguards to prevent the spread of fire and any arrangements in connection therewith provided in pursuance of the provisions of Part V (Means of escape in case of fire) of this Act or otherwise shall be kept and maintained in good condition and repair and in efficient working order by the owner of the building and no person shall do or permit or suffer to be done anything to impair the efficiency of any such means of escape safeguards or arrangements...

141. Power of owner and others to enter premises and execute work

(1) Notwithstanding any provision contained in any lease or contract it shall be lawful:

(a) for the owner of a building or structure to enter the building or structure or the curtilage thereof for the purpose of carrying out any work or providing any safeguard required to be carried out or provided by him in virtue of any of the provisions of Part V (Means of escape in case of fire) of this Act or for the purpose of maintaining in pursuance of section 133 (maintenance of means of escape etc.) of this Act all means of escape in case of fire and all safeguards for lessening danger from fire and any arrangements in connection therewith.

(3) In this section the expression "owner" in relation to any requirement in virtue of any of the provisions of Part V (Means of escape in case of fire) of this Act has the same meaning as in that Part of this Act.

SOUTHWARK & VAUXHALL WATER COMPANY v WANDSWORTH DISTRICT BOARD OF WORKS (1898)

Court of Appeal

In brief

- **Issues:** Building owner exercising lawful right – detrimental effects on neighbour – whether building owner liable in the absence of negligence – highway authority exercising statutory powers – building owner demolishing his building where adjoining owner has no easement of support – whether neighbour has a right to redress.

- **Facts:** The Highway Authority exercised its statutory powers to lower the surface of the street. This left the Water Company's pipes without sufficient cover to protect them from frost. The Water Company sought an injunction to prevent the works without the Highway Authority also exercising its power to reposition the pipes to maintain frost protection.

- **Decision:** The Highway Authority had a lawful right to lower the surface of the street. Their duty to the Water Company was simply to take reasonable care in undertaking this work. In the absence of negligence they had no further liability, notwithstanding the fact that the works, by their very nature, would have a detrimental effect on the Water Company. The situation was identical to the right of a building owner to demolish his building. In the absence of an easement of support in favour of his neighbour's building he was free to demolish. This remained the case, even where his neighbour's building would inevitably collapse as a consequence. He would be subject to a duty not to cause unnecessary damage during the course of the demolition but this could not deprive him of his lawful right to demolish.

- **Notes:** (1) See also *Dodd v Holme* (1834). (2) The clear statement of law in the present case has been complicated by the development of the principle that an occupier now owes a "general" (or "measured") duty of care to his neighbour in respect of hazards on his own land. See: *Leakey v National Trust for Places of Historic Interest or Natural Beauty* (1980); *Bradburn v Lindsay* (1983); *Holbeck Hall Hotel Ltd v Scarborough Borough Council* (2000); *Rees v Skerrett* (2001).

The facts

The Southwark and Vauxhall Water Company had laid some pipes under the surface of a street in pursuance of its statutory powers to do so. It had laid them to a depth of between 1 ft 4 ins and 1 ft 10 ins.

Subsequently the Highway Authority (the Wandsworth District Board of Works) lowered the surface of the street as part of works to remove unevenness in the surface of the highway. Their power to do this was contained in section 98 of the Metropolis Management Act 1855. The same section also included a power to alter the depth of existing pipes.

Despite this power they chose not to alter the depth of the Water Company's pipes. This resulted in the depth of cover being reduced to only 2 to 3 ins. As a consequence they were exposed to risk of frost.

The Water Company sought an injunction against the lowering of the street level without the simultaneous lowering of their pipes so that adequate protection from frost would be maintained. They argued that, in view of the statutory power to alter the depth of pipes, the Highway Authority were subject to a common law duty to do so in circumstances where the exercise of their other powers would cause injury to their neighbours.

Duty to neighbours when undertaking lawful works

An injunction was granted at first instance, whereupon the Highway Authority appealed to the Court of Appeal. The Court of Appeal held that the common law imposed no such duty. Its decision is summarised by Chitty LJ:

> "The effect of their works, which have been lawfully executed under their principal power, is to bring the surface of the street nearer to the pipes which remain *in situ*. The consequence is that the water in the pipes being but a few inches from the surface is more exposed to frost.
>
> In exercising the power the road authority has not been guilty of any negligence. I am unable to find in the section any express or implied duty cast upon the road authority, when they exercise their power of altering the level of the road, whether by raising or lowering it, to exercise at their own expense their power of altering the position of the pipes for the benefit of the company owning the pipes, much less any duty to place the pipes at a depth below the new surface corresponding with the depth at which they stood below the old surface. I think that no such duty is imposed upon the appellants.
>
> The real question is on whom the expense of altering the position of the pipes is to fall. It appears to me that it falls on the company."

Demolition of building where no easement of support

The judgment of Collins LJ is of particular interest in the context of party walls and neighbourly matters. In his judgment he makes reference to the earlier case of *Chadwick v Trower* (1839). That case concerned the demolition of a building which was providing support for a neighbouring property in circumstances where no easement of support existed.

As with the present case it addressed a situation where one party exercises a lawful right and this has some detrimental effect on a neighbour. In the present case, the Highway Authority had the right to lower the road surface notwithstanding the impact on the Water Company. In the earlier case the demolishing owner was entitled to demolish his property despite the resulting damage to the neighbouring property. In each case the duty imposed on the party exercising their right is to do no more than to take reasonable care in the exercise of their right.

Thus, where no right of support exists in favour of an adjoining building, an owner is free to demolish their own building, as long as they exercise all reasonable care, notwithstanding the fact that this causes the collapse of their neighbour's building.

> Collins LJ: "Here the plaintiffs have no right to any particular thickness of soil above their pipes ... I think they can have no higher claim to consideration at the hands of the defendants than the owner of a house, for which no right of support from the adjoining house had been acquired, would be entitled to claim against the adjoining owner, who in pulling his own house down withdraws support to which his neighbour is not entitled.
>
> I think it is clear in such case that, though the pulling-down owner must be careful to interfere as little as possible with the adjoining house, he is certainly not called upon to take active steps for its protection, as, for instance, by shoring it up. There is a broad distinction between exercising a right with reasonable care so as not to do avoidable damage, and taking active measures to insure the continuance of something that is not a right in the adjoining owner.
>
> *Chadwick v Trower* ... merely decides that, supposing there is a duty upon a person pulling down his own house to take care not to injure his neighbour's vault in so doing, where he knows of its existence, though it has acquired no right of support, there can be no such duty where he does not know of it. And it cannot be the law that the pulling-down owner is bound to find a substitute or equivalent for the support which he has a right to remove ...
>
> I think the result is that though the person pulling down is bound to do no unnecessary damage, he is not fixed with any obligation to take active steps to mitigate a mischief which follows inevitably upon the reasonable exercise of his own rights."

The decision

The Court of Appeal reversed the decision at first instance and entered judgment for the Highway Authority.

SPIERS & SON LTD v TROUP (1915)
King's Bench Division

In brief

- **Issues:** 1894 Act – service of dangerous structure notice – emergency demolition of party wall – rebuilding of party wall – party structure notice – required contents – validity of service – definition of "owner" – position of prospective purchaser – validity of party structure notice – adjoining owner's liability to contribute towards cost of works – effect of failure to serve account on adjoining owner within statutory time limit.

- **Facts:** Spiers was the prospective purchaser of a property which he intended to develop. Before entering into a contract he served a party structure notice on Troup who owned the adjoining property. It referred to the possibility of works to the party wall if these proved to be necessary following opening up works. The appointed surveyors failed to agree on an award. In the meantime the District Surveyor served a dangerous structure notice which required the party wall to be demolished. Spiers demolished the party wall and later rebuilt it as part of his new development. Over a month later he sent an account for half the expenses of demolishing and rebuilding the wall to Troup under the party wall provisions of the 1894 Act. Troup refused to pay.

- **Decision:** The party wall procedures were invalid because of procedural irregularities. A prospective purchaser was not an owner within the Act and, due to its provisional nature, the party structure notice had not contained the requisite information. Furthermore, time was of the essence for the service of an account under the Act. As Spiers had served his account outside the one-month time limit it was invalid. The procedural irregularities meant that Spiers was unable to recover any monies under the Act's party wall provisions. However, works in pursuance of a dangerous structure notice were exempt from the requirement to serve a party structure notice. These works had therefore been legitimately carried out notwithstanding the defects in content and service of the party structure notice. There was a common law right to recover monies compulsorily expended jointly on behalf of another. Spiers was therefore entitled to recover monies from Troup despite the failure to serve the account within the statutory time limit. This was restricted to the demolition works which were necessary to comply with the dangerous structure notice. Spiers was therefore entitled to recover the contribution towards the cost of these works but not those in respect of the rebuilding works.

- **Notes:** See also (on the required contents of a party structure notice): *Hobbs, Hart & Co v Grover* (1899) – (on service of an account for contribution towards the cost of works): *Reading v Barnard* (1827); *J. Jarvis & Sons Ltd v Baker* (1956) – (on the definition of "owner"): *Cowen v Phillips* (1863); *Fillingham v Wood* (1891); *List v Tharp* (1897); *Orf v Payton* (1904); *Crosby v Alhambra Co Ltd* (1907); *Solomons v Gertzenstein* (1954); *Lehmann v Herman* (1993) – (on the effect of procedural irregularities): *Whitefleet Properties Ltd v St Pancras Building Society* (1956); *Gyle-Thompson v Wall Street (Properties) Ltd* (1974); *Frances Holland School v Wassef* (2001).

The facts

Harris owned the freehold of numbers 36 and 37 Hatton Gardens, London EC1. The properties comprised two terraced houses, separated by a party wall. He had granted a long lease of number 36 to Troup and now wished to redevelop number 37.

Spiers and Son Ltd were the contractors who Harris wished to undertake the redevelopment works at number 37. Whilst he was still negotiating with them over the terms of a building agreement they served a party structure notice on Troup, under the 1894 Act, in respect of the party wall between the two properties. This was served in March 1912 and provided Troup with notice of their intention:

> "To pull down and rebuild such party structure if on survey it be found so far defective or out of repair as to make such operation necessary or desirable, and to perform all other necessary works incidental thereto."

A dispute was deemed to arise and, by the end of April 1912, both parties had appointed surveyors under the Act.

In May 1912 Spiers eventually entered into a building agreement with Harris under which they were to erect new buildings on the site of number 37. Once completed, Harris would grant a 99-year lease of the property to them. Pending the grant of this lease nothing in the agreement was to be construed as a lease of the premises.

Response to dangerous structure notice

Spiers then began removing the roof of number 37. This exposed the party wall which was found to be in a dangerous and dilapidated state. Negotiations between the appointed surveyors had stalled so the District Surveyor served a dangerous structure notice on both parties. Neither party complied with this so the magistrates made an order directing the demolition of the wall.

Spiers then demolished the wall and rebuilt it to an improved specification to accommodate the new buildings proposed for the site. In particular, the rebuilt wall was higher and thicker than the previous one. The rebuilding of the new wall was completed on 24 February 1913.

On 28 March 1913 Spiers sent an account for half the expenses of demolishing and rebuilding the wall to Troup. Troup refused to pay so Spiers issued proceedings against him to recover the amount due. In his defence, Troup alleged that both the party structure notice and the account for expenses were invalid due to various procedural irregularities under the 1894 Act.

Validity of party structure notice

Firstly, at the time that Spiers served the party structure notice in March 1912, they were not an owner within the terms of the Act. The party structure notice must therefore be invalid.

Section 5(29) defined "owner" in terms of either possession or occupation of land. There was authority that an occupier under a building agreement fell within this definition[1] providing there was no clause in the agreement to the effect that occupation was on the basis of a tenancy at will[2].

However, the court found that Spiers did not enter into occupation under the building agreement until May 1912. At the time that they served the party structure notice they had not signed the agreement, nor were they in occupation. The notice was therefore held to be invalid on this basis.

Secondly, Troup submitted that the party structure notice must also be invalid because of the provisional nature of the work described in it. It had been decided in *Hobbs, Hart & Co v Grover* (1898) that a party structure notice must be sufficiently clear and intelligible to enable the adjoining owner to see what counter notice he should give to the building owner. The notice which Spiers had served lacked the requisite clarity.

The court agreed with this and declared the notice invalid on this basis also. Scrutton J explained that:

> "The notice in this case was very vague and hypothetical ... There is nothing about a higher wall or a thicker wall, and what is to happen is made contingent on what is found when the wall is opened up. I cannot think that such a notice complies with the requirements of the Act, as explained by the Court of Appeal in the case referred to, and I think the notice was bad on this ground also."

---

[1] *List v Tharp* (1897).
[2] *Orf v Payton* (1905).

Validity of account for expenses

Finally, Troup submitted that the account for expenses was invalid as it had been submitted out of time. Section 96 of the Act required this to be delivered within one month after completion of the work[3]. As the work was completed on 24 February but the account was not delivered until 28 March this was outside the statutory time limit.

The court held that time was of the essence for the purpose of this provision and that the account was therefore invalid. Scrutton J explained the reasons for this:

> "It was ... argued that the requirement of section 96 as to one month was only directory, and not a condition precedent. But in my view the object of the provision was that the claim should be made promptly when the facts were fresh in everyone's memory, and that time was of the essence of the proceeding. The claim would fail, therefore in this respect also."

The court had therefore upheld all Troup's submissions about the invalidity of the procedures under the 1894 Act. It noted that, in normal circumstances, these would have prevented Spiers from recovering any of the expenses they had incurred.

Recovery of demolition costs

In fact, in the present case, the three procedural irregularities did not entirely defeat Spiers' claim because at least some of the works had been undertaken in pursuance of a dangerous structure notice.

Section 90 of the 1894 Act waived the requirement to serve a party structure notice when exercising rights under the Act where these were necessary due to the dangerous condition of the structure[4]. In the context of the demolition works, the invalidity of Spiers' party structure notice was therefore irrelevant, providing they had undertaken the works in their capacity as a building owner.

By the time they started the demolition works the building agreement was in force and they were in occupation pursuant to that agreement. At that point they were therefore a building owner within the meaning of the Act[5] and, being exempt from the requirement to serve a party structure notice, were free to proceed with the work under the Act.

---

[3] An equivalent provision appears in section 13 of the 1996 Act although the time limit has been extended to two months.
[4] An equivalent provision now appears in section 3 (3)(b) of the 1996 Act.
[5] See *List v Tharp* above.

The court also held that they were entitled to recover one half of the cost of the demolition works, notwithstanding that the account had been served out of time. This was because the demolition works had been carried out under Part IX of the Act, rather than Part VIII to which the strict time limit in section 96 applied.

After considering section 173, the court was of the view that the Act clearly "contemplated" contributions being made between building and adjoining owners where works were carried out under Part IX.

It found evidence of this in *Debenham v Metropolitan Board of Works* (1880) where the local authority had undertaken the works required by a magistrates' order due to the default of the building and adjoining owners. In that case the local authority had successfully recovered against both owners. Further evidence was to be found in *Hunt v Harris* (1865). In that case a building owner had recovered the expenses of demolishing and rebuilding a party wall from an adjoining owner where the wall had previously been condemned as dangerous.

The legal basis for the obligation to contribute lay, not in the Act, but at common law. This was explained by Scrutton J:

> "The claim is in respect of pulling down ... is for the expense of the work which each owner has been ordered to do under Part IX of the Act, and which one has done. This appears to me to come within the class of cases where, from the compulsory payment by one man of a debt for which he as well as another was liable, an implied request to pay on behalf of the other has been inferred."

Recovery of rebuilding costs

However, Spiers had done more than simply demolish the party wall, as required by the dangerous structure notice. They had also rebuilt it and in so doing had increased both its height and its thickness. This fell outside the waiver in section 90 and Spiers could not therefore recover a contribution towards the cost of this part of the work. Scrutton J noted:

> "The plaintiffs, in building a new party wall higher and thicker than the old one without any legal compulsion, must have been acting either under the powers of section 88(2)[6] (as to rebuilding); section 88(6)[7] (as to raising), or section 88(7)[8] (as to thickening), or without any legal authority at all.
>
> In the latter case they certainly cannot recover any share of the expenses of rebuilding from the adjoining owner. In the former case they can only do so if

---

[6] Section 2(2)(b) of the 1996 Act.
[7] Section 2(2)(a) of the 1996 Act.
[8] Section 2(2)(a) of the 1996 Act.

they comply with the provisions of Part VIII of the Act as to notice, arbitration, and expenses, which they have failed to do in the three ... respects above mentioned, as well as not obtaining an award of surveyors.

In my view the exception in section 90, as to dangerous party structures, only extends to work ordered to be done to the party wall under Part IX of the Act, and does not give the building owner a free hand in rebuilding and recovering expenses without regard to the restrictions of Part VIII of the Act. I hold, therefore, that the plaintiffs' claim for a proportion of the expenses of rebuilding fails."

The decision

The court made a declaration that Spiers were entitled to recover half the costs of demolishing the party wall but that they could recover nothing towards the costs of rebuilding it.

London Building Act 1894

PART I: INTRODUCTORY

5. Definitions

In this Act unless the context otherwise requires:

(29) The expression "owner" shall apply to every person in possession or receipt either of the whole or of any part of the rents or profits of any land or tenement or in the occupation of any land or tenement otherwise than as a tenant from year to year or for any less term or as a tenant at will.

PART VIII: RIGHTS OF BUILDING & ADJOINING OWNERS

88. Rights of building owner

The building owner shall have the following rights in relation to party structures (that is to say):

(2) A right to pull down and rebuild any party structure which is so far defective or out of repair as to make it necessary or desirable to pull it down.

(6) A right to raise and underpin any party structure permitted by this Act to be raised or underpinned or any external wall built against such party structure upon condition of making good all damage occasioned thereby to the adjoining premises or to the internal finishings and decorations thereof and of carrying up to the requisite height all flues and chimney stacks belonging to the adjoining owner on or against such party structure or external wall.

(7) A right to pull down any party structure which is of insufficient strength for any building intended to be built and to rebuild the same of sufficient strength for the above purpose upon condition of making good all damage occasioned thereby to the adjoining premises or to the internal finishings and decorations thereof.

90. Rules as to exercise of rights by building and adjoining owners

(1) A building owner shall not except with the consent in writing of the adjoining owner and of the adjoining occupiers or in cases where any wall or party structure is dangerous (in which cases the provisions of Part IX of this Act shall apply) exercise any of his rights under this Act in respect of any party fence wall unless at least one month or exercise any of his rights under this Act in relation to any party wall or party structure other than a party fence wall unless at least two months before doing so he has served on the adjoining owner a party wall or party structure notice stating the nature and particulars of the proposed work and the time at which the work is proposed to be commenced.

96. Account of expenses to be delivered to adjoining owner

Within one month after the completion of any work which a building owner is by this Part of this Act authorised or required to execute and the expense of which is in whole or in part to be borne by an adjoining owner the building owner shall deliver to the adjoining owner an account in writing of the particulars and expense of the work specifying any deduction to which such adjoining owner or other person may be entitled in respect of old materials or in other respects and every such work shall be estimated and valued at fair average rates and prices according to the nature of the work and the locality and the market price of materials and labour at the time.

PART IX: DANGEROUS & NEGLECTED STRUCTURES

PART XV: LEGAL PROCEEDINGS

173. Payment of expenses by owners

Where it is by any provision of this Act declared that expenses are to be borne by or may be recovered from the owner of any premises (including under the term "owner" the adjoining and building owners respectively) the following rules shall be observed with respect to the payment of those expenses...

STANDARD BANK OF BRITISH SOUTH AMERICA v STOKES (1878)

Chancery Division

In brief

- **Issues:** 1855 Act – existence of party wall – entitlement to underpin party wall – common law rights – statutory rights – relationship between common law rights and statutory rights – significance of statutory procedural requirements – approach to statutory interpretation.

- **Facts:** Stokes wished to underpin the wall between his building and the adjoining building owned by Standard Bank. He served a party structure notice on them but started work before the appointed surveyors had made an award.

- **Decision:** As there was common user of the wall it was presumed to be a party wall owned by the parties as tenants in common. Stokes would previously have had a common law right to underpin the wall but all common law rights had now been replaced by rights granted by the 1855 Act. The Act granted no express right to underpin a party wall but the court held that one should be implied from the wording of the statute. This right was subject to compliance with the Act's procedural requirements. Stokes therefore had no right to proceed with the work until the appointed surveyors had authorised this by an award.

- **Notes:** (1) See also: *Cubitt v Porter* (1828); *Selby v Whitbread & Co* (1917); *Lewis & Solome v Charing Cross Railway Co* (1906); *Bower v Peate* (1876). (2) An express right to underpin a party wall was included in the 1894 Act and has appeared in all subsequent party wall legislation.

The facts

The case concerned adjoining houses in Clements Lane, London EC4. One was owned by Stokes, the other by Standard Bank.

Stokes intended to construct a sub-basement below the existing basement to his house and this required him to underpin the party wall. The underpinning works would involve cutting away the existing concrete foundation, excavating the London clay beneath it, inserting a new concrete foundation at a lower level and carrying brickwork up from the new foundation to the base of the existing party wall.

In September 1877 he served a party structure notice on the Bank in respect of the proposed work under the 1855 Act. Surveyors were appointed and a third surveyor was eventually selected but no agreement was reached between the surveyors and no award was made by them. In the meantime Stokes started the works.

The Bank issued proceedings for an injunction to restrain the works on the basis that Stokes had no authority to undertake them.

The parties settled the matter before the hearing and agreed that the works would, in fact, be undertaken in accordance with an award. However they were unable to reach agreement on the liability for the costs of the proceedings and these were left for the court to decide.

In considering this issue the court addressed the substantive issues. Stokes' right to undertake the work, both at common law and under the 1855 Act, were examined.

Work to party walls at common law

The court had no evidence before it regarding the position of the boundaries between the two properties. However, *Cubitt v Porter* (1828) had held that the common user of a wall separating adjoining lands belonging to different owners was *prima facie* evidence that the wall belonged to the owners as tenants in common. The court therefore applied the presumption that the parties owned the wall on this basis.

It then considered whether Stokes, as one of the tenants in common, was entitled to undertake the current works which affected the whole of the party wall. It again referred to *Cubitt v Porter* in this context. Although the permanent demolition of the wall by a single tenant in common would amount to a trespass this case had held that a temporary demolition as part of the repair or improvement works was permissible.

Jessel MR therefore summarised the situation in the present case:

> "The result, therefore, is this: that what the defendant is doing, namely the removing of the foundation of the old wall, be it for the purpose of putting in as good a foundation or a better one – for he is doing it, as it is stated, by beautiful concrete – it is not a destruction by a tenant in common.
>
> All that has been done has been done in this case with the *bona fide* intention of supporting the wall, and will not entitle the other tenant in common to maintain an action, of course still less an injunction.
>
> I am not now speaking of a case of danger, which is not, as I understand, the case here. That being so, it appears to me that at common law there would have been no right of action as far as the wall is concerned."

Work to party walls under the Act

Having established that Stokes did have a right to undertake the work at common law, the court then agreed with the Bank's submission that this right no longer existed because of the effects of the 1855 Act.

The court noted that section 83[1] began with the following words: "The building owner shall have the following rights in relation to party structures" and that it then listed 11 separate rights. The court was of the view that this meant that these were the only rights that the building owner now had. The common law rights therefore no longer existed.

> Jessel MR: "Does not that mean he is to have no other? Is not that the definition of the rights he is to have, meaning those are all the rights he is to have? In my opinion that is the meaning of the section.
>
> The law of the metropolis now defines the rights of a building owner. He is a building owner within the definition of section 82, because he wants to do some work in respect of a party structure, and, being such a building owner, he has these rights, which, in my opinion, are exclusive, and he has no other rights."

This interpretation was supported by the court's examination of the right, contained in subsection 83(1) of the Act, to make good or repair a party structure which was out of repair. For the reasons already considered the court noted that, at common law, a tenant in common already had a right to undertake works of repair *or improvement* to the party wall.

If the purpose of the Act was solely to grant additional rights to the building owner then this particular subsection could have no useful purpose, because more extensive rights already existed at common law. The court therefore concluded that the effect of the subsection was not simply to grant additional rights. It was also intended to restrict the building owner's previous common law rights. Where section 83 did not refer to a right then no right existed – even if one had previously existed at common law.

The court also felt that its observations on the effect of subsection 83(1) were equally applicable to the right, in subsection 83(2), to demolish and rebuild a party structure which required rebuilding due to its defective state.

A right to underpin a party wall?

Neither of these two subsections, nor the former wider common law rights (which had now been replaced) could therefore provide the legal authority for Stokes to undertake the underpinning work. However, the court applied an extremely liberal interpretation to two further subsec-

---

[1] Broadly equivalent to section 2(2) of the 1996 Act.

tions within section 83, and considered that these provided sufficient authority.

The first of these was subsection 83(6) which granted a right to raise a party structure. The second was subsection 83(7) which contained a right to demolish and rebuild a party structure if this was not sufficiently strong for a building which a building owner intended to build against it.

Jessel MR explained the court's thinking in relation to subsection 83(6) in the following terms:

> "Is it necessary to limit the word "raise" to putting something on the wall on the top, and may not you raise or make it longer or build it up by something on the bottom? I do not think it is necessary so to hold, and, if it were absolutely impossible to underpin a wall except under this subsection, my impression is that the subsection would be wide enough to include it."

In relation to section 83(7) he adopted an equally liberal approach:

> "If you may pull it down and rebuild it, why may you not do something short of that, and underpin it under that subsection? It would be a very extraordinary reading of the subsection to say that, although you have the right of pulling it down altogether and putting it up again, yet in rebuilding it you may not do something less, that is to say, support it, and put a new wall underneath it. I think that would be a very narrow view of the subsection."

He also made passing reference to subsections 83(8), 83(9) and 83(11) in support of the view that underpinning was authorised by these provisions also:

> "Then there is the right to cut into a party structure, a right to cut away a footing, and then by section 83(11) there is a right 'to perform any other necessary works incident to the connexion of party structure with the premises adjoining thereto'. Now these are very large words. Why should I not read them to include again the case in question?
>
> I think, therefore, there are words here which are quite large enough to include this right, and my interpretation and decision is, that they are large enough to include them, and therefore that this right is a right within the Building Act when it becomes necessary, or reasonably necessary, to perform it, that the right is not limited to putting bricks upon the top of the wall, and that you may increase the wall by putting bricks below the wall so as to enlarge it in that way as much as you may by putting bricks above the wall to enlarge it in the other way."

Requirement for award

Having established that Stokes no longer had a right to undertake the work at common law but he did have a right under the Act, the court emphasised that this right could only be exercised in accordance with the Act's procedures. As a difference (dispute) had arisen under the Act this could only be resolved by a surveyors' award:

> Jessel MR: "...where there is a dispute or difference – for the words are identical in meaning – the work shall not be done except with the sanction of the other owner who is interested in the wall, or if two surveyors are appointed, one to be nominated by each party, then the two are to choose an umpire, and the work shall be done according to the direction of the majority of the surveyors or the umpire, and shall not be done in any other manner."

As this was the only way that authority could be granted for the works, Stokes clearly had no right to start the work pending the making of the award:

> Jessel MR: "As [the surveyors] have to determine the right to do and the time and manner of doing the work, it would really be reducing the Act to an absurdity to suppose that the building owner had a right to proceed with the work until they have so determined. It must mean that they are to determine the time, that is, the work is not to be done until the time is determined, and that, if the work is commenced before that time, the building owner is committing a breach of the Act of Parliament.
>
> Then there is a power of appealing, and the appeal may no doubt be delayed for a very considerable time; but if [the building owner's] right is, as I read it to be, a right merely to do that which the surveyor or the umpire may direct shall be done, the building owner has no right to do anything at all until he obtains such directions. The Defendant therefore had no right to proceed with his work until the direction of the surveyor had been obtained, and the Plaintiffs are entitled to come here to restrain his so proceeding."

The decision

Stokes had a right to undertake the underpinning work under the 1855 Act, but not at common law. His right to undertake the work under the Act could not be exercised until a surveyors' award had been made. In starting the work before an award had been made he had acted unlawfully. The court therefore ordered him to pay the costs of the suit.

Metropolitan Building Act 1855

PART III: PARTY STRUCTURES

PRELIMINARY

82. Definition of building owner and adjoining owner

In the construction of the following provisions relating to party structures, such one of the owners of the premises separated by or adjoining to any party structure as is desirous of executing any work in respect to such party structure shall be called the building owner, and the owner of the other premises shall be called the adjoining owner.

RIGHTS OF BUILDING AND ADJOINING OWNERS

83. Rights of building owner

The building owner shall have the following rights in relation to party structures; that is to say:

(1) A right to make good or repair any party structure that is defective or out of repair.

(2) A right to pull down and rebuild any party structure that is so far defective or out of repair as to make it necessary or desirable to pull down the same.

(6) A right to raise any party structure permitted by this Act to be raised, or any external wall built against such party structure, upon condition of making good all damage occasioned thereby to the adjoining premises or to the internal finishings and decorations thereof, and of carrying up to the requisite height all flues and chimney stacks belonging to the adjoining owner on or against such party structure or external wall.

(7) A right to pull down any party structure that is of insufficient strength for any building intended to be built, and to rebuild the same of sufficient strength for the above purpose, upon condition of making good all damage occasioned thereby to the adjoining premises, or to the internal finishings and decorations thereof.

(8) A right to cut into any party structure upon condition of making good all damage occasioned to the adjoining premises by such operation.

(9) A right to cut away any footing or any chimney breasts, jambs, or flues projecting from any party wall, in order to erect an external wall against such party wall, or for any other purpose, upon condition of making good all damage occasioned to the adjoining premises by such operation.

(11) A right to perform any other necessary works incident to the connection of party structure with the premises adjoining thereto.

RE STONE AND HASTIE (1903)
Court of Appeal

In brief

- **Issues:** 1894 Act – raising of party wall by predecessor in title – tenant's entitlement to receive contribution towards expenses where adjoining owner subsequently makes use of raised portion – surveyors' jurisdiction to adjudicate on entitlement to contribution – enforcement of arbitration award – challenge to invalid award after expiry of 14-day appeal period.

- **Facts:** L raised the party wall between his property and Stone's adjoining property. He subsequently granted a 21-year lease of his property to Hastie. Stone later increased the height of his own property and made use of the raised portion of the party wall. On completion of this work the appointed surveyors signed an award requiring Stone to pay a contribution towards the expenses of raising the wall to Hastie. Stone failed to pay so Hastie applied for leave to enforce the award as an arbitration award.

- **Decision:** The award was invalid and therefore unenforceable, notwithstanding that it had not been challenged within the statutory 14-day appeal period. The building owner who originally raised the wall (L) was entitled to payment of the contribution. This entitlement did not pass to Hastie on the granting of his lease. Furthermore this was not a matter on which the surveyors were competent to adjudicate. Their jurisdiction was confined to the difference which they had been appointed to settle. It did not extend to matters which had already arisen in respect of other work.

- **Notes:** (1) It is interesting that the surveyors' function was assumed to constitute an arbitration within the meaning of the Arbitration Act. This assumption would be unlikely to go unchallenged today. See also: *Re Metropolitan Building Act ex parte McBride* (1876). (2) The judgments place great emphasis on the fact that Hastie's lease was only for a term of 21 years. It is unclear whether the grant or assignment of a more substantial term would have resulted in rights under the Act being transferred to a tenant. Contrast the case with *Mason v Fulham Corporation* (1910) where identical rights were held to pass to a purchaser on the conveyance of a freehold estate.

The facts

Numbers 16 and 17 Queen Street, London W1, were adjoining terraced houses. Stone owned number 16 and Hastie had a 21-year lease of number 17.

At some stage prior to the grant of Hastie's lease his landlord[1] had increased the height of number 17 and this had involved raising the height of the party wall between the two properties.

Stone now wished to rebuild number 16 to an increased height and this entailed making use of the raised portion of the party wall. He therefore served a party structure notice on Hastie under the 1894 Act and all the works were undertaken in accordance with a surveyors' award.

On completion of the work the surveyors made an addendum award which required Stone to pay a sum of money to Hastie in respect of the use he was now making of the raised party wall, as contemplated by section 95(2)[2] of the Act.

Stone failed to pay the money so Hastie applied to the court for leave to enforce the award under section 12 of the Arbitration Act 1889[3]. His original application was successful but, following Stone's appeal, the court then ruled that the addendum award was invalid. It held that the surveyors had no jurisdiction to require Stone to pay monies to Hastie which were primarily due to his landlord.

Hastie then appealed to the Court of Appeal. He maintained that, as an "owner", he was one of the people entitled to receive payment under section 95(2). As the occupier of the premises he was the person most affected by the use of the raised party wall. The surveyors had come to this conclusion and awarded that payment be made to him. As Stone had not appealed against the award within the 14-day time limit it was now, by virtue of section 91(2), too late to question its validity. His appeal raised the following issues.

Entitlement to payment for use of raised wall

The first issue concerned the entitlement to receive payment, under section 95(2), for the subsequent use of a raised party wall. The court had to decide whether payment was due to the building owner who originally

---

[1] Hastie's landlord owned a long leasehold estate in the property.
[2] A similar provision now appears in section 11(11) of the 1996 Act.
[3] A substantially similar provision now appears as section 66(1) of the Arbitration Act 1996: "An award made by the tribunal pursuant to an arbitration agreement may, by leave of the court, be enforced in the same manner as a judgment or order of the court to same effect."

raised the wall, or to his tenant where a tenancy had subsequently been granted.

The court had no doubt that payment was due to Hastie's landlord, as the person who had originally paid for the wall to be raised, rather than to Hastie. Payment under section 95(2) was not a payment for the use of the wall but a contribution to the original expense of raising the wall. The retention of title provision in section 99 also supported this interpretation.

> Collins MR: "Does the statute contemplate that a portion of the original expense of the work shall be thus recouped to any one other than the owner who raised the wall? The provisions of the Act appear to me to contemplate that the recoupment shall be to the owner who raised the wall; and there does not appear to be any provision which contemplates any payment to a person, who, like the appellant, becomes a tenant for 21 years, after the raising of the wall.
>
> There is one other section which appears to point in the same direction, namely, section 99, which provides that, where the adjoining owner is liable to contribute to the expenses of building any party structure, then, until such contribution is paid, the building owner at whose expense the same was built shall stand possessed of the sole property in the structure.
>
> I think it is obvious when this code is looked at that it provides for the recoupment of the owner at whose expense the wall was raised, and not of any one else.
>
> A tenant like the appellant might, as it appears to me, just as well claim a right to a share of the money originally expended in building the old party wall as a right to a share of the money expended in raising it. The appellant never became a tenant till after both the old and the new portion of the wall had been built."

Do surveyors have jurisdiction to award payment?

The second issue addressed the nature of the surveyors' jurisdiction. The court decided that, quite apart from Hastie's substantive lack of entitlement to receive the payment, the surveyors had no jurisdiction to deal with such questions in their award.

The surveyors' role was confined to matters connected with the work which the building owner wished to undertake. They were concerned with the practicalities of adjusting the respective rights of the building and adjoining owner. In this context they were concerned with the administration of the earlier sections in Part VIII of the Act. Section 95(2) formed a separate part of the legislation and fell outside the surveyors' jurisdiction.

> Collins MR: "That is, as it seems to me, a separate provision parallel with but distinct from the earlier provisions which deal with what a building owner

shall have a right to do, and the reciprocal rights of an adjoining owner against him to which I have alluded.

In the present case the situation arose which is provided for by the earlier provisions of the Act to which I have referred – namely, those which provide as to the respective rights of the parties in view of the fact that the house of which Hastie was tenant adjoined the house of the building owner, Stone. In accordance with those provisions a notice was given by Stone...

... The jurisdiction of the arbitrators is then let in, but they have no jurisdiction whatever to decide anything beyond the dispute which is submitted to them by virtue of the statute. I do not think that the arbitration so provided for contemplates any question between the parties with regard to the initial expenses of raising the party wall in such a case as this...

... The case appears to me quite clear. I think the provisions of the Act restrict the questions with which the arbitrators have to deal in the statutory reference to the differences arising between a building owner and an adjoining owner as mentioned in section 91, and that such a reference does not embrace any claim on the part of a tenant in the position of the appellant to a share of the expenses originally incurred in raising the party wall.

It follows that the arbitrators had no jurisdiction to award to the appellant anything in respect of such a claim, and that the appeal must therefore be dismissed."

The decision

The 14-day time limit for appeals against valid awards had no relevance to challenges against invalid awards. The surveyors had no jurisdiction to require Stone to pay the contribution to Hastie. The award was therefore invalid and could not be enforced. Hastie's appeal was therefore dismissed.

London Building Act 1894

Part VIII: RIGHTS OF BUILDING AND ADJOINING OWNERS

91. Settlement of differences between building and adjoining owners

(1) In all cases (not specifically provided for by this Act) where a difference arises between a building owner and adjoining owner in respect of any matter arising with reference to any work to which any notice given under this Part of this Act relates unless both parties concur in the appointment of one surveyor they shall each appoint a surveyor and the two surveyors so

appointed shall select a third surveyor and such one surveyor or three surveyors or any two of them shall settle any matter from time to time during the continuance of any work to which the notice relates in dispute between such building and adjoining owner with power by his or their award to determine the right to do and the time and manner of doing any work and generally any other matter arising out of or incidental to such difference but any time so appointed for doing any work shall not unless otherwise agreed commence until after the expiration of the period by this Part of this Act prescribed for the notice in the particular case.

(2) Any award given by such one surveyor or by such three surveyors or by any two of them shall be conclusive and shall not be questioned in any court with this exception that either of the parties to the difference may appeal therefrom to the county court within 14 days from the date of the delivery of the award and the county court may subject as hereafter in this section mentioned rescind the award or modify it in such manner as it thinks fit.

95. Rules as to expenses in respect of party structures

(2) As to expenses to be borne by the building owner:

If at any time the adjoining owner make use of any party structure or external wall (or any part thereof) raised or underpinned as aforesaid or of any party fence wall pulled down and built as a party wall (or any part thereof) beyond the use thereof made by him before the alteration there shall be borne by the adjoining owner from time to time a due proportion of the expenses (having regard to the use that the adjoining owner may make thereof):

> (i) of raising or underpinning such party structure or external wall and of making good all such damage occasioned thereby to the adjoining owner and of carrying up to the requisite height all such flues and chimney stacks belonging to the adjoining owner on or against any such party structure or external wall as are by this part of this Act required to be made good and carried up;
>
> (ii) of pulling down and building such party fence wall as a party wall.

99. Structure to belong to building owner until contribution paid

Where the adjoining owner is liable to contribute to the expenses of building any party structure then until such contribution is paid the building owner at whose expense the same was built shall stand possessed of the sole property in the structure.

THOMPSON v HILL (1870)
Court of Common Pleas

In brief

- **Issues:** 1855 Act – laying open – definition of "owner" – tenant at will – duties to adjoining occupier – unnecessary inconvenience – duty of care.

- **Facts:** Thompson demolished and rebuilt the party wall between his property and that occupied by Hill. Hill was a tenant at will and therefore was not an "owner" within the Act. The work necessarily involved "laying open" Hill's property which consequently suffered damage due to rain penetration.

- **Decision:** Thompson was not liable to compensate Hill for the damage to his property. The court did not specifically address Hill's submission that Thompson owed him a common law duty to undertake the works reasonably as well as a statutory duty not to cause him unnecessary inconvenience. Whether this was the case or not, Thompson's duties did not include an obligation to erect hoardings nor other protection for Hill's exposed property.

- **Notes:** (1) The injustice created by this case was addressed by new provisions in the 1894 Act which survive in the current legislation. The obligation not to cause unnecessary inconvenience was expressly extended to include an adjoining occupier as well as an adjoining owner (section 7(1) of the 1996 Act). Additional provisions were also included which addressed the "laying open" of adjoining land or buildings. In these situations building owners were made subject to statutory obligations to provide protection (section 7(3) of the 1996 Act) and also to provide compensation for disturbance and inconvenience caused to an adjoining owner (section 11(6) of the 1996 Act). (2) See also (on the building owner's entitlement to undertake lawful works): *Southwark & Vauxhall Water Company v Wandsworth District Board of Works* (1898) – (on laying open): *Fillingham v Wood* (1891) – (on the meaning of unnecessary inconvenience): *Jolliffe v Woodhouse* (1894); *Barry v Minturn* (1913) – (on the status of a tenant at will): *Orf v Payton* (1905).

The facts

Thompson was a tailor who traded from a small shop near Tower Hill. He occupied the premises as tenant at will. Hill had recently purchased the adjoining houses in the terrace for redevelopment.

> **Metropolitan Building Act 1855**
>
> PART III: PARTY STRUCTURES
>
> RIGHTS OF BUILDING AND ADJOINING OWNERS
>
> **83. Rights of Building Owner**
>
> The building owner shall have the following rights in relation to party structures; that is to say:
>
> (2) A right to pull down and rebuild any party structure that is so far defective or out of repair as to make it necessary or desirable to pull down the same.
>
> (7) A right to pull down any party structure that is of insufficient strength for any building intended to be built, and to rebuild the same of sufficient strength for the above purpose, upon condition of making good all damage occasioned thereby to the adjoining premises, or to the internal finishings and decorations thereof.
>
> **85. Rules as to exercise of rights by building and adjoining owners**
>
> The following rules shall be observed with respect to the exercise by building owners and adjoining owners of their respective rights:
>
> (3) No building owner shall exercise any right hereby given to him in such manner or at such time as to cause unnecessary inconvenience to the adjoining owner.

Hill's proposals involved the demolition and rebuilding of his properties, including the party wall which he shared with the shop. Although he undertook the work to the party wall under the 1855 Act[1], Thompson was not entitled to notice as he was not an "owner" within the terms of the Act.

During the course of the works the shop was "laid open" during the temporary removal of the party wall. Hill made virtually no attempt to protect the shop during this work and the exposed rooms suffered damage

---

[1] Either under section 83(2) or section 83(7).

due to rain penetration. Thompson therefore issued proceedings against him to recover his losses.

Duties to adjoining occupier when laying open?

Thompson submitted that Hill was subject to a duty to provide protection for the exposed rooms, either under a common law duty to do so, or as part of his statutory obligation not to cause unnecessary inconvenience to the adjoining owner under section 85(3) of the Act.

Section 85(3) only made express mention of the adjoining owner, rather than an adjoining occupier. As Thompson was not an owner within the meaning of the Act his counsel phrased the statutory part of his claim in the following terms:

> "The Metropolitan Building Act 1855 authorised the pulling down of the wall, but only in a reasonable manner, and without causing unnecessary inconvenience, and it is not reasonable to pull down the wall of a house without in some way covering up and protecting the rooms thereby exposed to the weather. The rooms must be protected by someone; and if the court holds that it is not the duty of the person pulling down the wall, the expenses will be thrown on the occupier, who has no interest in the improvements, and not upon the owner, upon whom the Act clearly intends that the expense of the repairs should fall."

The decision

The court held that Hill was not subject to any obligation to protect Thompson's rooms. He was not therefore liable to compensate him for the losses which he had incurred.

Perhaps surprisingly, the fact that Thompson was not an adjoining owner did not feature in the court's deliberations. However, the court found other justifications for finding that Hill had not breached the statutory obligation not to cause unnecessary inconvenience.

> Willes J: "[Thompson] rested his claim under the third count on the statute, which provides that the pulling down of the wall shall be done so as to cause no unnecessary inconvenience. That shows the exact limit of the defendant's obligation, and it does not include the doing an independent act, such as putting up a hoarding, nor was there any implied duty on the defendants to do so."

> Bovill CJ: "The plaintiff is not entitled to recover on the third count. I think there was no duty imposed on the defendant, either by statute or at common

law, to board up or otherwise protect the plaintiff's rooms, while the wall was down. The inconvenience the plaintiff suffered is one to which persons who live in houses the walls of which are not reasonably safe, must be subject."

Keating J: "I should be glad if a count could be framed which would entitle the plaintiff to recover the damage he suffered through the exposure of his rooms, but I quite agree that no such count can be framed, as the defendants were under no legal obligation to protect the plaintiff's rooms."

The court therefore rejected Thompson's claim for damages.

THORNTON v HUNTER (1898)
Chancery Division

In brief

- **Issues:** 1894 Act – definition of "party wall" – status of two half brick skins in contact with each other – rebuilding of skin on building owner's land – right to place footings on adjoining owner's land – line of junction.

- **Facts:** Thornton and Hunter owned adjoining houses. The half brick flank wall of each property was in direct contact with the other. Thornton demolished his house. Hunter then wished to rebuild the flank wall of his house as this had become unstable due to the removal of Thornton's property.

- **Decision:** Hunter had no right to place the projecting footings of his new wall on Thompson's land. He had no right to do so at common law. He had no right to do so under the 1894 Act's line of junction notice provisions as the boundary was already built on. He had no right to do so under the party structure notice provisions as the two skins of brickwork did not together constitute a party wall.

- **Notes:** See also (on line of junction notices) *Orf v Payton* (1904) – (on the status of particular boundary structures under the legislation*):* *Wiltshire v Sidford* (1827); *Major v Park Lane Company* (1866); *Johnston v Mayfair Property Company* (1893); *Moss v Smith* (1977).

The facts

Thornton owned 63 High Street, Clapham. Hunter owned the adjoining property at number 65 (Figure 23). The two properties were in contact with each other but there was a difference of opinion as to whether the wall or walls which separated them constituted a party wall.

Each property had its own $4\frac{1}{2}$ in skin of brickwork which was in direct contact with the other's skin. However, the two skins were not bonded together, nor was there any mortar in the gaps between them.

Thornton demolished his building at number 63, including the skin of brickwork which was immediately in contact with number 65's skin. At the time of the demolition he erected shoring to support this skin but the following day part of it collapsed due to the effects of wind. Hunter immediately started rebuilding a new wall in its place and installed its footings on Thornton's side of the boundary line.

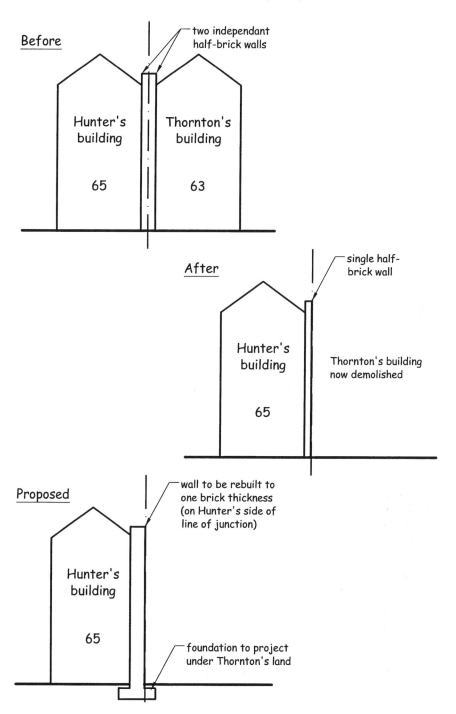

Figure 23: Thornton v Hunter

> # London Building Act 1894
>
> PART VIII: RIGHTS OF BUILDING AND ADJOINING OWNERS
>
> **87. Rights of owners of adjoining lands respecting erection of walls on line of junction**
>
> Where lands of different owners adjoin and are unbuilt on at the line of junction and either owner is about to build on any part of the line of junction the following provisions shall have effect:
>
> (6) Where ... the building owner proceeds to build an external wall on his own land he shall have a right at his own expense at any time after the expiration of one month from the service of the notice to place on the land of the adjoining owner below the level of the lowest floor the projecting footings of the external wall with concrete or other solid substructure thereunder making compensation to the adjoining owner or occupier for any damage occasioned thereby the amount of such compensation if any difference arise to be determined in the manner in which differences between building owners and adjoining owners are hereinafter directed to be determined.
>
> **88. Rights of building owner**
>
> The building owner shall have the following rights in relation to party structures (that is to say):
>
> (2) A right to pull down and rebuild any party structure which is so far defective or out of repair as to make it necessary or desirable to pull it down
>
> (7) A right to pull down any party structure which is of insufficient strength for any building intended to be built and to rebuild the same of sufficient strength for the above purpose upon condition of making good all damage occasioned thereby to the adjoining premises or to the internal finishings and decorations thereof

A party wall or two external walls?

Thornton issued proceedings against him. He sought an injunction to restrain the alleged trespass and damages for the losses which he had allegedly suffered. Hunter defended the action on the basis that he had a right to place footings on Thornton's land under the 1894 Act.

After hearing evidence from witnesses the court concluded that the structure previously separating the two properties consisted of two separate external walls which were in contact with each other. The judge said that the mere fact that two distinct walls are in contact with each other does not make them a party wall. It was always a question of whether, by usage, a party wall exists. As each property had its own skin of brickwork there was no party wall in the present case.

The wall which Hunter was rebuilding was being erected on his side of the boundary line in the same position as his old wall. He was therefore engaged in the rebuilding of his external wall.

Specifically, he was not building a new wall on land which was presently unbuilt upon (a line of junction notice situation), nor was he rebuilding an existing party wall (a party structure notice situation), either of which situations might have brought him within the Act[1].

The decision

On this basis the court found that Hunter had no entitlement to place his footings on Thornton's land. It therefore held that he had committed a trespass and that Thornton was at liberty to remove the footings from his own land. Damages were awarded to Thornton, to include the cost of removal.

---

[1] Under either section 87(6), 88(2) or 88(7).

UPJOHN v SEYMOUR ESTATES LTD (1938)
King's Bench Division

In brief

- **Issues:** 1930 Act – type 'a' party wall – work undertaken without surveyors' award – building owner's entitlement to demolish his half of the party wall at common law – Law of Property Act 1925 – rights of user over adjoining owner's half of a party wall – whether rights of user include a right of protection from other owner's half of the wall.

- **Facts:** Seymour demolished their building without waiting for the appointed surveyors to publish an award. This resulted in the collapse of their half of the party wall but Upjohn's half remained intact. However, the absence of Seymour's half of the wall exposed apertures in Upjohn's half. Construction debris entered through these apertures and damaged their stock.

- **Decision:** Seymour had committed no trespass as they had only interfered with their half of the party wall. However, Upjohn had rights of user in this half of the wall which had been interfered with by its removal. These included a right to the protection previously afforded by the wall. Seymour were therefore liable in nuisance for the damage to Upjohn's stock on account of their interference with this right.

- **Notes:** (1) The decision that a building owner retains his common law rights in respect of his own half of the wall runs contrary to accepted wisdom but may be consistent with the decision in *Major v Park Lane Company* (1866). However, it is difficult to reconcile this view with parts of the judgments in the following cases: *Standard Bank of British South America v Stokes* (1878); *Selby v Whitbread & Co* (1917). (2) The decision also suggests that rights of support and user in post-1926 party walls might include a right to be protected from the weather by one's neighbour's half of the wall. This would not challenge the well-known decision in *Phipps v Pears* (1964) which did not involve a party wall. It would also suggest that the decisions in the following cases were less of a departure from orthodoxy than is sometimes assumed: *Bradburn v Lindsay* (1983); *Marchant v Capital & Counties plc* (1983); *Rees v Skerrett* (2001).

The facts

Upjohn owned number 94 Camden Road. He lived there and carried on a tailoring business from the premises. Seymour Estates bought several houses in the same terrace, up to and including number 92 (Figure 24).

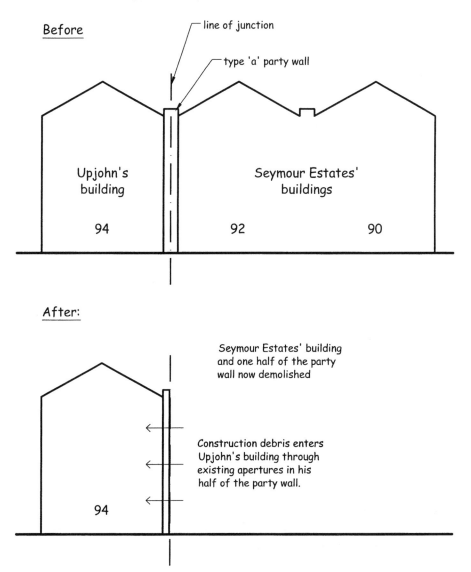

Figure 24: Upjohn v Seymour Estates Ltd

They intended to demolish them and to redevelop the site with flats. Numbers 92 and 94 were separated by a party wall.

Seymour served a party structure notice on Upjohn, under the 1930 Act, advising him of their intention to demolish number 92. A dispute arose under the Act and surveyors were appointed.

Seymour then started the demolition work without waiting for the surveyors to sign an award. This resulted in a bulge appearing in their side of the party wall and, a few days later, to the complete collapse of their half of the wall. Although Upjohn's side was unaffected, dirt and debris entered his building through apertures in his side of the wall which had previously been protected by Upjohn's half of the wall.

Subsequently, cracks and bulging also appeared in Upjohn's building due to the removal of support caused by the demolition of Seymour's building.

Upjohn issued proceedings against Seymour. He sought damages for the loss of his stock caused by the ingress of dirt and debris, and also for the subsequent damage to his building.

Rights and obligations in a party wall

The court considered whether Seymour were liable to him, notwithstanding the fact that no direct damage had initially been caused to Upjohn's half of the wall. This involved a consideration of the rights and obligations of parties in a party wall which were now described in section 38 of the new Law of Property Act 1925. That section now envisaged the ownership in a party wall as being severed down the centre line but also provided for mutual rights of support and user between the parties.

Goddard J's analysis of the legal situation was as follows:

> "[Counsel for the Defendants] argued that, as only the defendants' half had fallen, and as he had committed no trespass, his clients were not guilty of any

---

Law of Property Act 1925

38. Party structures

(1) Where under a disposition or other arrangement which, if a holding in undivided shares had been permissible, would have created a tenancy in common, a wall or other structure is or is expressed to be made a party wall or structure, that structure shall be and remain severed vertically as between the respective owners, and the owner of each part shall have such rights to support and user over the rest of the structure as may be requisite for conferring rights corresponding to those which would have subsisted if a valid tenancy in common had been created.

(2) Any person interested may, in case of dispute, apply to the court for an order declaring the rights and interests under this section of the persons interested in any such party structure, and the court may make such order as it thinks fit.

wrong towards the plaintiff. There is no evidence of any trespass by the defendants, but, by allowing their part of the wall to fall, they not only withdrew such support as it afforded to the plaintiff's part, and to the rest of the premises, but also deprived him of the right of user over the defendants' wall, which protected the apertures to which I have referred, and which were exposed, and left exposed, by its collapse, and to which user he was entitled by the effect of the statute."

The decision

The court therefore held (1) that Seymour had committed no trespass on the party wall as they had simply removed their half without interfering with Upjohn's half; (2) that Upjohn's premises and his half of the party wall were entitled to support (and protection) from Seymour's premises and their half of the party wall; and (3) that Seymour had withdrawn this support (and protection) and were therefore liable (presumably in nuisance) for the resulting damage which had occurred to the Upjohn's property.

VIDEO LONDON SOUND STUDIOS LTD v ASTICUS (GMS) LTD AND KELTBRAY DEMOLITION LTD (2001)

Technology and Construction Court

In brief

- **Issues:** 1939 Act – demolition of building owner's building – noise and dust – damage caused to adjoining owner's building by vibration – obligation to make good damage – validity of surveyors' award – demolition contractor's duty to undertake prior inspection of adjoining owner's building – negligence – building operations undertaken with reasonable care – liability in nuisance – situation where physical damage results.

- **Facts:** Keltbray were demolition contractors who were demolishing buildings on Asticus' site. The work displaced some rubble in Video London's chimney which fell into their property and damaged some expensive recording equipment. The appointed surveyors made an addendum award requiring Asticus to reimburse the cost of this under the 1939 Act's "making good" provisions. Asticus failed to pay. Video London therefore brought proceedings in nuisance, negligence and for a breach of the award.

- **Decision:** The surveyors' addendum award was invalid. The recording equipment was a chattel rather than a fixture. It therefore fell outside the making good provisions which only applied to the adjoining premises and their internal finishings and decorations. Asticus were not therefore liable for a breach of the award. There was also no liability in negligence for failing to undertake an adequate pre-demolition inspection of the chimney. Keltbray had undertaken a non-destructive inspection. They had also had a copy of the surveyors' schedule of condition and of an engineers' report relating to the design of shoring. There was no requirement to undertake destructive testing and, even if this had been undertaken, it would not have predicted the damage which actually occurred. However, both Asticus and Keltbray were liable in nuisance. Because Video London had suffered physical damage it was no defence that the works had been undertaken with reasonable care and skill. Furthermore, they could not escape liability on the ground that the damage suffered was too remote. The kind of damage which occurred was reasonably foreseeable. This was sufficient, even if the precise mechanism by which it occurred could not have been foreseen.

- **Notes:** (1) The court's decision in respect of the addendum award is surprising. The "making good" provisions in the 1939 Act included damage to internal finishings and decorations as well as to the premises themselves. The court held that damage to these items is only recoverable if they can be classified as fixtures. This would leave an adjoining owner who suffers damage to chattels without any redress at all in circumstances where there is no fault on the part of the building owner. It seems more likely that the reference to finishings and decorations was included with the specific intention of bringing chattels within the provision. This must be all the more so under the current Act which substitutes the word "furnishings" for the previous reference to "finishings". (2) See also *Andreae v Selfridge & Company Ltd* (1937); *Matania v National Provincial Bank Ltd and The Elevenist Syndicate* (1936).

The facts

Video London owned premises at 16 to 18 Ramillies Street, London W1. Their property comprised five storeys and a basement (Figure 25). It adjoined buildings owned by Asticus on two sides. On each side the properties were separated from one another by party walls.

Video London operated a sound studio from their property. Their business was concerned with the provision of post production sound facilities to the film and television industries. This involved the use of some extremely sensitive recording and editing equipment which was installed in the basement.

When they took up occupation in 1985 they had removed the chimney breast at second floor level and provided appropriate steel supports and brick sealing for the remainder of the stack. At some stage prior to their occupation the chimney breast in the basement had also been removed. Appropriate brick sealing had been provided at that time although no support had been provided for the stack. The remaining portion of the stack at basement level was also hidden from view by a false ceiling.

Asticus were in the process of redeveloping their site. This involved the demolition of their existing buildings and the erection of new ones. The first part of their site, on the south side of Video London's property, was cleared during 1994/1995 by Brutons Construction Ltd. In 1996 Keltbray were appointed to undertake the remainder of the demolition work.

Damage caused by demolition operations

In 1996, during the course of Keltbray's contract, a number of minor incidents occurred and there were complaints about noise and vibration from Video London's employees. In April about a "cup full" of dust fell into their premises from the chimney breast on the west elevation of their building at second floor level.

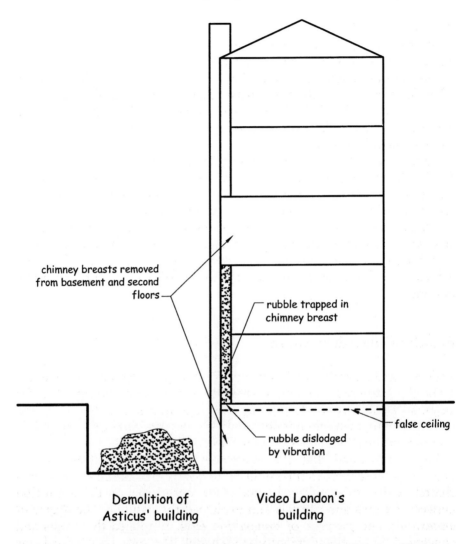

Figure 25: Video London Sound Studios Ltd v Asticus (GMS) Ltd and Keltbray Demolition Ltd

In September a more substantial quantity of debris fell out of a flue on the south elevation of their building. It fell on to Asticus' side where it became trapped between the party wall and the weather protection that had been installed following the demolition of the adjoining building on this elevation.

On 12 May 1997 the most serious incident occurred. Keltbray were removing concrete from the west side of the party wall at basement level. During the course of this operation a 70 lb jack hammer momentarily came into contact with the building's steel frame which was embedded in the concrete wall. The hammer blow transferred a vibration shock to the rigid steel frame in close proximity to the remaining portion of the chimney breast in Video London's premises.

This dislodged some rubble in the flue which had been left over from when the second floor chimney breast had been removed. This rubble fell with such force that it broke through the brick seal at basement level, causing a shower of dust, rubble and bricks to fall on to the recording and editing equipment below.

Damage was caused to the equipment to a value of approximately £225,000. As the demolition work was the subject of a party wall award under the 1939 Act, the surveyors made an addendum award which required Asticus to reimburse Video London with this sum.

Asticus failed to pay so Video London instituted the present proceedings against them and Keltbray. They sought to recover the awarded sum together with damages for business interruption. The claim was based on a breach of the terms of the addendum award and on negligence and nuisance.

Breach of addendum award

Asticus argued that the addendum award was ultra vires. Section 46 of the 1939 Act required building owners to make good damage "to the adjoining premises or to the internal finishings thereof". As the editing and recording equipment did not fall within this description the surveyors had no power to deal with it in an award under the Act.

The court noted that the equipment would only fall within the description in section 46 if it could be described as a fixture rather than a chattel. It therefore reviewed some of the authorities on the distinction between fixtures and chattels and explained the nature of the *degree of annexation* and *purpose of annexation* tests. It applied these tests and concluded that the equipment was a chattel. The surveyors' award was therefore indeed, ultra vires and Video London's claim on this basis must therefore fail. Wilcox J described the rationale for this decision:

> "I have had the benefit of a visit to the claimants' premises and have seen the type and nature of the equipment installed there. It is largely freestanding equipment and the attachment to the walls is for the purpose of power supply. This is not merely plugging in – there is cabling in conduits and channels affixed to the wall.
>
> Some of the equipment that I observed was replacement equipment. Some was repaired as was clear from the evidence before me. The technology in the field in which the claimants operate changes and they use state of the art equipment. By way of example, video taping and films are no longer used. Digital technology has superseded that. I have no doubt that in order to keep

up to date and remain competitive items of equipment will continue to be updated and replaced.

Those items particularised in the schedule are all items that could be removed from the premises with relative ease. It was contemplated, at one stage, that they should be removed from the premises whilst demolition work went on. Such an operation, although comparatively expensive, was technically easy. In my judgment the items contained in the schedule are not fixtures. They were never intended to be part of the realty. I am satisfied that they are not fixtures and do not fall within the addendum award."

Negligence

Video London alleged that Keltbray were negligent in failing to inspect the remains of the chimney breast in the basement above the false ceiling. Had they done so they would have been put on notice that the unsupported brick seal was vulnerable and they should then have provided some alternative support.

The court rejected this argument. In accordance with their contract, Keltbray had carried out an inspection of Video London's building prior to demolition, although this had been a non-destructive examination. They had also received the condition report from the party wall surveyors who had also considered it unnecessary to undertake a destructive examination above the false ceiling.

In addition, independent engineers had investigated the stability of Video London's building for the purpose of designing appropriate shoring. Keltbray had also had a copy of their report. In the circumstances the court considered that Keltbray had carried out a reasonable level of inspection. They also found that there was nothing that would have been apparent from this level of inspection that would have put a reasonable demolition contractor on notice as to any vulnerability of the chimney breast and flues at basement level.

Furthermore, even if Keltbray had conducted a destructive investigation above the false ceiling this would not have made them aware of the presence of the brick plug in the chimney, nor would the condition of the brick seal have put him on enquiry either. The manner in which the damage actually occurred was therefore not reasonably foreseeable, even if this additional level of inspection had been undertaken.

Video London's claim in negligence therefore failed.

Nuisance

Asticus and Keltbray both argued that they could not be liable in nuisance as they had taken all reasonable care in the conduct of the demolition.

They relied, in particular, on the decision in *Andreae v Selfridge & Company Ltd* (1937) in support of this position. That case had decided that, for reasons of public policy, everyone had to put up with a certain amount of discomfort caused by demolition and building operations, provided that these were carried out in a reasonable manner.

They also relied on the more recent case of *Wildtree Hotels Ltd v Harrow London Borough Council* (2000) in this context. That case had considered a claim for compensation for injurious affection, under section 10 of the Compulsory Purchase Act 1965, caused by noise, dust and vibration. Asticus and Keltbray referred to Lord Hoffman's judgment in that case:

> "In my view it will be almost impossible for any claim for damage caused by noise, dust or vibrations to satisfy [the legal requirements for an award of compensation]. Being 'things productive of sensible personal discomfort' within the meaning of Lord Westbury's dichotomy in *St Helen's Smelting Company v Tipping* (1865), the claim is subject to the principle that a reasonable use of land with due regard to the interest of neighbours is not actionable.
>
> The implications for building operations were spelled out by Sir Wilfred Greene MR in *Andreae v Selfridge & Company Ltd* ... Actionability at common law therefore depends upon showing that the building works were conducted without reasonable consideration for the neighbours."

In considering these arguments the court made reference to the definition of private nuisance in *Clerk & Lindsell on Torts*[1]:

> "A private nuisance may be and usually is caused by a person doing, on his own land, something which he is lawfully entitled to do. His conduct only becomes a nuisance when the consequences of his act are not confined to his own land but extend to the land of his neighbour by:
> (1) Causing an encroachment on his neighbour's land when it closely resembles trespass;
> (2) Causing physical damage to his neighbour's land or buildings or works or vegetation upon it;
> (3) Unduly interfering with his neighbour in the comfortable and convenient enjoyment of his land."

The court stressed that Sir Wilfred Greene's observations in *Andreae* had been confined to interference falling within the third category of interference. The court in the *Wildtree Hotels* case, which Asticus and Keltbray relied on, had also been concerned with this category. In contrast, it was

---

[1] Dugdale, A. (2000) *Clerk & Lindsell on Torts*, (18th edn) Sweet & Maxwell.

misleading to think of the damage which had been suffered by Video London in these terms:

> Wilcox J: "The electronic equipment was mostly affected by the dust after the fall of the brick plug in the chimney on 12 May 1997. It is tempting therefore when considering the analysis in *Clerk and Lindsell* earlier referred to, to consider this nuisance within Class 3. But that is to overlook the fact that there was directly caused physical damage to the claimant's property as a result of the activity carried out by the second defendants to the party wall. The brick seal shattered and itself fell together with the brick plug and dust above it. In falling, dust was released and spread. In my judgment this is not a case falling within the third category in the analysis of *Clerk and Lindsell*."

Nuisance causing physical damage

As Video London had suffered physical damage Asticus and Keltbray's liability could not be limited by their having taken all reasonable steps in the conduct of the demolition.

The court also made reference to another recent injurius affection case, *Clift v Welsh Office* (1998) in support of this finding. In that case dust and mud, generated by construction operations, had caused damage to external decorations. The court had awarded compensation and Sir Christopher Slade had explained the policy issues in the following terms:

> "... we see no sufficient reason why, as a matter of policy, the law should expect the neighbour, however patient, to put up with actual physical damage to his property in such circumstances. Where there is physical damage, the loss should, in our judgment, fall on the doer of the works rather than his unfortunate neighbour. No authority has been cited to us in which it has been held that the rule in *Andreae* applies in respect of physical damage caused by nuisance. In the absence of such authority, we hold that it does not..."

Keltbray could not therefore escape liability on the basis that they had taken all reasonable care. However, they could only be held liable if the physical damage suffered by the claimant was not too remote. Under the rule in *Overseas Tankship (UK) Ltd v Miller Steamship Co Pty, The Wagon Mound* (No. 2) (1966) the damage suffered must have been of the kind which was reasonably foreseeable. The court considered that the damage to Video London's property was reasonably foreseeable by the second defendants.

> Wilcox J: "In my judgment it is clear that the second defendant's demolition works caused physical damage to Video London's property. Reasonable user does not provide a defence to nuisance where there is physical damage to a

claimant's property. However it is necessary for the claimant to establish that such damage was reasonably foreseeable...

... the second defendants' site manager accepted that on several occasions following complaints he went to the claimants' premises and saw debris and dust albeit in small quantities, that had fallen on to the claimants' equipment...

He was in charge of the operations on the second defendants' demolition site. A reasonable person in his position would have realised that activity in relation to the party wall could cause debris to fall within the claimants' building and cause damage thereby. On 12 May 1997 the second defendant could not reasonably foresee the precise mechanism by which damage was caused.

[The site manager] was aware that some debris from time to time remains in chimney flues and can be dislodged and fall. It was also foreseeable that, if there was physical activity involving force in relation to the party wall, that could affect the adjoining building and cause some physical damage to the fabric of the building. In my judgment the claimants succeed on their claim in nuisance."

The decision

The claims in negligence and for breach of the addendum award failed. However, as Video London had suffered physical damage to their property, both Asticus and Keltbray were liable in nuisance. Judgment was entered accordingly.

WATSON v GRAY (1880)
Chancery Division

In brief

- **Issues:** Freestanding boundary wall – meaning of "party wall" at common law – tenancy in common – rights of owners – rights to raise wall – rights to remove a trespass from the wall.

- **Facts:** A freestanding boundary wall separated yards belonging to Watson and Gray. Watson raised the wall so as to build a lean-to shed against it. Gray argued that he owned the wall entirely, subject only to easements in Watson's favour. He therefore removed the increased courses. Watson argued that they owned the wall as tenants in common and that he was therefore entitled to raise it.

- **Decision:** The wall was a party wall, owned by the parties as tenants in common. Nevertheless, Watson's plans to raise it represented an ouster of Gray's rights and therefore amounted to a trespass. Gray had therefore been justified in removing the raised portion.

- **Notes:** (1) Contrast this decision with that in *Cubitt v Porter* (1828) where no ouster was involved and a building owner was held to be entitled to raise the party wall. (2) See also: *Matts v Hawkins* (1813); *Wiltshire v Sidford* (1827); *Standard Bank of British South America v Stokes* (1878).

The facts

The parties owned adjoining terraced houses at 7 and 9 Queen's Terrace, Middlesborough (Figure 26). Both properties had formerly been in single ownership but Watson's house had been sold off with a separate title in 1855 and Gray's in 1864.

Each of the two conveyances contained agreements and declarations relating to the boundaries. The front and rear walls were declared to fall within the ownership of the respective properties. The separating wall between the two houses was stated to be a party wall. This latter provision also applied to a $4\frac{1}{2}$ in brick wall separating the rear yards of each property.

Watson decided to build a lean-to shed in his yard against the $4\frac{1}{2}$ in brick party wall. To facilitate this he first raised the height of the wall for its full thickness by a further 3 ft 4 ins. Gray objected to this and demolished the additional courses at the top of the wall. Watson therefore issued proceedings for trespass, claiming damages and an injunction to restrain him from further interfering with his raising of the wall.

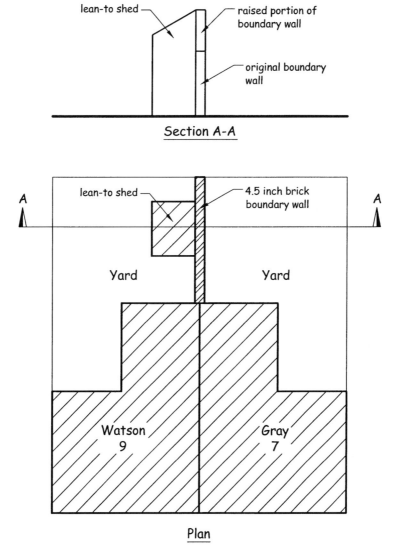

Figure 26: Watson v Gray

Watson argued that, when his house had been conveyed as a separate title in 1855, the entire thickness of the wall must have been conveyed to his predecessor in title, subject to certain easements in favour of the adjoining property.

He maintained that, where a wall is built by the original owner of both properties, there could be no presumption that adjoining owners become tenants in common of the wall. On the contrary, the presumption must be that the original owner conveyed the whole thickness of the wall by the first conveyance.

Gray, in turn, argued that the normal presumption should apply that the parties were tenants in common of the wall. As Watson had attempted to exclude him from the wall, he was justified in removing the obstruction and could not therefore be liable for trespass.

Definition of "party wall"

Fry J delivered his now famous judgment in which he defined four different categories of party wall:

> "What is the meaning of the term 'party wall' as there used? The words appear to me to express a meaning rather popular than legal, and they may, I think, be used in four different senses.
>
> They may mean, first, a wall of which the two adjoining owners are tenants in common, as in *Wiltshire v Sidford* (1827) and *Cubitt v Porter* (1828). I think that the judgments in those cases show that that is the most common and the primary meaning of the term.
>
> In the next place the term may be used to signify a wall divided longitudinally into two strips, one belonging to each of the neighbouring owners, as in *Matts v Hawkins* (1813).
>
> Then, thirdly, the term may mean a wall which belongs entirely to one of the adjoining owners, but is subject to an easement or right in the other to have it maintained as a dividing wall between the two tenements. The term is so used in some of the Building Acts[1].
>
> Lastly, the term may designate a wall divided longitudinally into two moieties, each moiety being subject to a cross easement in favour of the owner of the other moiety[2]."

In the present case the court was of the view that the wall fell into the first of these categories. The parties owned the wall jointly as tenants in common. It based this decision on two factors. Firstly, this was the ordinary meaning of the term. Secondly, in addressing all the boundaries to the property, the 1855 conveyance had expressly contrasted party walls with walls belonging entirely to one owner. On this basis Watson's suggestion that the side walls belonged to him in addition to those at the front and rear appeared unconvincing.

---

[1] Including a type 'b' party wall within the 1996 Act.
[2] Fry J's first category of party wall was abolished by Law of Property Act 1925, section 39 and Schedule 1, Part V, which deemed it to be a party wall within this fourth category. In terms of common law rights the fourth category therefore now describes the "most common and the primary meaning" of the term "party wall".

Owners' rights to interfere with party wall

Once it had concluded that the parties owned the wall as tenants in common the court then had to decide whether Watson was justified in raising the wall, and if not, whether Gray was then justified in demolishing the raised portion.

It decided that Watson had no right to raise the wall. In an apparent contradiction of Bayley J's views in *Cubitt v Porter* it considered that raising a party wall amounted to an ouster of the other party's entitlement to possession.

In coming to this conclusion it placed reliance on Crompton J's judgment in *Stedman v Smith* (1857). In that case the defendant had raised a freestanding garden wall in order to build a lean-to wash house against it. The work had involved removing the copings from the top of the wall, raising the wall, and then replacing the copings. However, the roof of the wash house was then installed in such a way that it occupied the whole width of the top of the wall. The defendant had also set a stone into the wall, with an inscription on it stating that the wall and the land on which it stood belonged to him.

> Fry J: "On these facts it was held that a jury might find an actual ouster by the defendant of the plaintiff from possession of the wall which would constitute a trespass. In the course of the argument Crompton J said, 'You certainly had no longer the use of the same wall; you could not put flower pots on it, for instance' ... And in his judgment he said: 'The plaintiff is excluded from the top of the wall; he might have wished to train fruit trees there, or to amuse himself by running along the top of the wall'. Just so in the present case, the plaintiff has excluded the defendant from the top of the wall."

Having decided that Watson had violated Gray's rights the court then addressed the Gray's response to this. Somewhat ironically, having rejected the first part of Bayley J's analysis in *Cubitt v Porter*, the court now placed reliance on other parts of his judgment to support its conclusions in the present case. Fry J decided that Gray had been justified in demolishing the raised portion of the wall and explained this in the following terms:

> "In *Cubitt v Porter*, Bayley J said: 'One tenant in common has upon that which is the subject matter of the tenancy in common laid bricks and heightened the wall. If that be done further than it ought to have been done, what is the remedy of the other party? He may remove it. That is the only remedy he can have.'
>
> That is the precise remedy to which the defendant has had recourse in the present case ... I hold, therefore, the plaintiff is not entitled to any damages in

respect of the throwing down of the wall, and that the injunction asked for cannot be granted."

The decision

Watson had no right to raise the whole thickness of the party wall. Gray was therefore entitled to remove the raised portion. The court dismissed Watson's claim for damages and an injunction.

WESTON v ARNOLD (1873)
Court of Appeal in Chancery

In brief

- **Issues:** Bristol Improvement Acts 1840 and 1847 – definition of type 'b' party wall – wall performing separating function for only part of its height – interference with easement of light.

- **Facts:** Weston demolished his three-storey building with a view to rebuilding it. Before he could do so Arnold started erecting a large warehouse on his adjacent land which interfered with easements of light in favour of the windows in Weston's building. Weston sought an injunction to restrain the erection of the warehouse. Arnold had previously built a single-storey lean-to building against one of the walls of Weston's building. Because the lower part of Weston's wall therefore performed a separating function he argued that the whole of the wall was a party wall within the Bristol Acts. Windows were not permitted in party walls under the Acts so Weston had no right to install windows in this wall when reinstating his building. In these circumstances Arnold could not be held liable for interfering with an easement of light.

- **Decision:** An injunction was granted to restrain the erection of the warehouse. In the absence of an express statutory provision requiring walls to be party walls for their whole height, there was a presumption that they were only party to the extent that they performed a separating function. There was no such provision in the present case. Weston's wall was therefore an external wall above ground floor level and there was no restriction on installing windows in it from that point upwards.

- **Notes:** See also: *Knight v Pursell* (1879); *Drury v Army & Navy Auxilliary Co-operative Supply Ltd* (1896).

The facts

Weston owned the White Lion Inn in Bristol. This was a three-storey building. The south side of the Inn adjoined land owned by Arnold. Some single-storey outbuildings were situated on Arnold's land, one of which was built up against, and made use of, the south wall to the Inn (Figure 27).

There were a large number of windows in the Inn's south wall at first and second floor level, which were entitled to a prescriptive easement of light. These windows overlooked Arnold's land, over the top of the single-storey outbuildings.

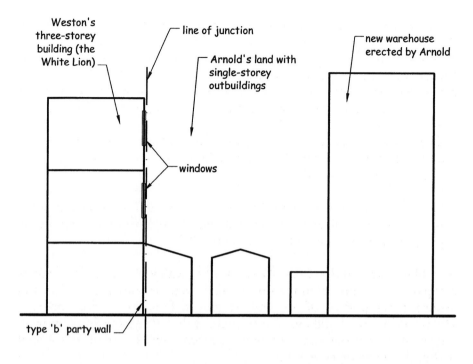

Figure 27: Weston v Arnold

The Inn's south wall was in poor repair and required rebuilding. Weston therefore served notice on Arnold, under section 27 of the Bristol Improvement Act 1847, of his intention to undertake this work. This Act made provision for work to party walls in similar terms to those contained in the London Building Acts. Service on Arnold was appropriate because the work would affect the outbuilding on his side of the wall. In due course surveyors were appointed and they authorised the work.

Weston then commenced extensive works to the White Lion which involved the demolition and rebuilding of virtually the whole of the building. The proposed rebuilding of the south wall was to include the reinstatement of windows in the same location as those existing previously. Whilst this work was proceeding Arnold began erecting a large warehouse on his land which obscured the prescriptive light to Weston's property.

Definition of party wall

Weston obtained an *ex parte* injunction restraining Arnold from interfering with his easement of light by erecting the warehouse. Arnold appealed against this decision.

He argued that the south wall was a party wall within the meaning of the Bristol Improvement Acts 1840 and 1847. Section 25 of the 1840 Act prohibited the construction of openings in party walls for any purpose except for communication from one building to another. He therefore contended that as Weston had no right to construct windows in the south wall in any event, he could have suffered no infringement of the easement of light to them.

The court had to consider the status of the Inn's south wall. The Bristol Improvement Acts contained no clear definition of a party wall. Section 24 of the 1847 Act provided that:

> "... every wall, arch, fence wall, or partition between separate houses, warehouses, or other buildings, should be deemed to be respectively a party wall, party arch, party fence wall, or partition..."[1]

Arnold maintained that the wall was a party wall within this definition as it separated the Inn from the outbuilding. Weston only accepted this in respect of the portion of the wall, at ground floor level, which actually separated the two buildings. He contended that above this level, as it no longer separated the two buildings, the wall ceased to be a party wall within the meaning of the Act.

Can part of a wall be a party wall?

The court considered the question of whether a wall could be a party wall for only part of its height. It concluded that it could. Furthermore, where the evidence suggested that only part of the wall was a party wall this should be presumed to be the case. This presumption could only be rebutted by a clear contrary intention in the words of the statute.

> Sir W M James LJ: "Unless there is something in the Act of Parliament which shows a clear intention to the contrary, it appears to me that a wall may be in part of its length a party wall, and in part of its length an external wall, and there is no distinction between height and length.
>
> A wall may be a party wall up to part of its height, and may be an external wall for the rest of its height ... There is nothing in fact or in law to make it impossible or improbable that a wall should be a party wall up to a certain height, and above that height be the separate property of one of the owners."

---

[1] As with the type 'b' party wall in the 1996 Act, the ownership of the wall is irrelevant to this definition. A wall becomes a party wall by virtue of its function in separating two buildings rather than because it is in shared ownership.

The court found no clear requirement in the Bristol Improvement Acts that a wall had to be a party wall for the whole of its height. It therefore favoured what it described as the "natural construction" of the statutes which had been proposed by Weston.

> Sir G Mellish LJ: "Section 24, as it appears to me, is equally consistent with the natural construction, namely, that for the purposes of the Act ... every wall between separate houses shall be deemed to be a party wall.
>
> The wall in this case is not a wall between separate houses. What is wanted in order that the defendants may make out their construction is a section which says that where any party wall continues in height after it ceases to separate two buildings, then the whole of that wall shall be a party wall. There are no such words in these Acts..."

Presumption against interference with property rights

The presumption which the court had applied in this case was further supported by a more general presumption of statutory interpretation, namely that Parliament does not intend to interfere with private property rights without providing full compensation in return. In this context the court was unconvinced by Arnold's suggestion that the Bristol Improvement Acts could be used to deprive Weston of his easement of light.

> Sir G Mellish LJ: "... and when the contention is that the private rights of an individual have been taken away from him and given to another by a public Act of Parliament, and that without compensation – for the compensation clause here clearly does not apply to this case – the intention of the Legislature ought to appear in perfectly clear and plain terms. Here it does not so appear; on the contrary, the construction of the Acts supports the plaintiff's case, and not the defendants'."

> Sir W M James LJ: "It is clear in this case that there was a party wall, and there was a separate wall, and that the windows were in the separate wall. The owner of that wall has acquired a right to those windows, and there is nothing in the Acts of Parliament that can deprive him of his right for the benefit of his neighbour, or that has given his neighbour a right to raise his building so as to darken those windows. There is nothing in the words or spirit if the Acts which can justify such transfer of a valuable property from one person to another, without any reference to the public good, in the city of Bristol, or in any other place.

The decision

The court therefore found that the wall, above ground floor level, was not a party wall within the meaning of Bristol Improvement Acts. The Acts did not therefore deprive Weston of his easement of light. The court made the injunction permanent against Arnold's erection of the warehouse.

WHITE v PETO BROTHERS (1888)
Chancery Division

In brief

- **Issues:** 1855 Act – damage to adjoining owner's building – damage a "necessary consequence" or "inevitable result" of the works – damage caused by negligence – definition of "party wall".

- **Facts:** Peto Brothers were building contractors. They were involved in the redevelopment of a site adjacent to White's property. They demolished their clients' building and excavated its site. They caused extensive damage to White's building whilst underpinning its flank wall in accordance with an award under the 1855 Act. Peto Brothers denied liability on the basis that the work had been authorised under the Act.

- **Decision:** Peto Brothers were liable for the damage. No liability arises where damage occurs to an adjoining owner's property as a necessary consequence/inevitable result of work authorised by the Act. In these circumstances the damage must be addressed by the surveyors under the making good and compensation provisions within the Act. However, the Act does not authorise acts of negligence. The damage to White's property had been caused by Peto Brothers' negligence and they were therefore liable for it. Although it made no difference to the decision the court also held that the wall which was underpinned was not a party wall. There was therefore no authority to underpin it. Peto Brothers would therefore still have been liable, even in the absence of negligence.

- **Notes:** See also: *Thornton v Hunter* (1898); *Adams v Marylebone Borough Council* (1907); *Louis v Sadiq* (1997).

The facts

White owned a shop and living accommodation at 10 Russell Street, London WC2. He had let the property to Ward. The property adjoined the Hummums Hotel, at 11 Russell Street, which was owned by Smith.

Smith wished to demolish the hotel and to rebuild it on a much grander scale. He therefore appointed Peto Brothers, a firm of contractors, to undertake the work. He also served a party structure notice on Ward[1], under the 1855 Act, in respect of the works to what was mutually assumed to be a party wall between the two properties.

---

[1] No notice was served on White, as freeholder, but he raised no objection to this in the proceedings.

The notice referred to the demolition, rebuilding, and raising of the wall and to cutting in works associated with the installation of new chimney breasts. No mention was made of the excavation works for the construction of a new basement as there were no adjacent excavation provisions in the 1855 Act. Nevertheless, the works also involved the underpinning of the 'party wall' and this work seems to have been referred to in the notice[2].

Damage due to adjacent excavations

In due course a surveyors' award was published and Peto Brothers undertook the work. Unfortunately the excavation and underpinning works caused extensive cracking damage to the full height of the "party wall". White's property became uninhabitable and he was unable to relet it on the expiry of Ward's tenancy. White therefore issued proceedings against Peto Brothers to recover his losses from them on the basis of their negligence in undertaking the work below ground.

Peto Brothers argued that they could not be liable as the works had been undertaken under the Act. These had been authorised in accordance with the Act's procedures and any damage that resulted was a necessary consequence of carrying them out. On this basis, if White had any remedy, it must be against Smith as he was the building owner under the Act[3]. Peto Brothers, on the other hand, had behaved lawfully throughout.

The court did not accept Peto Brothers' argument and found against them on two grounds.

Liability for negligent party wall works

Firstly, the Act did not authorise acts of negligence. Kay J made a distinction between damage which occurs as a "necessary consequence" or an "inevitable result" of work which has been properly authorised and that which occurs due to the negligent performance of such work.

> "The law applicable to the case is stated in *Addison on Torts* (5th edition), p. 671, as follows: 'If an action is brought against contractors and workmen who are personally engaged in the execution of public works under the order or

---

[2] There was no express right to underpin a party wall in the 1855 Act. Nevertheless, *Standard Bank of British South America v Stokes* (1878) had held that this right existed due to its similarity with some of the other party wall rights contained in the Act. Kay J's judgment, in the present case, confirms that the underpinning work was properly authorised on the basis of the notice which Smith had served.

[3] Presumably on the basis of his obligation, as building owner, to make good.

authority of trustees, or a board of public works, and the damage of which the plaintiff complains is the inevitable result of the execution of a public work under statutory authority, the action will fail; but if the damage arises from the negligent execution of the work, and might have been avoided by the exercise of proper skill and care, the contractors and workmen will be personally answerable for the damage done.' The authority for this is *Jones v Bird* (1822), which fully supports that statement of the law.

From the evidence he was satisfied that the damage had been caused by Peto Brothers' negligence and was not simply a necessary consequence of the works. Cracking a wall from top to bottom was not a necessary consequence of underpinning works. On this basis Peto Brothers were therefore liable for the damage.

> Kay J: "The first question is, was this work negligently done, or was the damage the necessary consequence of what the contractors did?
> ... There were cracks in the south wall which went to the top of the house, and they were admittedly caused by the acts of the contractors, which I assume were duly done under the Metropolitan Building Acts. The notices to the occupier were good, and authorised the contractors to do the work. Now, what they did, so far as concerned the plaintiff's house, was this:
> They dug out the foundations of the Hummums hotel 9 ft lower than they were before, and they dug 9 ft lower than the foundations of the plaintiff's house, and in doing so they underpinned the plaintiff's house. They did one of the most common operations and one of everyday occurrence in London, that of underpinning an adjoining house.
> I am asked to say that underpinning a house by a thoroughly experienced firm, such as Peto Brothers, involves as a necessary consequence to the house underpinned that it should be cracked from top to bottom. If all the builders in London came and said that that is a necessary consequence of underpinning I would not believe them. Nothing is more common than underpinning a house in London, and I have myself seen it done without the house being let down the slightest fraction of an inch. An experienced builder can perform an operation of that kind without damaging the adjoining house.
> On the evidence before me I am satisfied that there has been negligence, and very considerable negligence, in the way in which these works have been carried out."

A party wall or two external walls?

Secondly, the court held that the wall was not, as a matter of evidence, a party wall within the Act. Peto Brothers therefore had no right to underpin it.

Throughout the case there had been some confusion as to whether the wall was a party wall or whether White's wall and the hotel wall were really independent structures. Smith's party structure notice had referred to the wall as a party wall and White's pleadings had also referred to it as such. On this basis Peto Brothers had always assumed that there was no dispute as to the status of the wall.

White had called two witnesses who gave evidence that it was not a party wall but Peto Brothers' counsel had failed to cross-examine them on the point. When the significance of their evidence became apparent, he requested permission to call further witnesses but Kay J declined permission.

He considered that it would not further the ends of justice to allow further evidence on the point. The damage had been caused by Peto Brothers' negligence and the Act had been found to provide no defence to liability in such circumstances. The precise status of the wall would therefore have made no difference to the outcome of the case.

Nevertheless, on the basis of the witnesses' testimony, which had not been challenged by Peto Brothers, Kay J found that the wall was not a party wall. Peto Brothers therefore had no authority under the Metropolitan Building Act to underpin it in any event.

The decision

The Act conferred no authority to undertake work negligently. The damage to White's property had been caused by Peto Brothers' negligence. The court granted judgment for White.

WHITEFLEET PROPERTIES LTD v ST PANCRAS BUILDING SOCIETY (1956)

Chancery Division

In brief

- **Issues:** 1939 Act – appointment of surveyors – contents of party structure notice – service by agent – service on agent – contents of award.

- **Facts:** A dangerous structure notice was served on the parties in respect of a party wall between their two properties. They each appointed consultants to deal with the matter. A party structure notice was served and the consultants published an award relating to the proposed remedial works. St Pancras subsequently dismissed their consultant and argued that they were not bound by the award on account of its invalidity.

- **Decision:** The award was valid. St Pancras' surveyor had been validly appointed under the Act. They had instructed him to deal with matters arising out of the dangerous structure notice, and the party wall procedures were an obvious part of this. Although they had initially instructed him by telephone, they had subsequently confirmed their instructions by letter. Service of the party structure notice on St Pancras' consultant was valid as he had authority to accept service. Furthermore, both the party structure notice and the award were valid notwithstanding the fact that they referred to Whitefleet's managing agents and made no reference to Whitefleet at all.

- **Notes:** It cannot be assumed that courts will generally interpret the legislation as liberally as this. In circumstances where a building owner seeks to interfere with an adjoining owner's property in order to improve his own building, the courts are likely to insist on a more literal interpretation of the Act's procedural requirements. See *Gyle-Thompson v Wall Street (Properties) Ltd* (1974).

The facts

Whitefleet Properties owned 21 Bride Lane, off Fleet Street, London EC4. JRC were agents who managed the property on Whitefleet's behalf. St Pancras Building Society owned the adjoining property at 20 Bride Lane.

Early in 1955 the local authority served a dangerous structure notice on both owners in respect of the party wall between the two properties. JRC therefore appointed an architect (Consultant A) to deal with matters arising from the notice. Consultant A then wrote to St Pancras to inform them of his involvement and to enquire how they would like to resolve the matter.

St Pancras then telephoned a chartered surveyor (Consultant B) to instruct him to deal with the matter on their behalf. This telephone conversation was then confirmed by letter (also enclosing a copy of Consultant A's letter) dated 19 July 1955.

Consultant A then served a party structure notice on Consultant B, under the 1939 Act, in respect of the proposed remedial works[1]. The notice was stated to be served on behalf of JRC. Consultant A and Consultant B then agreed the necessary remedial works and published an award on 25 August 1955. This authorised the works and provided that their costs should be shared beteen the parties.

Before the works could start, a dispute appears to have arisen between St Pancras and Consultant B (their own surveyor). On 29 November 1955, Consultant B informed Consultant A that he was withdrawing from the matter. On 2 January 1956 St Pancras also informed Consultant A that Consultant B had never had authority to enter into the award. The award was therefore invalid.

Whitefleet applied to the court for a declaration that St Pancras were bound by the award. St Pancras submitted that the award was invalid on five separate counts. The court considered each of these in turn.

Appointment of surveyor

St Pancras maintained that they had never appointed Consultant B as their surveyor under the Act. He had simply been instructed to resolve the various issues with Consultant A rather than to perform the statutory function of appointed surveyor. He therefore had no authority to make an award.

The court held that Consultant B had been validly appointed. It was obvious that the procedure under the Act would form an essential feature of the process for resolving the problem with the defective party wall. In these circumstances, when St Pancras instructed Consultant B to deal with the matter they obviously also intended to appoint him as their surveyor under the Act.

> Danckwerts J: "It seems to me that where you have a party wall in the City and it gets into a dangerous condition, it is obvious that work will have to be done

---

[1] It is unclear why this was thought to be necessary. Under section 47(4) of the 1939 Act works which are carried out in compliance with a dangerous structure notice did not require the prior service of a party structure notice. The same provision now appears in section 3(3)(b) of the 1996 Act.

on it by the owners and adjoining owners and the Act provides procedure in case they do not agree in all respects. There was nothing unexpected in appointing a surveyor to act throughout, even in arbitration proceedings."

Requirement for appointment in writing

St Pancras made the further submission that, even if they had intended to appoint Consultant B as their surveyor, his appointment must be invalid as it had not been made in writing, as required by section 55(h) of the Act[2].

The court held that the appointment had been in writing. Although the initial appointment had been made by telephone this had then been confirmed in writing by the letter of 19 July 1955. Taken together, there had been a valid appointment in writing.

Requirement for service on adjoining owner

St Pancras also argued that the party structure notice had been invalidly served. Section 47(1) of the Act required the notice to be served on them as the adjoining owner and section 124(1) prescribed the method of service[3]. The notice had not been served in accordance with the Act but had, instead, been served on Consultant B.

The court held that the party structure notice had been validly served. Section 124 laid down a number of ways in which service could be effected on the adjoining owner. However, these were not exclusive and a notice could be served by other means. In particular the section did not prevent notices being served on someone having authority to accept service. The court held, on the facts, that Consultant B had authority to accept service on behalf of St Pancras[4]. By sending the notice to him Whitefleet had therefore effectively served the notice on St Pancras.

Contents of party structure notice

St Pancras also submitted that the party structure notice itself was invalid. It was stated to be served on behalf of JRC (the managing agents). Section 47(1) of the Act required service by the building owner[5] but there was no reference to Whitefleet in the notice.

---

[2] The same requirement is now contained in section 10(2) of the 1996 Act.
[3] The relevant provisions are now contained in sections 3(1) and 15 of the 1996 Act.
[4] Presumably on the basis of the wording in the letter of 19 July 1955.
[5] The relevant provision is now contained in section 3(1) of the 1996 Act.

The court did not regard this as a reason for the notice being invalid. In the circumstances, service by managing agents on behalf of the building owner was acceptable.

Parties to the award

Finally, St Pancras maintained that the award must be invalid as it too made no reference to Whitefleet, as the building owner. Once again, JRC were named as the building owner.

Once again, the court held that this did not affect the validity of the procedures:

> Danckwerts J: "The question of an undisclosed principle has nothing to do with the case. [Consultant A] put in the managers quite properly because they were the people who were giving him his instructions. There was nothing contrary to the provisions of the Act in regarding the managers as the building owners."

The decision

The notice, the appointment of surveyors and the award were all valid. The court therefore made the declaration, requested by Whitefleet, that St Pancras were bound by the award.

London Building Acts (Amendment) Act 1939

PART VI: RIGHTS ETC. OF BUILDING AND ADJOINING OWNERS

RIGHTS ETC. OF OWNERS

47. Party structure notices

(1) Before exercising any right conferred on him by section 46 (rights of owners of adjoining lands where junction line built on) of this Act a building owner shall serve on the adjoining owner notice in writing (in this Act referred to as a "party structure notice") stating the nature and particulars of the proposed work the time at which it will be begun and those particulars shall where the building owner proposes to construct special foundations include plans sections and details of construction of the special foundations with reasonable particulars of the loads to be carried thereby.

DIFFERENCES BETWEEN OWNERS

55. Settlement of differences

Where a difference arises or is deemed to have arisen between a building owner and an adjoining owner in respect of any matter connected with any work to which this Part of this Act relates the following provisions shall have effect:

(a) Either:

   (i) both parties shall concur in the appointment of one surveyor (in this section referred to as an "agreed surveyor"; or

   (ii) each party shall appoint a surveyor and the two surveyors so appointed shall select a third surveyor (all of whom are in this section together referred to as "the three surveyors").

(h) All appointments and selections made under this section shall be in writing.

PART XI: LEGAL PROCEEDINGS

NOTICES ETC.

124. Service of notices etc. by council and others

(1) Subject to the provisions of this section and section 125 (service of documents relating to dangerous or neglected structures) of this Act any notice order or other document (in this section referred to as a "document") authorised or required by or under the London Building Acts or any byelaws made in pursuance of those Acts to be served by or on behalf of the Council or a local authority or by the superintending architect or the district surveyor or any other person on any person shall be deemed to be duly served:

   (c) where the person to be served is a public body or a corporation society or other body if the document is addressed to the clerk secretary treasurer or other head officer of that body corporation or society at its principal office or (if there is no office) at the premises to which the document relates and is either:

      (i) sent by post in a prepaid letter; or

      (ii) delivered at that office or the premises as the case may be.

WILLIAMS v BULL (1890)
Chancery Division

In brief

- **Issues:** 1855 Act – building owner raising a party wall – definition of "party wall" – status of raised portion which straddles boundary but performs no separating function – adjoining owner subsequently making use of raised portion – whether adjoining owner liable to contribute to the cost of raising – whether adjoining owner entitled to make use of raised portion.

- **Facts:** Williams raised the party wall separating his property from Bull's property. Bull subsequently increased the height of his property and enclosed on the raised portion. Williams demanded payment of a contribution towards the original cost of raising the wall. Bull refused to pay. Williams argued that the raised portion was therefore his sole property and sought to restrain Bull from trespassing on it.

- **Decision:** The 1855 Act only expressly included a type 'b' party wall. Nevertheless, the raised portion straddled the boundary line and was therefore a party wall. Consequently Bull was entitled to enclose on it under the Act. However, in the absence of any statutory provision requiring him to contribute towards the original cost of raising it, Williams was unable to recover this from him. Bull was not therefore liable for trespass.

- **Notes:** (1) The anomalies within this decision were addressed in the 1894 Act and the changes made in that statute are now incorporated in the current Act. The definition of party wall was widened to expressly include walls straddling the boundary in addition to those which perform a separating function between buildings (section 20 of the 1996 Act). A provision was also included to make provision for the retrospective contribution towards expenses where an adjoining owner later makes use of work undertaken by a building owner at his sole expense (section 11(11) of the 1996 Act). (2) See also: *Re Stone and Hastie* (1903); *Orf v Payton* (1905); *Mason v Fulham Corporation* (1910).

The facts

The parties owned adjoining terraced houses in London. The two houses were separated by a party wall which straddled the boundary line between the two plots (Figure 28). In 1884 Williams had rebuilt his house and increased its height. This had involved increasing the height of the party

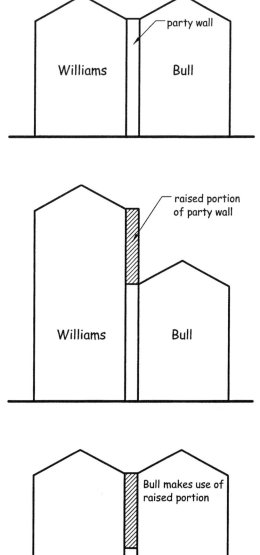

Figure 28: Williams v Bull

wall. The work had been carried out under the party wall provisions in the 1855 Act.

The following year Bull decided to increase the height of his house and to build the new storey against the raised portion of the party wall. He served a party structure notice on Williams of his intentions and, in due course, he completed the work in accordance with the Act's procedures. Williams then demanded payment for half the cost of raising the party wall. Bull refused to pay this so Williams issued proceedings against him for trespass.

Definition of party wall

Williams claimed that he was unwilling to allow Bull to use the raised portion of the wall without also making a contribution towards its erection. The raised portion was not being used to separate the two buildings as required by the definition of "party wall" in section 3 of the Act. This portion was therefore an "external wall" rather than a "party wall". Consequently Bull had no rights to interfere with it under the Act.

Williams cited *Weston v Arnold* (1873) and *Knight v Pursell* (1879) to demonstrate that a wall could be a party wall for only part of its height or length. He maintained that, although the wall as a whole straddled the boundary line between the two properties, the raised portion had been erected by him and therefore remained his property. As Bull had no right to enclose against the wall under the Act and had no proprietary right in it either, his actions constituted a trespass against his property for which he now sought redress.

Liability for cost of raising party wall

Bull argued that, as the wall was built astride the boundary line it must fall within the definition of "party wall" notwithstanding the more restrictive wording in the Act's definition of the term. He therefore had the right to undertake work to the wall in pursuance of the Act.

Despite this, he denied that he had any responsibility to contribute towards the cost of raising the wall. An adjoining owner's liabilities to contribute were restricted to situations where (under section 88) work was necessary to remedy some defect in a structure, or where (under section 93) he had required the building owner to undertake the work.

Where, as in the present situation, work has been undertaken by a building owner entirely for his own benefit, there was, under the 1855 Act, no requirement that an adjoining owner should contribute, even if he later derived some benefit from the work.

The decision

The court favoured Bull's argument. For reasons which are not explained in the published law report, the court appears to have accepted that the wall was a party wall within the meaning of the Act.

Nevertheless, in the absence of any provision, in the 1855 Act, requiring an adjoining owner to contribute to expenses where use is subsequently made of works undertaken by the building owner, no such liability could exist. Bull was therefore entitled to enclose upon the raised portion of the party wall but was not liable to contribute towards Williams' expenses of raising it. Judgment was therefore entered for Bull.

Metropolitan Building Act 1855

PRELIMINARY

3. Interpretation of certain terms in this act

In the construction of this Act (if not inconsistent with the context) the following terms shall have the respective meanings hereinafter assigned to them; (that is to say):

"External Wall" shall apply to every outer wall or vertical enclosure of any building not being a party wall.

"Party Wall" shall apply to every wall used or built in order to be used as a separation of any building from any other building, with a view to the same being occupied by different persons.

PART III: PARTY STRUCTURES

RIGHTS OF BUILDING AND ADJOINING OWNERS

84. Rights of adjoining owner

Whenever the building owner proposes to exercise any of the foregoing rights with respect to party structures the adjoining owner may require the building owner to build on any such party structure certain chimney jambs, breasts or flues, or certain piers or recesses, or any other like works for the convenience of such adjoining owner; and it shall be the duty of the building owner to comply with such requisition in all cases where the execution of the required works will not be injurious to the building owner or cause to him unnecessary inconvenience or unnecessary delay in the exercise of his right; and any difference that arises between the building owner and adjoining owner in respect of the execution of such works as aforesaid shall be determined in manner in which differences between building owners and adjoining owners are hereinafter directed to be determined.

88. Rules as to expenses in respect of party structure

The following rules shall be observed as to expenses in respect of any party structure; (that is to say):

As to expenses to be borne jointly by the building owner and adjoining owner:

(1) If any party structure is defective or out of repair the expenses of making good or repairing the same shall be borne by the building owner and adjoining owner in due proportion, regard being had to the use that each owner makes of such structure.

(2) If any party structure is pulled down and rebuilt by reason of its being so far defective or out of repair as to make it necessary or desirable to pull down the same, the expense of such pulling down and rebuilding shall be borne by the building owner and adjoining owner in due proportion, regard being had to the use that each owner makes of such structure.

As to expenses to be borne by the building owner:

(6) If any party structure or external wall built against the same is raised in pursuance of the power hereinafter vested in any building owner, the expense of raising the same, and of making good all such damage, and of carrying up to the requisite height all such flues and chimneys as are hereinbefore required to be made good and carried up, shall be borne by the building owner.

(7) If any party structure which is of proper materials and sound, or not so far defective or out of repair as to make it necessary or desirable to pull down the same, is pulled down and rebuilt by the building owner, the expense of pulling down and rebuilding the same, and of making good all such damage as is hereinbefore required to be made good, shall be borne by the building owner.

93. As to expenses on requisition of adjoining owner

Where any building owner has incurred any expenses on the requisition of an adjoining owner, the adjoining owner making such requisition shall be liable for all such expenses, and in default of payment the same may be recovered from him as a debt.

WILTSHIRE v SIDFORD (1827)
Court of King's Bench

In brief

- **Issues:** Definition of "party wall" at common law – entitlement to demolish half thickness of structure between two buildings – entitlement to enclose on remaining half – status of party wall at common law – absence of clear evidence of ownership – presumption of tenancy in common.

- **Facts:** Wiltshire and Sidford owned adjoining semi-detached houses. Sidford demolished his house and half the thickness of the dividing structure between the houses. He then enclosed his new building on the remaining half of the structure. Wiltshire argued that this was a trespass on a wall which was owned solely by him.

- **Decision:** As a matter of evidence the structure was a party wall rather than two separate skins of brickwork. There was no evidence as to the precise quantities of land contributed by each party for the erection of the wall so it could not be a longitudinally divided party wall. In these circumstances the wall must be presumed to be owned by both parties as tenants in common. Sidford was therefore entitled to undertake the work to the wall and no trespass had been committed.

- **Notes:** See also (on the demolition of half the thickness of a dividing structure between buildings): *Thornton v Hunter* (1898); *Upjohn v Seymour Estates Ltd* (1938) – (on common law rights in party walls): *Matts v Hawkins* (1813); *Cubitt v Porter (1828);* *Watson v Gray* (1880) *Standard Bank of British South America v Stokes* (1878).

The facts

The parties were the respective owners of two adjoining semi-detached houses. Sidford demolished and rebuilt his house. The works appear to have included the removal of a skin of brickwork from his side of the wall between the two houses which he did not then replace as part of the rebuilding process (Figure 29).

Wiltshire issued proceedings against him for trespass by placing the reconstructed house directly against his wall. He maintained that the structure between the two properties had previously consisted of two entirely separate walls. Sidford had therefore demolished his own wall and was now purporting to enclose on his (Wiltshire's) wall, which he had no right to do.

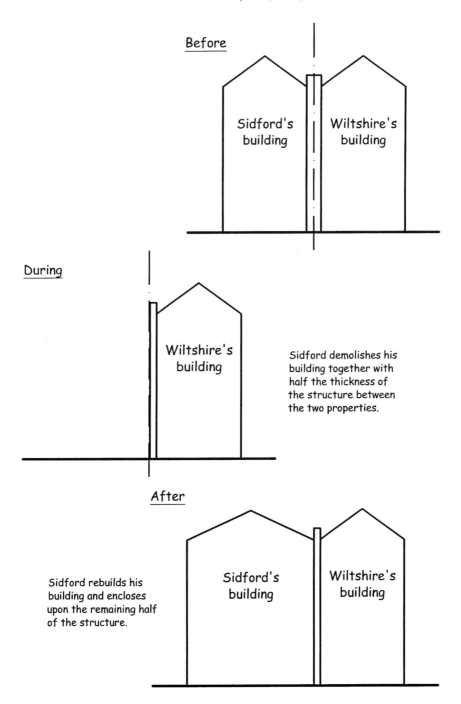

Figure 29: Wiltshire v Sidford

Sidford maintained that the structure between the two houses had previously consisted of a single party wall which the parties owned jointly as tenants in common. He was therefore within his rights in treating the fabric of the wall as if it were his own, providing that he did not deprive Wiltshire of the use of the wall.

Because of the conflicting evidence, both of the parties and of surveyors, as to the previous nature of the structure, the matter was left to the jury. They decided, on the facts, that the wall had previously been a party wall rather than two separate walls.

Wall in common use presumed to be party wall

Wiltshire then argued that, even if the structure was a party wall, it did not necessarily mean that the two parties were tenants in common.

He relied on *Matts v Hawkins* (1813) as demonstrating that the parties had retained their own rights in their own side of the wall, even though the wall was a party wall. Therefore, although Sidford might have been within his rights in removing his own side of the wall, he was not entitled to enclose his new building on the part of the wall which belonged to Wiltshire.

The court distinguished the present case from *Matts v Hawkins*. In that case, the precise quantity of land contributed by each party for the building of the party fence wall had been known. It was therefore possible to state with certainty precisely where the boundary line between the two properties lay and to identify the portions of the wall which lay on either side of it.

In the present case this was not known. In such situations, there was a presumption arising from the common use of the wall that, prima facie, the wall, and the strip of land on which it was built, were owned by the parties as tenants in common. The rationale for this presumption was described by Bayley J:

> "When the builder of two houses grants off one, it is more reasonable to presume he grants the whole wall in undivided moieties, than that he should leave to either party the power of cutting the wall in half. That would be the case if the houses were built by one and the same person.
>
> If two persons built at the same time, the probability is that they would take a conveyance of an undivided moiety of the ground on which the wall was to be erected, in order that the property might afterwards be kept in the same state."

The decision

The wall was therefore presumed, in the present case, to be owned by the parties as tenants in common. Sidford had, therefore, acted within his rights and Wiltshire's claim for trespass was dismissed. Judgment was entered for Sidford.

WOODHOUSE v CONSOLIDATED PROPERTY CORPORATION LTD (1993)

Court of Appeal

In brief

- **Issues:** 1939 Act – validity of surveyor's award – meaning of surveyors' jurisdiction to settle "any matter which ... may be in dispute" – meaning of surveyors' power to determine "any other matter arising out of or incidental to the difference" – surveyors' authority to adjudicate on matters arising prior to their appointment.

- **Facts:** Consolidated Property commenced works to their building. They subsequently served a party structure notice on Woodhouse in respect of works which would affect the party wall between their two properties. Before surveyors could be appointed the party wall collapsed, causing damage to Woodhouse's property. Surveyors were subsequently appointed and the third surveyor made an award. This found that Consolidated were to blame for the collapse and required them to pay compensation for the damage.

- **Decision:** The third surveyor's award was invalid. The apparent breadth of the words in the Act had to be read in the context of the surveyors' primary role under the legislation. This was to adjudicate on the circumstances under which the building owner was entitled to undertake work under the Act. Surveyors had no authority to adjudicate on other disputes between the parties. The dispute about the collapse of the party wall had arisen prior to the surveyors' appointment and the surveyors therefore had no authority to settle it.

- **Notes:** (1) See also *Leadbetter v Marylebone Corporation [No.1]* (1904). (2) Some minor changes have been made to the wording of the relevant sections in the 1996 Act[1]. As the primary role of the appointed surveyors remains unchanged, it seems unlikely that this case would have been decided any differently if it had arisen under this Act.

The facts

Consolidated Property owned 68a Neal Street, London WC2. This was a former mission hall comprising ground floor, first floor and basement. The basement floor was some 2.3 metres below the external ground level

---

[1] Sections 10(10) and 10(12).

and the rear wall of the basement acted as a retaining wall for land owned by Woodhouse (Figure 30).

Woodhouse's property comprised a timber yard at 61 Endell Street. A timber storage shed on his land adjoined Consolidated's mission hall building. The shed consisted of three brick pillars which were in contact with the mission hall wall (a party wall) and two further brick pillars on its far side. The pillars supported a monopitch roof.

Consolidated planned to undertake extensive works to their property. These included the demolition and rebuilding of the upper sections of the party wall and the lowering of the basement floor. The excavation of the basement floor would also require the underpinning and reinforcing of the lower sections of the party wall so that it could continue to act as a retaining wall for the excavated basement.

Damage to adjoining property

In the autumn of 1990 Consolidated started their construction operations, including some work in the basement. In January 1991 they served a party structure notice on Woodhouse, under the 1939 Act, in respect of the work to the party wall. At the time they claimed that this had not yet started. However, before the 14-day notice period had expired, the party wall collapsed and seriously damaged the shed.

Once the notice period had expired, surveyors were appointed and a third surveyor was selected. The two surveyors failed to agree on the responsibility for the collapse and therefore referred the matter to the third surveyor. He made an award. This provided that the collapse had been caused by Consolidated's work and required them to pay for the damage to the shed.

Consolidated failed to pay this so Woodhouse issued proceedings against them in nuisance, trespass and negligence. He sought damages for the cost of rebuilding the shed, and for loss of profits caused by the damage to his property.

In interlocutory proceedings he later sought to rely on the third surveyor's finding of fact about the cause of the damage. He claimed that Consolidated were estopped from denying that they were responsible for this due to the statement to this effect in the third surveyor's award. The matter was referred to the Court of Appeal (together with a number of other issues which are not considered here).

Surveyors' jurisdiction

Consolidated submitted that the third surveyor had no jurisdiction to make the award concerning the damage to the shed. The basis for their

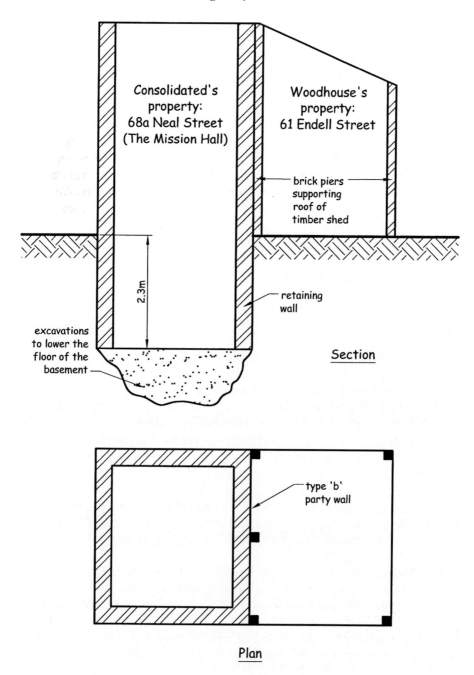

Figure 30: Woodhouse v Consolidated Property Corporation Ltd

submission is not set out in the published law report but presumably relied on the fact that the damage predated the appointment of the surveyors and was therefore an entirely different matter from the difference which was actually referred to them. If the surveyors had no authority then the award must be invalid. If the award was invalid then it could not estop Consolidated in any way.

The court considered the statutory basis for the surveyors' authority in section 55 of the Act. Although the words of subsection 55(i) (which defined the surveyors' jurisdiction) appeared quite wide, these should be read in their proper context. Subsection 55(k) (which defined the surveyors' powers) made it clear that their primary role was to determine the

---

London Building Acts (Amendment) Act 1939

PART VI: RIGHTS &c. OF BUILDING AND ADJOINING OWNERS

DIFFERENCES BETWEEN OWNERS

55. Settlement of differences

Where a difference arises or is deemed to have arisen between a building owner and an adjoining owner in respect of any matter connected with any work to which this Part of this Act relates the following provisions shall have effect:

(a) Either:

(i) both parties shall concur in the appointment of one surveyor (in this section referred to as an "agreed surveyor"; or

(ii) each party shall appoint a surveyor and the two surveyors so appointed shall select a third surveyor (all of whom are in this section together referred to as "the three surveyors").

(i) The agreed surveyor or as the case may be the three surveyors or any two of them shall settle by award any matter which before the commencement of any work to which a notice under this Part of this Act relates or from time to time during the continuance of such work may be in dispute between the building owner and the adjoining owner.

(k) The award may determine the right to execute and the time and manner of executing any work and generally any other matter arising out of or incidental to the difference

building owner's right to undertake the proposed work and the time and manner of its execution.

Their further power, in this subsection, to determine "any other matter arising out of or incidental to the difference" was concerned with further issues arising out of the terms on which the building owner was authorised to undertake the work. For example (although this was not expressly stated in the judgment) if work was authorised subject to the condition that damage to an adjoining owner's property be made good, the surveyors would also have power to adjudicate on this issue.

This was the limit of the surveyors' authority. Specifically, they did not have a general jurisdiction to determine other disputes between the parties which had not been referred to them under the Act. Glidewell LJ explained the court's decision in the following terms:

> Glidewell LJ: "Although section 55(i) requires the surveyor to settle by his award 'any matter which ... from time to time during the continuance of such work may be in dispute...', this must, in my view, be read in its context. The context in particular includes the provisions of section 55(k), which commences: 'the award may determine the right to execute and the time and manner of executing any work...'
>
> In my judgment, the provisions of section 55 relate only to the resolution of differences between adjoining owners as to whether one of them shall be permitted under the Act to carry out works, the subject of a section 47 notice, and if so, the terms and conditions under which he is permitted to carry out such works.
>
> A matter which arises during the carrying out of the works, about which there is a dispute, must therefore be a matter which relates to the consent for the works to be carried out, e.g. whether the building owner is complying with a particular requirement in the consent.
>
> Section 55 does not permit or authorise the surveyor appointed under this part of the 1939 Act to determine other disputes arising between the parties. It follows, in my judgment, that under the 1939 Act [the third surveyor] had no jurisdiction to make the award which he purported to make. [The third surveyor's] award in this action is evidence, but no more."

The decision

The third surveyor's award addressed a dispute which was separate from the difference which had been referred to the surveyors under the Act. He had no jurisdiction to make the award, which was therefore invalid. His findings provided evidence about the cause of the damage but no more. Consolidated were not estopped from denying the truth of these findings.

Appendix

Party Wall etc. Act 1996

An Act to make provision in respect of party walls, excavation and construction in proximity to certain buildings or structures; and for connected purposes.

CONSTRUCTION AND REPAIR OF WALLS ON LINE OF JUNCTION

1. New buildings on line of junction

(1) This section shall have effect where lands of different owners adjoin and:

(a) are not built on at the line of junction; or
(b) are built on at the line of junction only to the extent of a boundary wall (not being a party fence wall or the external wall of a building),

and either owner is about to build on any part of the line of junction.

(2) If a building owner desires to build a party wall or party fence wall on the line of junction he shall, at least one month before he intends the building work to start, serve on any adjoining owner a notice which indicates his desire to build and describes the intended wall.

(3) If, having been served with notice described in subsection (2), an adjoining owner serves on the building owner a notice indicating his consent to the building of a party wall or party fence wall:

(a) the wall shall be built half on the land of each of the two owners or in such other position as may be agreed between the two owners; and
(b) the expense of building the wall shall be from time to time defrayed by the two owners in such proportion as has regard to the use made or to be made of the wall by each of them and to the cost of labour and materials prevailing at the time when that use is made by each owner respectively.

(4) If, having been served with notice described in subsection (2), an adjoining owner does not consent under this subsection to the building of a party wall or party fence wall, the building owner may only build the wall:

(a) at his own expense; and

(b) as an external wall or a fence wall, as the case may be, placed wholly on his own land,

and consent under this subsection is consent by a notice served within the period of fourteen days beginning with the day on which the notice described in subsection (2) is served.

(5) If the building owner desires to build on the line of junction a wall placed wholly on his own land he shall, at least one month before he intends the building work to start, serve on any adjoining owner a notice which indicates his desire to build and describes the intended wall.

(6) Where the building owner builds a wall wholly on his own land in accordance with subsection (4) or (5) he shall have the right, at any time in the period which:

(a) begins one month after the day on which the notice mentioned in the subsection concerned was served, and

(b) ends twelve months after that day,

to place below the level of the land of the adjoining owner such projecting footings and foundations as are necessary for the construction of the wall.

(7) Where the building owner builds a wall wholly on his own land in accordance with subsection (4) or (5) he shall do so at his own expense and shall compensate any adjoining owner and any adjoining occupier for any damage to his property occasioned by:

(a) the building of the wall;

(b) the placing of any footings or foundations placed in accordance with subsection (6).

(8) Where any dispute arises under this section between the building owner and any adjoining owner or occupier it is to be determined in accordance with section 10.

2. Repair etc. of party wall: rights of owner

(1) This section applies where lands of different owners adjoin and at the line of junction the said lands are built on or a boundary wall, being a party fence wall or the external wall of a building, has been erected.

(2) A building owner shall have the following rights:

(a) to underpin, thicken or raise a party structure, a party fence wall, or an external wall which belongs to the building owner and is built against a party structure or party fence wall;

(b) to make good, repair, or demolish and rebuild, a party structure or party fence wall in a case where such work is necessary on account of defect or want of repair of the structure or wall;

(c) to demolish a partition which separates buildings belonging to different owners but does not conform with statutory requirements and to build instead a party wall which does so conform;

(d) in the case of buildings connected by arches or structures over public ways or over passages belonging to other persons, to demolish the whole or part of such buildings, arches or structures which do not conform with statutory requirements and to rebuild them so that they do so conform;

(e) to demolish a party structure which is of insufficient strength or height for the purposes of any intended building of the building owner and to rebuild it of sufficient strength or height for the said purposes (including rebuilding to a lesser height or thickness where the rebuilt structure is of sufficient strength and height for the purposes of any adjoining owner);

(f) to cut into a party structure for any purpose (which may be or include the purpose of inserting a damp proof course);

(g) to cut away from a party wall, party fence wall, external wall or boundary wall any footing or any projecting chimney breast, jamb or flue, or other projection on or over the land of the building owner in order to erect, raise or underpin any such wall or for any other purpose;

(h) to cut away or demolish parts of any wall or building of an adjoining owner overhanging the land of the building owner or overhanging a party wall, to the extent that it is necessary to cut away or demolish the parts to enable a vertical wall to be erected or raised against the wall or building of the adjoining owner;

(j) to cut into the wall of an adjoining owner's building in order to insert a flashing or other weather-proofing of a wall erected against that wall;

(k) to execute any other necessary works incidental to the connection of a party structure with the premises adjoining it;

(l) to raise a party fence wall, or to raise such a wall for use as a party wall, and to demolish a party fence wall and rebuild it as a party fence wall or as a party wall;

(m) subject to the provisions of section 11(7), to reduce, or to demolish and rebuild, a party wall or party fence wall to:

(i) a height of not less than two metres where the wall is not used by an adjoining owner to any greater extent than a boundary wall; or
(ii) a height currently enclosed upon by the building of an adjoining owner;

(n) to expose a party wall or party structure hitherto enclosed subject to providing adequate weathering.

(3) Where work mentioned in paragraph (a) of subsection (2) is not necessary on account of defect or want of repair of the structure or wall concerned, the right falling within that paragraph is exercisable:

> (a) subject to making good all damage occasioned by the work to the adjoining premises or to their internal furnishings and decorations; and
> (b) where the work is to a party structure or external wall, subject to carrying any relevant flues and chimney stacks up to such a height and in such materials as may be agreed between the building owner and the adjoining owner concerned or, in the event of dispute, determined in accordance with section 10;

and relevant flues and chimney stacks are those which belong to an adjoining owner and either form part of or rest on or against the party structure or external wall.

(4) The right falling within subsection (2)(e) is exercisable subject to:

> (a) making good all damage occasioned by the work to the adjoining premises or to their internal furnishings and decorations; and
> (b) carrying any relevant flues and chimney stacks up to such a height and in such materials as may be agreed between the building owner and the adjoining owner concerned or, in the event of dispute, determined in accordance with section 10;

and relevant flues and chimney stacks are those which belong to an adjoining owner and either form part of or rest on or against the party structure.

(5) Any right falling within subsection (2)(f), (g) or (h) is exercisable subject to making good all damage occasioned by the work to the adjoining premises or to their internal furnishings and decorations.

(6) The right falling within subsection (2)(j) is exercisable subject to making good all damage occasioned by the work to the wall of the adjoining owner's building.

(7) The right falling within subsection (2)(m) is exercisable subject to:

> (a) reconstructing any parapet or replacing an existing parapet with another one; or
> (b) constructing a parapet where one is needed but did not exist before.

(8) For the purposes of this section a building or structure which was erected before the day on which this Act was passed shall be deemed to conform with statutory requirements if it conforms with the statutes regulating buildings or structures on the date on which it was erected.

3. Party structure notices

(1) Before exercising any right conferred on him by section 2 a building owner shall serve on any adjoining owner a notice (in this Act referred to as a "party structure notice") stating:

(a) the name and address of the building owner;
(b) the nature and particulars of the proposed work including, in cases where the building owner proposes to construct special foundations, plans, sections and details of construction of the special foundations together with reasonable particulars of the loads to be carried thereby; and
(c) the date on which the proposed work will begin.

(2) A party structure notice shall:

(a) be served at least two months before the date on which the proposed work will begin;
(b) cease to have effect if the work to which it relates:

(i) has not begun within the period of twelve months beginning with the day on which the notice is served; and
(ii) is not prosecuted with due diligence.

(3) Nothing in this section shall:

(a) prevent a building owner from exercising with the consent in writing of the adjoining owners and of the adjoining occupiers any right conferred on him by section 2; or
(b) require a building owner to serve any party structure notice before complying with any notice served under any statutory provisions relating to dangerous or neglected structures.

4. Counter notices

(1) An adjoining owner may, having been served with a party structure notice serve on the building owner a notice (in this Act referred to as a "counter notice") setting out:

(a) in respect of a party fence wall or party structure, a requirement that the building owner build in or on the wall or structure to which the notice relates such chimney copings, breasts, jambs or flues, or such piers or recesses or other like works, as may reasonably be required for the convenience of the adjoining owner;
(b) in respect of special foundations to which the adjoining owner consents under section 7(4) below, a requirement that the special foundations:

(i) be placed at a specified greater depth than that proposed by the building owner; or

(ii) be constructed of sufficient strength to bear the load to be carried by columns of any intended building of the adjoining owner,

or both.

(2) A counter notice shall:

(a) specify the works required by the notice to be executed and shall be accompanied by plans, sections and particulars of such works; and
(b) be served within the period of one month beginning with the day on which the party structure notice is served.

(3) A building owner on whom a counter notice has been served shall comply with the requirements of the counter notice unless the execution of the works required by the counter notice would:

(a) be injurious to him;
(b) cause unnecessary inconvenience to him; or
(c) cause unnecessary delay in the execution of the works pursuant to the party structure notice.

5. Disputes arising under sections 3 and 4

If an owner on whom a party structure notice or a counter notice has been served does not serve a notice indicating his consent to it within the period of fourteen days beginning with the day on which the party structure notice or counter notice was served, he shall be deemed to have dissented from the notice and a dispute shall be deemed to have arisen between the parties.

ADJACENT EXCAVATION AND CONSTRUCTION

6. Adjacent excavation and construction

(1) This section applies where:

(a) a building owner proposes to excavate, or excavate for and erect a building or structure, within a distance of three metres measured horizontally from any part of a building or structure of an adjoining owner; and
(b) any part of the proposed excavation, building or structure will within those three metres extend to a lower level than the level of the bottom of the foundations of the building or structure of the adjoining owner.

(2) This section also applies where:

(a) a building owner proposes to excavate, or excavate for and erect a building or structure, within a distance of six metres measured horizontally from any part of a building or structure of an adjoining owner; and

(b) any part of the proposed excavation, building or structure will within those six metres meet a plane drawn downwards in the direction of the excavation, building or structure of the building owner at an angle of forty-five degrees to the horizontal from the line formed by the intersection of the plane of the level of the bottom of the foundations of the building or structure of the adjoining owner with the plane of the external face of the external wall of the building or structure of the adjoining owner.

(3) The building owner may, and if required by the adjoining owner shall, at his own expense underpin or otherwise strengthen or safeguard the foundations of the building or structure of the adjoining owner so far as may be necessary.

(4) Where the buildings or structures of different owners are within the respective distances mentioned in subsections (1) and (2) the owners of those buildings or structures shall be deemed to be adjoining owners for the purposes of this section.

(5) In any case where this section applies the building owner shall, at least one month before beginning to excavate, or excavate for and erect a building or structure, serve on the adjoining owner a notice indicating his proposals and stating whether he proposes to underpin or otherwise strengthen or safeguard the foundations of the building or structure of the adjoining owner.

(6) The notice referred to in subsection (5) shall be accompanied by plans and sections showing:

(a) the site and depth of any excavation the building owner proposes to make;
(b) if he proposes to erect a building or structure, its site.

(7) If an owner on whom a notice referred to in subsection (5) has been served does not serve a notice indicating his consent to it within the period of fourteen days beginning with the day on which the notice referred to in subsection (5) was served, he shall be deemed to have dissented from the notice and a dispute shall be deemed to have arisen between the parties.

(8) The notice referred to in subsection (5) shall cease to have effect if the work to which the notice relates:

(a) has not begun within the period of twelve months beginning with the day on which the notice was served; and
(b) is not prosecuted with due diligence.

(9) On completion of any work executed in pursuance of this section the building owner shall if so requested by the adjoining owner supply him with particulars including plans and sections of the work.

(10) Nothing in this section shall relieve the building owner from any liability to which he would otherwise be subject for injury to any adjoining owner or any adjoining occupier by reason of work executed by him.

RIGHTS ETC.

7. Compensation etc.

(1) A building owner shall not exercise any right conferred on him by this Act in such a manner or at such time as to cause unnecessary inconvenience to any adjoining owner or to any adjoining occupier.

(2) The building owner shall compensate any adjoining owner and any adjoining occupier for any loss or damage which may result to any of them by reason of any work executed in pursuance of this Act.

(3) Where a building owner in exercising any right conferred on him by this Act lays open any part of the adjoining land or building he shall at his own expense make and maintain so long as may be necessary a proper hoarding, shoring or fans or temporary construction for the protection of the adjoining land or building and the security of any adjoining occupier.

(4) Nothing in this Act shall authorise the building owner to place special foundations on land of an adjoining owner without his previous consent in writing.

(5) Any works executed in pursuance of this Act shall:

(a) comply with the provisions of statutory requirements; and
(b) be executed in accordance with such plans, sections and particulars as may be agreed between the owners or in the event of dispute determined in accordance with section 10;

and no deviation shall be made from those plans, sections and particulars except such as may be agreed between the owners (or surveyors acting on their behalf) or in the event of dispute determined in accordance with section 10.

8. Rights of entry

(1) A building owner, his servants, agents and workmen may during usual working hours enter and remain on any land or premises for the purpose of executing any work in pursuance of this Act and may remove any furniture or fittings or take any other action necessary for that purpose.

(2) If the premises are closed, the building owner, his agents and workmen may, if accompanied by a constable or other police officer, break open any fences or doors in order to enter the premises.

(3) No land or premises may be entered by any person under subsection (1) unless the building owner serves on the owner and the occupier of the land or premises:

> (a) in case of emergency, such notice of the intention to enter as may be reasonably practicable;
> (b) in any other case, such notice of the intention to enter as complies with subsection (4).

(4) Notice complies with this subsection if it is served in a period of not less than fourteen days ending with the day of the proposed entry.

(5) A surveyor appointed or selected under section 10 may during usual working hours enter and remain on any land or premises for the purpose of carrying out the object for which he is appointed or selected.

(6) No land or premises may be entered by a surveyor under subsection (5) unless the building owner who is a party to the dispute concerned serves on the owner and the occupier of the land or premises:

> (a) in case of emergency, such notice of the intention to enter as may be reasonably practicable;
> (b) in any other case, such notice of the intention to enter as complies with subsection (4).

9. Easements

Nothing in this Act shall:

> (a) authorise any interference with an easement of light or other easements in or relating to a party wall; or
> (b) prejudicially affect any right of any person to preserve or restore any right or other thing in or connected with a party wall in case of the party wall being pulled down or rebuilt.

RESOLUTION OF DISPUTES

10. Resolution of disputes

(1) Where a dispute arises or is deemed to have arisen between a building owner and an adjoining owner in respect of any matter connected with any work to which this Act relates either:

(a) both parties shall concur in the appointment of one surveyor (in this section referred to as an "agreed surveyor"); or

(b) each party shall appoint a surveyor and the two surveyors so appointed shall forthwith select a third surveyor (all of whom are in this section referred to as "the three surveyors").

(2) All appointments and selections made under this section shall be in writing and shall not be rescinded by either party.

(3) If an agreed surveyor:

(a) refuses to act;

(b) neglects to act for a period of ten days beginning with the day on which either party serves a request on him;

(c) dies before the dispute is settled; or

(d) becomes or deems himself incapable of acting,

the proceedings for settling such dispute shall begin *de novo*.

(4) If either party to the dispute:

(a) refuses to appoint a surveyor under subsection (1)(b), or

(b) neglects to appoint a surveyor under subsection (1)(b) for a period of ten days beginning with the day on which the other party serves a request on him,

the other party may make the appointment on his behalf.

(5) If, before the dispute is settled, a surveyor appointed under paragraph (b) of subsection (1) by a party to the dispute dies, or becomes or deems himself incapable of acting, the party who appointed him may appoint another surveyor in his place with the same power and authority.

(6) If a surveyor:

(a) appointed under paragraph (b) of subsection (1) by a party to the dispute; or

(b) appointed under subsection (4) or (5),

refuses to act effectively, the surveyor of the other party may proceed to act *ex parte* and anything so done by him shall be as effectual as if he had been an agreed surveyor.

(7) If a surveyor:

(a) appointed under paragraph (b) of subsection (1) by a party to the dispute; or

(b) appointed under subsection (4) or (5),

neglects to act effectively for a period of ten days beginning with the day on which either party or the surveyor of the other party serves a request on him, the surveyor of the other party may proceed to act *ex parte* in respect of the subject matter of the request and anything so done by him shall be as effectual as if he had been an agreed surveyor.

(8) If either surveyor appointed under subsection (1)(b) by a party to the dispute refuses to select a third surveyor under subsection (1) or (9), or neglects to do so for a period of ten days beginning with the day on which the other surveyor serves a request on him:

- (a) the appointing officer; or
- (b) in cases where the relevant appointing officer or his employer is a party to the dispute, the Secretary of State,

may on the application of either surveyor select a third surveyor who shall have the same power and authority as if he had been selected under subsection (1) or subsection (9).

(9) If a third surveyor selected under subsection (1)(b):

- (a) refuses to act;
- (b) neglects to act for a period of ten days beginning with the day on which either party or the surveyor appointed by either party serves a request on him; or
- (c) dies, or becomes or deems himself incapable of acting, before the dispute is settled,

the other two of the three surveyors shall forthwith select another surveyor in his place with the same power and authority.

(10) The agreed surveyor or as the case may be the three surveyors or any two of them shall settle by award any matter:

- (a) which is connected with any work to which this Act relates, and
- (b) which is in dispute between the building owner and the adjoining owner.

(11) Either of the parties or either of the surveyors appointed by the parties may call upon the third surveyor selected in pursuance of this section to determine the disputed matters and he shall make the necessary award.

(12) An award may determine:

- (a) the right to execute any work;
- (b) the time and manner of executing any work; and
- (c) any other matter arising out of or incidental to the dispute including the costs of making the award;

but any period appointed by the award for executing any work shall not unless otherwise agreed between the building owner and the adjoining owner begin to run until after the expiration of the period prescribed by this Act for service of the notice in respect of which the dispute arises or is deemed to have arisen.

(13) The reasonable costs incurred in:
- (a) making or obtaining an award under this section;
- (b) reasonable inspections of work to which the award relates; and
- (c) any other matter arising out of the dispute,

shall be paid by such of the parties as the surveyor or surveyors making the award determine.

(14) Where the surveyors appointed by the parties make an award the surveyors shall serve it forthwith on the parties.

(15) Where an award is made by the third surveyor:

(a) he shall, after payment of the costs of the award, serve it forthwith on the parties or their appointed surveyors; and
(b) if it is served on their appointed surveyors, they shall serve it forthwith on the parties.

(16) The award shall be conclusive and shall not except as provided by this section be questioned in any court.

(17) Either of the parties to the dispute may, within the period of fourteen days beginning with the day on which an award made under this section is served on him, appeal to the county court against the award and the county court may:

(a) rescind the award or modify it in such manner as the court thinks fit; and
(b) make such order as to costs as the court thinks fit.

EXPENSES

11. Expenses

(1) Except as provided under this section expenses of work under this Act shall be defrayed by the building owner.

(2) Any dispute as to responsibility for expenses shall be settled as provided in section 10.

(3) An expense mentioned in section 1(3)(b) shall be defrayed as there mentioned.

(4) Where work is carried out in exercise of the right mentioned in section 2(2)(a), and the work is necessary on account of defect or want of repair of the

structure or wall concerned, the expenses shall be defrayed by the building owner and the adjoining owner in such proportion as has regard to:

(a) the use which the owners respectively make or may make of the structure or wall concerned; and
(b) responsibility for the defect or want of repair concerned, if more than one owner makes use of the structure or wall concerned.

(5) Where work is carried out in exercise of the right mentioned in section 2(2)(b) the expenses shall be defrayed by the building owner and the adjoining owner in such proportion as has regard to:

(a) the use which the owners respectively make or may make of the structure or wall concerned; and
(b) responsibility for the defect or want of repair concerned, if more than one owner makes use of the structure or wall concerned.

(6) Where the adjoining premises are laid open in exercise of the right mentioned in section 2(2)(e) a fair allowance in respect of disturbance and inconvenience shall be paid by the building owner to the adjoining owner or occupier.

(7) Where a building owner proposes to reduce the height of a party wall or party fence wall under section 2(2)(m) the adjoining owner may serve a counter notice under section 4 requiring the building owner to maintain the existing height of the wall, and in such case the adjoining owner shall pay to the building owner a due proportion of the cost of the wall so far as it exceeds:

(a) two metres in height; or
(b) the height currently enclosed upon by the building of the adjoining owner.

(8) Where the building owner is required to make good damage under this Act the adjoining owner has a right to require that the expenses of such making good be determined in accordance with section 10 and paid to him in lieu of the carrying out of work to make the damage good.

(9) Where:

(a) works are carried out, and
(b) some of the works are carried out at the request of the adjoining owner or in pursuance of a requirement made by him,

he shall defray the expenses of carrying out the works requested or required by him.

(10) Where:
(a) consent in writing has been given to the construction of special foundations on land of an adjoining owner; and

(b) the adjoining owner erects any building or structure and its cost is found to be increased by reason of the existence of the said foundations,

the owner of the building to which the said foundations belong shall, on receiving an account with any necessary invoices and other supporting documents within the period of two months beginning with the day of the completion of the work by the adjoining owner, repay to the adjoining owner so much of the cost as is due to the existence of the said foundations.

(11) Where use is subsequently made by the adjoining owner of work carried out solely at the expense of the building owner the adjoining owner shall pay a due proportion of the expenses incurred by the building owner in carrying out that work; and for this purpose he shall be taken to have incurred expenses calculated by reference to what the cost of the work would be if it were carried out at the time when that subsequent use is made.

12. Security for expenses

(1) An adjoining owner may serve a notice requiring the building owner before he begins any work in the exercise of the rights conferred by this Act to give such security as may be agreed between the owners or in the event of dispute determined in accordance with section 10.

(2) Where:

(a) in the exercise of the rights conferred by this Act an adjoining owner requires the building owner to carry out any work the expenses of which are to be defrayed in whole or in part by the adjoining owner; or
(b) an adjoining owner serves a notice on the building owner under subsection (1),

the building owner may before beginning the work to which the requirement or notice relates serve a notice on the adjoining owner requiring him to give such security as may be agreed between the owners or in the event of dispute determined in accordance with section 10.

(3) If within the period of one month beginning with:
(a) the day on which a notice is served under subsection (2); or
(b) in the event of dispute, the date of the determination by the surveyor or surveyors,

the adjoining owner does not comply with the notice or the determination, the requirement or notice by him to which the building owner's notice under that subsection relates shall cease to have effect.

13. Account for work carried out

(1) Within the period of two months beginning with the day of the completion of any work executed by a building owner of which the expenses are to be wholly or partially defrayed by an adjoining owner in accordance with section 11 the building owner shall serve on the adjoining owner an account in writing showing:

    (a) particulars and expenses of the work; and
    (b) any deductions to which the adjoining owner or any other person is entitled in respect of old materials or otherwise;

and in preparing the account the work shall be estimated and valued at fair average rates and prices according to the nature of the work, the locality and the cost of labour and materials prevailing at the time when the work is executed.

(2) Within the period of one month beginning with the day of service of the said account the adjoining owner may serve on the building owner a notice stating any objection he may have thereto and thereupon a dispute shall be deemed to have arisen between the parties.

(3) If within that period of one month the adjoining owner does not serve notice under subsection (2) he shall be deemed to have no objection to the account.

14. Settlement of account

(1) All expenses to be defrayed by an adjoining owner in accordance with an account served under section 13 shall be paid by the adjoining owner.

(2) Until an adjoining owner pays to the building owner such expenses as aforesaid the property in any works executed under this Act to which the expenses relate shall be vested solely in the building owner.

MISCELLANEOUS

15. Service of notices etc.

(1) A notice or other document required or authorised to be served under this Act may be served on a person:

    (a) by delivering it to him in person;
    (b) by sending it by post to him at his usual or last-known residence or place of business in the United Kingdom; or
    (c) in the case of a body corporate, by delivering it to the secretary or clerk of the body corporate at its registered or principal office or sending it by post to the secretary or clerk of that body corporate at that office.

(2) In the case of a notice or other document required or authorised to be served under this Act on a person as owner of premises, it may alternatively be served by:

    (a) addressing it "the owner" of the premises (naming them), and
    (b) delivering it to a person on the premises or, if no person to whom it can be delivered is found there, fixing it to a conspicuous part of the premises.

16. Offences

(1) If:

    (a) an occupier of land or premises refuses to permit a person to do anything which he is entitled to do with regard to the land or premises under section 8(1) or (5); and
    (b) the occupier knows or has reasonable cause to believe that the person is so entitled,

the occupier is guilty of an offence.

(2) If:

    (a) a person hinders or obstructs a person in attempting to do anything which he is entitled to do with regard to land or premises under section 8(1) or (5); and
    (b) the first-mentioned person knows or has reasonable cause to believe that the other person is so entitled,

the first-mentioned person is guilty of an offence.

(3) A person guilty of an offence under subsection (1) or (2) is liable on summary conviction to a fine of an amount not exceeding level 3 on the standard scale.

17. Recovery of sums

Any sum payable in pursuance of this Act (otherwise than by way of fine) shall be recoverable summarily as a civil debt.

18. Exception in case of temples etc.

(1) This Act shall not apply to land which is situated in inner London and in which there is an interest belonging to:

    (a) the Honourable Society of the Inner Temple,
    (b) the Honourable Society of the Middle Temple,
    (c) the Honourable Society of Lincoln's Inn, or
    (d) the Honourable Society of Gray's Inn.

(2) The reference in subsection (1) to inner London is to Greater London other than the outer London boroughs.

19. The Crown

(1) This Act shall apply to land in which there is:

(a) an interest belonging to Her Majesty in right of the Crown,
(b) an interest belonging to a government department, or
(c) an interest held in trust for Her Majesty for the purposes of any such department.

(2) This Act shall apply to:

(a) land which is vested in, but not occupied by, Her Majesty in right of the Duchy of Lancaster;
(b) land which is vested in, but not occupied by, the possessor for the time being of the Duchy of Cornwall.

20. Interpretation

In this Act, unless the context otherwise requires, the following expressions have the meanings hereby respectively assigned to them:

"adjoining owner" and "adjoining occupier" respectively mean any owner and any occupier of land, buildings, storeys or rooms adjoining those of the building owner and for the purposes only of section 6 within the distances specified in that section;

"appointing officer" means the person appointed under this Act by the local authority to make such appointments as are required under section 10(8);

"building owner" means an owner of land who is desirous of exercising rights under this Act;

"foundation", in relation to a wall, means the solid ground or artificially formed support resting on solid ground on which the wall rests;

"owner" includes:

(a) a person in receipt of, or entitled to receive, the whole or part of the rents or profits of land;

(b) a person in possession of land, otherwise than as a mortgagee or as a tenant from year to year or for a lesser term or as a tenant at will;

(c) a purchaser of an interest in land under a contract for purchase or under an agreement for a lease, otherwise than under an agreement for a tenancy from year to year or for a lesser term;

"party fence wall" means a wall (not being part of a building) which stands on lands of different owners and is used or constructed to be used for separating such adjoining lands, but does not include a wall constructed on the land of one owner the artificially formed support of which projects into the land of another owner;

"party structure" means a party wall and also a floor partition or other structure separating buildings or parts of buildings approached solely by separate staircases or separate entrances;

"party wall" means:

> (a) a wall which forms part of a building and stands on lands of different owners to a greater extent than the projection of any artificially formed support on which the wall rests; and
> (b) so much of a wall not being a wall referred to in paragraph (a) above as separates buildings belonging to different owners;

"special foundations" means foundations in which an assemblage of beams or rods is employed for the purpose of distributing any load; and

"surveyor" means any person not being a party to the matter appointed or selected under section 10 to determine disputes in accordance with the procedures set out in this Act.

21. Other statutory provisions

(1) The Secretary of State may by order amend or repeal any provision of a private or local Act passed before or in the same session as this Act, if it appears to him necessary or expedient to do so in consequence of this Act.

(2) An order under subsection (1) may:

> (a) contain such savings or transitional provisions as the Secretary of State thinks fit;
> (b) make different provision for different purposes.

(3) The power to make an order under subsection (1) shall be exercisable by statutory instrument subject to annulment in pursuance of a resolution of either House of Parliament.

GENERAL

22. Short title, commencement and extent

(1) This Act may be cited as the Party Wall etc. Act 1996.

(2) This Act shall come into force in accordance with provision made by the Secretary of State by order made by statutory instrument.

(3) An order under subsection (2) may:

   (a) contain such savings or transitional provisions as the Secretary of State thinks fit;
   (b) make different provision for different purposes.

(4) This Act extends to England and Wales only.

Glossary

Abatement: a self-help remedy whereby the victim of a nuisance can lawfully remove the subject matter of the nuisance from his/her neighbour's land.
Act: the Party Wall etc. Act 1996 (the 1996 Act) or, according to the context, one of the earlier London Building Acts.
Addendum award: an award which is made by the appointed surveyor(s) which is supplementary to the primary award.
Adjoining owner: an owner of land which adjoins that of a building owner. Defined in section 20 of the 1996 Act.
Adjoining occupier: a person, who is not an owner, who occupies land which adjoins that of a building owner. Defined in section 20 of the 1996 Act.
Adverse possession: occupation of land which is inconsistent with the rights of the true owner and which, over time, can lead to the extinguishment of the true owner's title.
Agreed surveyor: a surveyor who is appointed jointly by the parties to act alone in determining a dispute under the Act. Contrast with party-appointed surveyor.
Appointed surveyor: a surveyor who is appointed by one or both of the parties to determine a dispute under the Act. See the definition of surveyor in section 20 of the 1996 Act.
Arbitration: a dispute resolution mechanism whereby one or more independent third parties, rather than the courts, are empowered to determine the dispute. Although it remains a minority view, some would argue that an appointed surveyor performs an arbitral function.
Award: the decision of the appointed surveyor(s) which determines a dispute under the Act. As this is invariably included in a written document the term is also used to describe this document.
Boundary wall: a generic term for any type of wall which acts as a physical boundary, including a party wall, a fence wall and an external wall.
Bristol Acts: the statutes which have historically regulated construction operations (including party wall work) in Bristol.
Building owner: An owner who wishes to undertake construction work which involves exercising rights under the Act. Defined in section 20 of the 1996 Act.
Causation: the principle that liability in tort is dependent on the claimant's damage being caused or contributed to by the defendant's tortious act.
Civil Procedure Rules (CPR): the rules which govern the process of litigation in the main civil courts in England and Wales. They came into effect on 26 April

1999 and embody the reforms proposed by Lord Woolf in his Access to Justice reports.

Compensation: Financial recompense for loss or damage caused by works lawfully carried out under the Act. Payable to an adjoining owner and/or adjoining occupier in accordance with an award made by the appointed surveyor(s). Contrast with: damages.

Co-ownership: the simultaneous ownership of the same title to land by two or more persons. Two forms of co-ownership exist – the joint tenancy and the tenancy in common.

Counter notice: strictly, a notice served under the Act by an adjoining owner which requires the building owner to undertake work proposed by a party structure notice in a particular manner. More generally, any written response by an adjoining owner to a notice served under the Act.

Damages: Financial recompense, ordered by the courts, for loss or damage caused by unlawful acts. Includes loss or damage suffered by an adjoining owner and/or adjoining occupier as a result of trespass, nuisance or negligence. Contrast with: compensation.

De minimis: the principle that some infringements of the law are too trivial to justify the intervention of the courts.

Difference: the term used in the London Building Acts to describe a dispute.

Dispute: a failure by the parties to agree on matters relating to works under the Act. In the absence of written consent to the proposed works, a dispute is deemed to arise fourteen days after service of a party structure notice or of a notice of adjacent excavation.

District surveyor: an officer, appointed under the London Building Acts, having responsibility for building control, dangerous structures and means of escape in case of fire.

Dominant owner: the owner of land having the benefit of an easement.

Dominant tenement: land which has the benefit of an easement.

Easement: a right which confers a benefit on one piece of land (the dominant tenement) and imposes a burden on another (the servient tenement). Easements can take many forms but include: a right of way; a right of support; a right to light. Unlike natural rights, an easement will not exist unless it has been acquired in some way, for example by prescription.

Estoppel: a rule of law and evidence which prevents a party from later denying the truth of his or her earlier statements or of the facts contained therein.

Ex parte: a decision made by a tribunal in the absence of one party to a dispute. A single party-appointed surveyor can proceed on this basis where the other surveyor has refused to act effectively, or has failed to comply with a ten-day request.

External wall: an outer wall of a building which is not a party wall.

Fence wall: a freestanding wall which acts as a physical boundary.

Independent contractor: a person who undertakes work as a contractor or subcontractor, rather than as an employee.

Injunction: a court order requiring a party to carry out a certain act or to restrain from a particular course of action.

Joint tenancy: a form of co-ownership in which all owners have an identical interest in the whole of the land. It is characterised by the right of survivorship whereby, on the death of one owner, ownership of the land passes automatically to the surviving owners. Contrast with tenancy in common.

Laying open: the process of opening up and exposing land or buildings. Includes the act of demolishing a party wall thereby exposing the interior of an adjoining building.

Legal boundary: the plane at which two pieces of land in different ownerships adjoin.

Line of junction: the term used in the Act to describe a legal boundary. See section 1 of the 1996 Act.

Line of junction notice: a notice served under the Act prior to building on a line of junction.

Longitudinally divided party wall: a common law term which describes any wall which straddles a legal boundary. The term therefore includes a party fence wall as well as a type 'a' party wall. Contrast with the party wall, owned as a tenancy in common, which is addressed by a number of the earlier legal cases.

London Building Acts: the statutes which have historically regulated construction operations (including party wall work) in inner London.

Making good: work carried out by a building owner to remedy damage which he/she has caused to an adjoining property.

Mitigation of loss: the responsibility of an injured party to minimise their losses when claiming damages.

Natural rights: rights to which an owner of land is entitled, simply by virtue of his/her ownership of the land. They include a right to have the land supported by the soil of neighbouring land. Contrast with: easement.

Negligence: a tort which is committed when one party breaches a duty of care which is owed to another party. Liability may arise where a building owner causes injury or damage to an adjoining owner or adjoining occupier through failing to take reasonable care in the execution of construction works.

Non-delegable duty: a class of duty whose breach will always result in personal liability of the duty holder, notwithstanding that he/she has delegated performance of the duty to some other person. For example, a building owner owes a non-delegable duty to an adjoining owner to exercise reasonable care in carrying out work to a party wall. Negligence by the contractor in undertaking the works will therefore lead to liability by the building owner, notwithstanding the absence of vicarious liability.

Notice of adjacent excavation: a notice served under the Act prior to excavating to certain depths within certain distances of a neighbouring building or structure. See section 6 of the 1996 Act.

Nuisance: a tort which arises from an indirect interference with the use or enjoyment of neighbouring land. For example a nuisance may be committed where construction operations interfere with neighbouring land through the production of excessive noise, fumes, smells or vibrations. The tort is similar to the concept of unnecessary inconvenience under the Act.

Ouster: the act of wrongfully dispossessing someone from their property.

Owner: a person who is entitled to one of the various qualifying interests in land described in the Act. Defined in section 20 of the 1996 Act.

Parties: the building owner and the adjoining owner.

Party-appointed surveyor: a surveyor who is appointed by one of the parties to determine a dispute under the Act in collaboration with a surveyor appointed by the other party.

Party fence wall: a fence wall which straddles a line of junction. Defined in section 20 of the 1996 Act.

Party structure: any structure (including a type 'a' or type 'b' party wall) which performs a separating function between adjoining premises. The term can sometimes include structures (for example floors or non-structural partitions) which separate different premises within a single building. Defined in section 20 of the 1996 Act.

Party structure agreement: see party wall agreement.

Party structure notice: a notice served under the Act prior to carrying out work to a party structure or to certain other boundary structures.

Party wall: any wall forming a physical boundary which is subject to rights by the owners of both adjacent plots. The nature of these rights was described in Watson v Gray (1880). The term includes a longitudinally-divided party wall; a type 'a' party wall; a type 'b' party wall and a party fence wall.

Party wall agreement: a contract between a building owner and an adjoining owner which authorises works to a party wall or party structure. It may also require the works to be carried out in a particular way and may make provision for making good or the payment of compensation should damage be caused to the adjoining property.

Physical boundary: a physical boundary feature (wall, fence, hedge, etc.) marking the general location of the legal boundary between two properties.

Prescription: the acquisition of an easement by the long enjoyment of a right, usually over a period of twenty years.

Presumption: a rule of evidence which requires a particular inference of fact to be drawn from certain other established facts.

Prima facie: at first appearance.

Primary award: the main (and usually the first) award made by the appointed surveyor(s).

Quiet enjoyment: the landlord's obligation, contained in every lease, not to interfere with the tenant's right to peacefully hold and enjoy the property.

Receiver: a person appointed to manage property on behalf of someone other than the owner, for example on behalf of a creditor to ensure that a debt is repaid.

Remoteness: the rules which determine the extent of a defendant's liability for consequential losses arising out of his/her unlawful act.

Rylands v Fletcher: a tort which is committed, without proof of fault, when dangerous things escape from land and cause damage to neighbouring land.

Servient owner: the owner of land which is subject to an easement.

Servient tenement: land which is subject to an easement.

Statutory consent: a term which is sometimes used to describe a party wall agreement which has been negotiated by surveyors whilst acting as agents for the parties. Some surveyors advocate its use, in lieu of an award, following the service of a party structure notice.

Ten-day notice: see ten-day request.

Ten-day request: (1) a request by one of the parties for the other to appoint a surveyor within ten days. A failure to comply within ten days entitles the requesting party to make the appointment on the other's behalf. (2) a request by one party-appointed surveyor that the other should act effectively within ten days. A failure to comply within ten days entitles the requesting surveyor to act ex parte.

Tenancy at will: a tenancy that can be terminated by either of the parties at any time.

Tenancy in common: a form of co-ownership in which each owner has a separate interest in the land which he/she is free to dispose of. Hence, in contrast to the situation with a joint tenancy, an owner is free to dispose of his/her interest by will and the right of survivorship does not operate. Before 1st January 1926 most party walls were owned by the parties as tenants in common. From that date any wall falling within this category became a longitudinally divided party wall.

Third surveyor: a surveyor who is selected by the two party-appointed surveyors to act as umpire in the event of them being unable to determine the dispute.

Trespass: a tort which results from the direct and unauthorised interference with land. Examples include unauthorised physical entry onto adjoining land and the use of neighbouring airspace for crane oversailing where no licence has first been obtained.

Tribunal: generally, any legal entity having power to determine claims or disputes. The term is wide enough to include the "practical tribunal" comprising either an agreed surveyor or two party-appointed surveyors and a third surveyor.

Type 'a' party wall: a wall forming part of a building which straddles a line of junction, whether or not it also performs a separating function between terraced or semi-detached properties. Defined in section 20 of the 1996 Act.

Type 'b' party wall: a wall which performs a separating function between terraced or semi-detached properties but which does not also straddle the line of junction between them. Defined in section 20 of the 1996 Act.

Ultra vires: an act by a legal entity which is outside its jurisdiction or goes beyond the powers which have been conferred on it. Hence an award will be ultra vires (and therefore invalid) if it relates to matters which the appointed surveyor(s) have no power or jurisdiction to deal with under the Act, or where it results from some procedural irregularity.

Unnecessary inconvenience: the carrying out of works, otherwise authorised by the Act, in a manner which fails to adequately limit the inconvenience suffered by an adjoining owner or adjoining occupier. The works are thereby rendered unlawful. The building owner is thus deprived of the protection afforded by the Act and may be liable in damages to those who have been affected.

Vicarious liability: an employer's liability for torts which are committed by an employee during the course of his/her employment. There is no vicarious liability for torts committed by an independent contractor.

Table of Cases

Cases in bold type are included within the Digest of Cases. A page number accompanied by an 'n' indicates that the case is referred to in a footnote to the listed page. The following abbreviations are used for case references:

A & E	Adolphus & Ellis, Reports (King's/Queen's Bench)
AC	Law Reports, Appeal Cases (House of Lords) [1891] onwards
All ER	All England Law Reports
App Cas	Law Reports, Appeal Cases (House of Lords) (1875–90)
B & C	Barnewall & Cresswell, Reports (King's Bench)
B & S	Best & Smith Reports (Queen's Bench)
Barn & Ald	Barnewall & Alderson, Reports (King's Bench)
Beav	Beavan
Bing NC	Bingham, New Cases in the Court of Common Pleas
BLR	Building Law Reports
CBNS	Common Bench Reports, New Series
Ch	Law Reports, Chancery Division [1891] onwards
Ch App	Law Reports, Chancery Appeal Cases
Ch D	Law Reports, Chancery Division (1875–90)
Co Rep	Coke's Reports
Con LR	Construction Law Reports
DLR	Dominion Law Reports
E & B	Ellis & Blackburn, Reports (Queen's Bench)
EB & E	Ellis, Blackburn & Ellis, Reports (Queen's Bench)
EG	Estates Gazette
EGLR	Estates Gazette Law Reports
ER	English Reports
EWCA Civ	England & Wales Court of Appeal, Civil Division
Ex	Exchequer Reports
H & C	Hurlstone & Coltman, Exchequer Reports
HL Cas	House of Lords Cases
JP	Justice of the Peace Reports
K & J	Kay & Johnson, Reports (Chancery)
KB	Law Reports, King's Bench Division [1901]–[1952] 1
LGR	Knight's Local Government Reports
LJKB	Law Journal Reports, Kings Bench (1832–1946)
Lloyd's Rep	Lloyd's Law Reports

LRCP	Law Reports, Court of Common Pleas (1865–75)
LR Eq	Law Reports, Equity Cases (1865–75)
LR HL	Law Reports, Appeal Cases (House of Lords) (1866–75)
LRQB	Law Reports, Court of Queen's Bench (1865–75)
LT	Law Times Reports
M & W	Meeson & Welsby, Reports (Exchequer)
Moo & Malk	Moody & Malkin, Reports (Nisi Prius)
P & CR	Property, Planning and Compensation Reports
P Wms	Peere Williams, Reports (Chancery)
QB	Queen's Bench Reports (1841–52)
QB	Law Reports, Queen's Bench Division [1891]–[1900] and [1952] 2 onwards
QBD	Law Reports, Queen's Bench Division (1875–90)
TLR	Times Law Reports
Saund	Saunders, Reports (King's Bench)
Swans	Swanston, Reports (Chancery)
Taunt	Taunton, Reports (Common Pleas)
WLR	Weekly Law Reports
WN	Weekly Notes

Adams v Marylebone Borough Council [1907] 2 KB 822
................8, 9, 10, 12, 14, 17, 19, 20, **23**, 88, 239, 257, 261, 423
Alcock v Wraith and Swinhoe (1991) 59 BLR 16
..........6, 10, 17, **34**, 58, 64, 68n, 106, 169, 183, 184, 190, 276, 295
Aldin v Latimer, Clark, Muirhead & Co [1894] 2 Ch 437............146n
Anchor Brewhouse Developments Ltd v Berkley House (Docklands
  Developments) Ltd [1987] 2 EGLR 173........................241n
Andreae v Selfridge & Company Ltd [1938] Ch 1; [1937] 3 All ER 255
........................10, 40, 125, 127, 276, 279n, 407, 411
Apostal v Simons [1936] 1 All ER 207 ....**4, 15, 47**
Attorney General v Horner (1884) 14 QBD 245 ....62n
Attorney General v Roe [1915] 1 Ch 235 342
Attorney-General v Tod Heatley [1897] 1 Ch 560...................78

Backhouse v Bonomi (1861) 9 HL Cas 503; (1861) EB & E 646 .... 293, 294
Balfour v Barty King [1957] 1 QB 496; [1957] 1 All ER 156...........182
Bank of England v Vagliano Brothers [1891] AC 107 ....245n
Bar Gur v Bruton (1993), Court of Appeal, 29 July (unreported)
  .... 165, 183, 332–3
Barry v Minturn [1913] AC 585 .....6, 10, 13, **52**, 80, 97, 190, 295, 324, 394
Belvedere Fish Guano Co Ltd v Rainham Chemical Works Ltd [1920] 2 KB 487
 37
Bennett v Harrod's Stores Ltd (1907) The Builder, Dec 7, p. 624.......6, 58
Birmingham, Dudley and District Banking Co v Ross (1888) 38 Ch D 295
 146–7

Bland v Moseley (1587) 9 Co Rep 58a 310
Bond v Nottingham Corporation [1940] Ch 429; [1940] 2 All ER 12
.... 4, 5, 17, **61**, 69, 75, 76, 164, 165, 166, 317, 328, 332, 340
Bonomi v Backhouse (1858) EB & E 622 .... 65, 66, 112
Bower v Peate (1876) 1 QBD 321 .... 17, 34, 38, **64**, 110, 113, 169, 172, 174, 174n, 184, 190, 191, 225, 228, 276, 282, 292, 317, 347, 356, 382
Brace v South East Regional Housing Association Ltd [1984] 1 EGLR 144
.... 4, 6, 17, 58, 61, **69**
Bradburn v Lindsay [1983] 2 All ER 408; (1983) 268 EG 152
.... 4, 5, 15, 17, 18, 20, 61, 69, 75, 115, 163, 165, 179, 183, 194, 214, 266, 307, 328, 332–3, 340, 371, 402
Bridlington Relay Ltd v Yorkshire Electricity Board [1965] Ch 436; [1965] 1 All ER 264. .... 126n
Broder v Saillard (1876) 2 Ch D 692. 342
Brooke v Bool [1928] 2 KB 578 37
Brown v Minister of Housing & Local Government [1953] 2 All ER 1385
.... 138–9
Burlington Property Company Limited v Odeon Theatres Limited [1939] 1 KB 633; [1938] 3 All ER 469 ...... .6, 13, 52, 80, 97, 151, 154, 204, 295

Carlish v Salt [1906] 1 Ch 335 .... 15, 85, 271, 300
Chadwick v Trower (1839) 6 Bing NC 1. .... 372, 373
Chartered Society of Physiotherapy v Simmonds Church Smiles [1995] 1 EGLR 155 .... 11, 14, 88
Clift v Welsh Office [1998] 4 All ER 852 412
Colebeck v Girdlers Co (1876) 1 QBD 234 342
Cowen v Phillips (1863) 33 Beav 18
.... 9, **94**, 101, 128, 130, 132, 135, 235, 304, 364, 375
Crofts v Haldane (1867) 2 LRQB 194. .... 5, 14, **97**, 249, 287, 289n, 359
Crosby v Alhambra Company Ltd [1907] 1 Ch 295
.... 9, **94**, 101, 128, 135, 219, 221, 235, 304, 364, 375
Cubitt v Porter (1828) 8 B & C 256
   3, 18, **106**, 169, 186, 188, 190, 191, 284, 295, 382, 383, 414, 416, 417, 438

Dalton v Angus (1881) 6 App Cas 740 .... 19, 34, 38, 64, 68n, **110**, 172, 174, 225, 226, 276, 282, 292, 317, 322n, 330n, 342
Darley Main Colliery Co v Mitchell (1886) 11 App Cas 127; (1886) 2 TLR 301
.... 293
Debenham v Metropolitan Board of Works (1880) 6 QBD 112 378
Dodd v Holme (1834) 1 A & E 492 .... 20, **115**, 317, 347, 356, 371
Doltis (J) Ltd v Isaac Braithwaite & Sons [1957] 1 Lloyd's Rep 52 183
Donoghue v Stevenson [1932] AC 562 322
Drury v Army & Navy Auxiliary Co-operative Supply Ltd [1896] 2 QB 271
.... 7, **119**, 143, 144, 200, 243, 245n, 419

Table of Cases

Emms v Polya (1973) 227 EG 1659 . . . . 8, 10, **125**, 239, 257, 276

Fillingham v Wood [1891] 1 Ch 51
. . . . 9, 94, 101, 102, **128**, 135, 235, 304, 364, 375, 394
Filliter v Phippard (1847) 11 QB 347 180
Frances Holland School v Wassef [2001] 2 EGLR 88
. . . . 9, 12, 94, 101, 128, **134**, 235, 304, 359, 364, 375
Frederick Betts Ltd v Pickfords Ltd [1906] 2 Ch 87 . . . . 17, **143**

Giles v Walker (1890) 24 QBD 656. . . . . .215n
Goldman v Hargrave [1967] 1 AC 645; [1966] 2 All ER 989
. . . . 165, 183, 215–18
Gray v Pullen (1864) 5 B & S 970. . . . . 68n
Gyle-Thompson v Wall Street (Properties) Ltd [1974] 1 WLR 123; [1974]
1 All ER 295
. . . . 6, 9, 11, 12, 13, 14, 52, 80, 88, 97, 135, **151**, 249, 253, 295, 375, 427

H & N Emanuel Ltd v Greater London Council [1971] 2 All ER 835 182
Hall v Hardy (1733) 3 P Wms 187 358
Harrison v Southwark & Vauxhall Water Co [1891] 2 Ch 409 279
Hiscox v Outhwaite (No. 1) [1991] 3 All ER 124 . . . . .139n
Hobbs, Hart & Co v Grover [1899] 1 Ch 11. . . . . 8, **161**, 375, 376
Hodgson v York Corporation (1873) 28 LT 836. . . . . .215n
Holbeck Hall Hotel Ltd v Scarborough Borough Council [2000] 2 WLR 1396;
[2000] 2 All ER 705
. . . . 20, 61, 115, **163**, 179, 183, 194, 214, 292, 317, 328, 332–3, 371
Honeywill & Stein Ltd v Larkin Brothers Ltd [1934] 1 KB 191
. . . . 36, 67n, 184, 281–3
Horsfall v Thomas (1862) 1 H & C 90. 87
Hughes v Lord Advocate [1963] AC 837; [1963] 2 WLR 779; [1963]
1 All ER 705. 166
Hughes v Percival (1883) 8 App Cas 443
. . . . 9, 10, 17, 18, 34, 38, 64, 67n, **169**, 190, 191, 225, 276
Hunt v Harris (1865) 19 CBNS 13 . . . . 129, 130, 378

India (Republic of) v India Steamship Co. Ltd (The Indian Endurance) (No. 2)
[1996] 3 All ER 641 . . . . .139n

J. Jarvis & Sons Ltd v Baker (1956) 167 EG 123. . . . . 16, **175**, 324, 375
Job Edwards Ltd v Birmingham Navigations [1924] 1 KB 341. 78
Johnson (T/A Johnson Butchers) v BJW Property Developments Ltd (2002)
Technology and Construction Court, 30 January (unreported)
. . . . 18, **179**, 214
Johnston v Mayfair Property Company [1893] WN 73
. . . . 7, **186**, 263, 295, 398

Jolliffe v Woodhouse (1894) 10 TLR 553 ...... 4, 8, 9, 10, 18, 52, **190**, 394
Jones v Bird (1822) 5 Barn & Ald 837 425
Jones v Pritchard [1908] 1 Ch 630
.............................5, 15, 75, 76, 163–4, **194**, 328, 340, 342

Knight v Pursell [1879] 11 Ch D 412 ...7, 119, 120, **200**, 243, 245, 419, 434

Ladd v Marshall [1954] 1 WLR 1489; [1954] 3 All ER 745 91
Lamb v Walker (1878) 3 QBD 389......................293
Leadbetter v Marylebone Corporation (No. 1) [1904] 2 KB 893
...................8, 13, 17, 80, 84, **204**, 210, 239, 257, 269, 356, 442
Leadbetter v Marylebone Corporation (No. 2) [1905] 1 KB 661
.... 8, 11, 204, **210**, 359, 361n
Leakey v National Trust for Places of Historic Interest or Natural Beauty [1980]
  QB 485; [1980] 2 WLR 65; [1980] 1 All ER 17
   6, 18, 20, 75, 77, 163, 164, 165, 166, 183, 194, **214**, 317, 328, 332–5, 371
Lehmann v Herman [1993] 1 EGLR 172
.... 8, 94, 101, 128, 135, **219**, 235, 249, 252, 304, 364, 375
Lemaitre v Davis (1881) 19 Ch D 281..................... 17, 61, 69, **225**
Lewis & Solome v Charing Cross, Euston, & Hampstead Railway Company
  [1906] 1 Ch 508 .... 7, 8, **229**, 263, 347, 356, 382
List v Tharp [1897] 1 Ch 260
................9, 94, 101, 103, 128, 135, **235**, 304, 364, 375, 376n
London & Manchester Assurance Company Ltd v O & H Construction Ltd
  [1989] 2 EGLR 185....................... 8, 17, 18, **239**, 257
London, Gloucester & North Hants Dairy Company v Morley & Lanceley
  [1911] 2 KB 257.................................. 7, 143, **243**
Loost v Kremer (1997) West London County Court, 12 May (unreported)
.... 7, 11, 14, 97, **249**, 359
Lord, Re (1854) 1 K & J 90 289
Louis v Sadiq [1997] 1 EGLR 136; (1998) 59 Con LR 127
................8, 10, 13, 16, 17, 18, 125, 127n, 239, **257**, 344, 423
Luttrel's Case (1601) 4 Co Rep 86a............................321

McKinnon Industries Ltd v Walker [1951] 3 DLR 577 .... 126n
Macpherson v London Passenger Transport Board (1946) 175 LT 279 ... 165
Major v Park Lane Company (1866) LR 2 Eq 453
.... 8, 229, 260, **263**, 398, 402
Marchant v Capital & Counties plc [1983] 2 EGLR 156
...........................5, 13, 17, 183, **266**, 307, 328, 402
Mareva Compania Naviera SA v International Bulkcarriers SA [1975]
  2 Lloyd's Rep 509....................................302n
Mason v Fulham Corporation [1910] 1 KB 631
.... 15, 58, 85, **271**, 300, 304, 347, 350–51, 389, 432

Matania v National Provincial Bank Ltd and The Elevenist Syndicate Ltd [1936]
2 All ER 633.................10, 17, 34, 38, 64, 125, 184, **276**, 407
Matts v Hawkins (1813) 5 Taunt 20
............3, 47, 58, 106, 107, **284**, 295, 414, 416, 438, 440
Metropolitan Building Act *ex parte* McBride, Re (1876) 4 Ch D 200
................................5, 11, 14, **287**, 389
Midland Bank plc v Bardgrove Property Services Ltd and John Willmott (WB)
Ltd [1992] 2 EGLR 168, CA....................19, 110, **292**, 317
Moss v Smith (1977) 76 LGR 284........6, 47, 58, 106, 186, 263, **295**, 398
Musgrove v Pandelis [1919] 2 KB 43.........................183

Nicklin v Williams (1854) 10 Ex 259 ....293, 294

Observatory Hill Ltd v Camtel Investments SA [1997] 1 EGLR 140
....15, 85, 271, **300**
Orf v Payton (1905) 69 JP 103
.... 9, 15, 18, 94, 101, 128, 135, 235, **304**, 364, 375, 376n, 394, 398, 432
Overseas Tankship (UK) Ltd v Miller Steamship Co Pty, The Wagon Mound
(No. 2) [1967] 1 AC 617.........................183, 217, 412

Percival v Hughes (1882) 9 QBD 441 356
Phipps v Pears [1965] 1 QB 76; [1964] 2 WLR 996; [1964] 2 All ER 35
.....................5, 71, 75, 79, 266, **307**, 328, 329–34, 402
Pickard v Smith (1861) 10 CBNS 470....................172
Pomfret v Ricroft (1669) 1 Saund 321........................342
Powell v McFarlane (1977) 38 P & CR 452 313
Pratt v Hillman (1825) 4 B & C 269.........................356
Prudential Assurance Co Ltd v Waterloo Real Estate Inc [1999] 2 EGLR 85
....7, 311

R v Fulling [1987] 2 WLR 925 ....245n
Ray v Fairway Motors (Barnstable) Ltd (1968) 20 P & CR 261
..........................19, 20, 110, 115, 292, **317**
Reading v Barnard (1827) 1 Moo & Malk 71........16, 52, 175, **324**, 375
Redland Bricks Ltd v Morris [1970] AC 652; [1969] 2 WLR 1437;
[1969] 2 All ER 576..................................294
Rees v Skerrett [2001] EWCA Civ 760; [2001] 1 WLR 1541..........5, 17,
18, 20, 61, 69, 75, 115, 163, 179, 183, 194, 214, 266, 307, **328**, 340, 371, 402
Riley Gowler Ltd v National Heart Hospital Board of Governors [1969]
3 All ER 1401...................................14, 151, 159, **336**
Robinson v Kilvert (1889) 41 Ch D 88 ....126n
Rylands v Fletcher [1868] LR 3 HL 330 .... 18, 36, 179, 180, 182, 184, 470

Sack v Jones [1925] Ch 235 ....5, 164, 165, **340**

St Helens Smelting Company v Tipping (1865) 11 HL Cas 642;
(1865) 11 ER 1483 411
Salisbury v Woodland [1970] 1 QB 324 37
Saunders v Williams [2002] EWCA Civ 673 (unreported) 17, **344**
Sedleigh-Denfield v O'Callaghan [1940] AC 880
.... 78, 165, 166, 183, 215, 334
Selby v Whitbread & Co [1917] 1 KB 736
... 5n, 7, 11, 12, 14, 15, 61, 69, 73n, 85, 88, 260, 271, 300, **347**, 382, 402
Shirlaw v Southern Foundries (1926) Ltd [1939] 2 All ER 113 ...... 73, 73n
Sims v The Estates Company (1866) 14 LT 55 .... 14, 97, 249, **359**
Solomons v R. Gertzenstein Ltd [1954] 2 QB 243; [1954] 3 WLR 317;
[1954] 2 All ER 625 ... 9, 94, 101, 128, 135, 219, 222, 235, 304, **364**, 375
Southwark & Vauxhall Water Company v Wandsworth District Board of Works
[1898] 2 Ch 603 .... 10, 16, 61, 69, 115, 317, 322n, **371**, 394
Spiers & Son Ltd v Troup (1915) 84 LJKB 1986
.... 8, 9, 15, 16, 94, 101, 128, 135, 161, 175, 235, 304, 324, 364, **374**
Standard Bank of British South America v Stokes (1878) 9 Ch D 68
.... 4, 7, 11, 34,
73n, 106, 190, 229, 231, 257, 260, 284, 347, 356, **382**, 402, 414, 424n, 438
Stedman v Smith (1857) 8 E & B 1 417
Stone and Hastie, Re [1903] 2 KB 463
11, 13, 14, 15, 85, 88, 151, 156, 245n, 271, 273, 287, 300, 347, 356, **389**, 432

Tarry v Ashton (1876) 1 QBD 314 .... 68n, 172
Thompson v Hill (1870) 5 LRCP 564 .... 9, 10, 17, 18, 52, 190, **394**
Thornton v Hunter (1898) The Builder, Feb 4, p. 182
.... 7, 19, 263, 304, **398**, 423, 438
Titterton v Conyers (1813) 5 Taunt 465 .... 289n

Upjohn v Seymour Estates Ltd [1938] 1 All ER 614
.... 5, 8, 17, 263, 307, 438, **402**

Video London Sound Studios Ltd v Asticus (GMS) Ltd and Keltbray Demolition Ltd (2001) Technology and Construction Court, 6 March (unreported)
.... 10, 14, 17, 18, 40, 276, **406**

Wagon Mound (No. 2), The, *see* Overseas Tankship (UK) Ltd v Miller Steamship Co Pty
Watson v Gray (1880) 14 Ch D 192 .... 3, 47, 58, 106, 284, **414**, 438
Webb v Bird (1861) 10 CBNS 268 310
Wells v Ody (1836) 1 M & W 452 .... 289n
West Leigh Colliery Co v Tunnicliffe & Hampson Ltd [1908] AC 27 293
Weston v Arnold (1873) 8 Ch App 1084
.... 6, 7, 119, 120, 122, 143, 200, 202, 243, 245, 289n, **419**, 434
White v Peto Brothers (1888) 58 LT 710 ..... 7, 10, 16, 18, 19, 23, 257, **423**

Whitefleet Properties Ltd v St Pancras Building Society (1956) 167 EG 262
.... 9, 151, 375, **427**
Wildtree Hotels Ltd v Harrow London Borough Council [2000] 3 WLR 165
.... 411
Williams v Bull (1890) The Times, Feb 15 .... 7, **432**
Wiltshire v Sidford (1827) 8 B & C 259n
............................3, 106, 107, 284, 398, 414, 416, **438**
Wood v Griffith (1818) 1 Swans 43............................... 358
Woodhouse v Consolidated Property Corporation Ltd [1993] 1 EGLR 174
.... 13, 204, **442**

Table of Statutory Extracts

Party Wall etc., Act 1996
Appendix

London Building Acts (Amendment) Act 1939
s. 1   223, 254, 369
s. 33(1)   369
s. 44   254
s. 46(1)(a)   159
s. 46(1)(k)   159
s. 47(1)   223, 430
s. 47(3)   159
s. 55(a)   160, 431, 445
s. 55(e)   140
s. 55(h)   160, 431
s. 55(i)   92, 160, 445
s. 55(j)   160
s. 55(k)   160, 445
s. 55(m)   92, 160, 338
s. 55(n)   92, 160, 338
s. 58(1)   177
s. 58(2)   177
s. 58(3)   177
s. 124(1)   431
s. 133(2)   370
s. 141(1)   370
s. 141(3)   370

London Building Act 1930
s. 5   105, 140, 224, 254, 369
s. 114(7)   83
s. 117(1)   83

Land Registration Act 1925
s. 54(1)   302

Law of Property Act 1925
s. 1(6)   50
s. 38(1)   50, 298, 404
s. 38(2)   51, 298, 404
s. 39   51
sch. 1, Part V   51

London Building Act 1894
s. 5(16)   123, 149, 246
s. 5(29)   105, 237, 305, 379
s. 5(32)   105
s. 54(3)   123, 149
s. 58   246
s. 59(1)   124, 150
s. 75   124
s. 77(3)(c)   124, 150
s. 87(1)   305
s. 87(2)   305
s. 87(3)   305
s. 87(6)   31, 400
s. 88(1)   56
s. 88(2)   379, 400
s. 88(6)   31, 380
s. 88(7)   31, 380, 400
s. 89(1)   208
s. 90(1)   208, 380
s. 90(3)   32, 56
s. 90(4)   212
s. 91(1)   32, 209, 357, 392
s. 91(2)   393
s. 91(3)   212
s. 93(3)   32
s. 93(4)   32
s. 94   357
s. 95(2)   33, 209, 274, 393
s. 96   380

s. 99	274, 305, 393	s. 84	435
s. 173	381	s. 85(1)	133, 362
s. 201(8)	233	s. 85(3)	192, 395
		s. 85(9)	362

Charing Cross, Euston, and Hampstead Railway Act 1893

s. 5	233
s. 31	233

Metropolitan Building Act 1855

s. 3	96, 104, 132, 188, 202, 435
s. 82	105, 132, 387
s. 83(1)	387
s. 83(2)	387, 395
s. 83(6)	99, 387
s. 83(7)	99, 192, 387, 395
s. 83(8)	188, 387
s. 83(9)	387
s. 83(11)	387

s. 88(1)	436
s. 88(2)	436
s. 88(6)	436
s. 88(7)	436
s. 93	436
s. 97(2)	133

Common Law Procedure Act 1854

s. 12	290

Railway Clauses Consolidation Act 1845

s. 16	233
s. 45	233

Subject Index

adjacent excavations, 19–20, 64, 110, 115, 163, 214, 225, 292, 317, 423, *see also* support, rights of
   and non-delegable duties, *see* independent contractor, liability for torts of
   legal status of, 65–6
adjoining occupier, 396–7, 466
adjoining owner, 8–9, 466, *see also* building owner
   co-owners as, 103–4
   in occupation of only part of a building, 131–2
   prospective purchaser, status of, 376
   receiver, status of, 364–8
   statutory tenants, status of, 138–9
   whether more than one entitled to service of notice, 101–4, 129–31
   with agreement for a lease, 94–6
   with building agreement, 235–8
   with equitable interest, 95–6, 238
adverse possession, 466
   intention to possess, 313–16
   requirement for discrete feature, 316
   requirements for, 312–13
agreement
   effect on time limit for commencement of works, 211–13
   party wall, 60, 73–4, 272
   to line of junction works, 304–6
   to withdraw claim for cost of repairs to party wall, 49–50
arbitration, 89–91, 288–9, 390, 466
architect, 147–8
award, 466
   addendum, 135–6, 354–5, 390, 409–10, 466
   adjoining owner's rights under, 352–4
   appeals against, 14
   challenges to awards after 14 days, 156, 355–6, 392
   role of, 91–3, 355–6
   building owner's liabilities under, 350–52
   delivery
      requirement for, 158–9, 385–6
      time of, 337–9
   disclosure to purchaser, 85–7
   effect on purchaser, *see* change of ownership
   effect on time limit for commencement of works, 211–13
   ex parte, 137, 141–2
   interim, 152–3
   interpretation of, 268–9
   nature of, 89–91
   parties to, 430
   primary, 135–6, 469
   registration as a caution, 301–3
   relevant considerations when making, 56–7
   retrospective, 24, 260–62
   service of, *see* delivery
   specific performance of, 358
   third surveyor's, 135–6
   validity of, *see* surveyors, powers *and* jurisdiction

Bristol Acts, 295–9, 419–22, 466
building owner, 8, 466, *see also* adjoining owner
   co-owners as, 219–23, 251–2, 467
   duties of, 9–10
      duty of care, 9–10, *see also* negligence
      limits of, 372–3
      non-delegable, 468, *see also* independent contractor, liability for torts of

unnecessary inconvenience, *see* unnecessary inconvenience
landlord as, 251–2
receiver, status of, *see* adjoining owner
tenant at will, excluded from definition of, 304–6

change of ownership, 14–15, 85–7, 271–5, 300-3, 347–54, 390–91
Civil Procedure Rules, 345–6, 466
common law rights
  construction rights, 6, 7, 38, 173–4
    demolition of half thickness of party wall, 440–41
    demolition and rebuilding of party wall, 108, 191–3
    enclosing on a party wall, 440–41
    no right to place projecting footings on adjoining owner's land, 401
    raising of party wall, 108–9, 285, 417–18
    repairs, 197
      to roof covering above party wall, 34, 38
    underpinning of party wall, 383
  fabric rights, 4–6, 196–7, 404–5, *see also* support, rights of; weather protection, rights of
    to use a flue on neighbour's side of party wall, 198–9
  relationship with statutory rights, 7–8, 191–3, 231–2, 334–5, 356, 383–4, 402–5
compensation and making good, 16–17, 467, 468
  and associated rights of access, 300–1
  for adjacent excavations, 25
  for line of junction works, 24–5
  for loss of trade, 24, 30
  for unnecessary inconvenience, 26–9
  for works to a party wall, 26, 265
  not available for damage to chattels, 409–10
  when laying open, 25–6
consent, *see* agreement

damages, 17–18, 467
  for cost of repairs to party wall, 48–9
  for nuisance, 45–6
  caused by unauthorised works, 258–9
  mitigation of loss, 345–6, 468
  not available for infringement of easement of support where works remoteness, 259, 344–6
  undertaken under the Act, 356
dangerous structure notice, 48, 375–6
demolition of building owner's building, 61, 69, 75, 88, 151, 190, 225, 229, 263, 266, 307, 328, 347, 359, 371, 394, 398, 402, 406, 419, 423, 438
derogation from grant, 146–7
dispute, 355, 467
district surveyor, 120, 144–6, 375–6, 467

easements, *see* support, rights of; light, rights of; common law rights
estoppel, 139–41, 443–6, 467
expenses, 15–16
  adjoining owner's liability for contractor's account, 177–8
  adjoining owner's obligation to contribute
    emergency demolition works, 377–8
    line of junction works, 306
  service of account on adjoining owner
    incorrect form, 325
    outside statutory time limit, 177–8, 377
  where adjoining owner makes subsequent use of party wall works, 272–5, 390–91, 434–5

fire
  liability for spread of, 180
    at common law, 180–82
    in negligence and nuisance, 183–4
    under Rylands v Fletcher, 182–3
  means of escape in case of, 365
    civil liability for failure to maintain, 368

independent contractor, liability for torts of employer authorising commission of torts, 65–6, 171–3
  exceptions to general rule, 36, 65–6, 171–2, 281

extra-hazardous acts exception, 36–8,
    173–4, 281–3
  general rule, 36, 65–6, 171–2, 227–8,
    281
  party wall exception, 173–4, 183–4,
    191–3
  where a serious risk of nuisance, 281–3
  withdrawal of support exception, 38–9,
    67–8, 113–14, 227–8
injunction, 468, *see also* notices, failure to
    serve; light, rights of
quia timet, 293–4

laying open, 394–7, 468
light, rights of
  derogation from grant of, 146
  implied grant of, 146–7
  injunction restraining interference with,
    420–22
  surveyors having no authority to
    authorise infringement, *see*
    surveyors, jurisdiction to
    determine matters of law
line of junction works, 304–6, 398–401,
  468

making good, *see* compensation and
  making good

negligence, 34–6, 75–6, 169–71, 344–6,
    468
  causation, 117–18, 466
  causing spread of fire, 183–4
  during adjacent excavations, 116,
    322–3, 424–5
  measured duty of care, *see* nuisance
  standard of care, 116, 410
notices, 8
  contents of, 161–2, 376, 429–30
  entitlement to service of, *see* adjoining
    owner
  failure to serve, 24, 126, 205, 239–40,
    242, 257–8, 259–60, *see also*
    trespass
  line of junction, *see* line of junction
    works
  requirement for, 205–6, 231–2, *see also*
    statutory rights, construction
    rights

retrospective, *see* award, retrospective
service of, 157, 429, *see also* adjoining
    owner; building owner
time limit for commencement of works,
    211–13, 361–3
withdrawal of, 359–63
nuisance, 34–6, 468–9
  abatement, 77–8, 466
  abnormally sensitive adjoining owner,
    115–18, 126–7
  causation, 259
  caused by excessive hours of working,
    125–7
  caused by spread of fire, 183–4
  caused by inadequate supervision of
    workforce, 125–7
  caused by landslip, 163–8, 214–18
  caused by noise and/or dust from
    construction operations, 40–45,
    125–7, 278, 279–80, 410–12
  caused by vibration, 410–13
  caused by withdrawal of support, 64–5,
    69–70, 75–6, 320–22, 330–31,
    341–3, 405
  requirement for actual damage, 293–4
  damages for, *see* damages
  future damage, probability of, 293–4
  measured duty of care, 163–8, 183–4,
    214–18, 331–4

owner, *see* adjoining owner/building owner

party wall, 469
  agreement, *see* agreement
  definition of, 3–4, 416
    status of raised portion, 432–5
    two external walls distinguished from,
      400–1, 425–6, 438–40
  existence of, *see* presumptions;
    surveyors, jurisdiction to
    determine matters of law
  liability for cost of repairs to, 48–9,
    76–7, 198–9
  longitudinally divided, 4, 48–9, 196–7,
    284–6, 402–5, 468
    effect of Act upon, 285–6
  type 'a', 470
    adverse possession of, 311–16
    entitlement to exercise statutory rights

irrespective of ownership, 252,
  285–6, 435
type 'b', 470
  entitlement to exercise statutory rights
    irrespective of ownership, 201,
    266–70,
  for part of wall's height, 119–23,
    143–6, 246–8, 421–2
  for part of wall's length, 186–9,
    201–2
  owned as tenancy in common, 3, 106–9,
    416–18, 470
  presumptions, 107–8, 122–3, 383,
    421–2, 440

quiet enjoyment, breach of covenant for,
  279, 469

retention of title, *see* statutory rights
Rylands v Fletcher, 470
  liability for spread of fire under, 182–3

statutory consent, 470, *see also* agreement,
  party wall
statutory duty, breach of, 368
statutory interpretation
  approach to, 245
  examples of, 230–32, 365–8
  presumption against appropriation of
    property, *see* statutory rights
  purpose of legislation, 99–100, 138–9
statutory rights
  absence of reciprocal common law duty
    to avoid harmful consequences of,
    372–3, 396–7, 424–5
  absence of statutory fabric rights, *see*
    support, rights of; weather
    protection, rights of
  construction rights, 6–7
    appropriation of adjoining owner's
      property, 56–7, 82–3, 99–100,
      154–5, 298–9, 325–7, 422
    demolition and rebuilding of party
      wall, 80–81, 82–3, 191–3,
      324–7
    exposing a party wall, 229–32, 264–5
    placing projecting footings on
      adjoining owner's land, 400–1
    raising height of party wall, 58–60,
      97–100, 205, 245, 272, 297–8,
      432–4
    reducing height of party wall, 154–5
    repairs to party wall, 55, 324–7
    underpinning party wall, 384–5
  relationship with common law rights, *see*
    common law rights
  to demolish, 62–3, 78, 229–32
  to retain title in party wall works, 273–5,
    306
support, rights of, 19–20, *see also* nuisance;
  negligence
  absence of, 116–17, 372–3
  acquired as easements, 112–13, 226–7,
    321–2, 330–31, 340–41, 356
  contained in award, 348
  duty to provide alternative support when
    demolishing servient tenement,
    62, 78–9, 341, 404–5
  measured duty of care, and, 163–8,
    331–3
  natural rights, 111–12, 164, 292–3,
    320–21, 468
  nature of, 70–71, 330–31, 341
  provision of alternative support by
    adjoining owner, 293–4
  underground water, removal of, 71–2
  whether a duty to repair servient
    tenement, 62, 76–7, 342–3
surveyors, 10–14
  agreed, 466
  authority to accept service, 157, 429
  appointment of, 157–8, 428–9
  jurisdiction, 12–13, 83–4, 151–9, 206–7,
    443–6
    to determine matters of law, 98–9,
      255, 289–90, 361
    to make addendum awards, 354–5
    to make *ex parte* awards, 141–2
    to require payment for subsequent use
      of party wall works by adjoining
      owner, 391–2
  party-appointed, 469
  powers, 13–14, 151–9
    and matters arising out of or incidental
      to a dispute, 207–8, 443–6
    to impose continuing obligations,
      269–70
    to require payment of compensation

and making good of damage, 26, 409–10
project architect as, 253
role of, 10–11, 26–9, 157
third, 255–6, 288–91, 470

ten-day notice/ten-day request, 137, 141–2, 361–3, 470
tenancy at will, 470, *see also* building owner
trespass, 34–6, 106–7, 240–41, 284, 285–6, 295–7, 360–61, 400–1, 405, 417–18, 470
and derogation from grant, 146–7
by crane oversailing, 241
by enclosing on an external wall, 148–9, 186–9
injunction for, 241–2
tribunal, 470
practical, 11, 27

ultra vires awards, 470, *see also* surveyors, powers *and* jurisdiction

unnecessary inconvenience, 9–10, 471, *see also* nuisance
common law obligation to avoid, 191–3
compensation for, *see* compensation and making good
failure to perform work expeditiously, 191–3
limits of duty to avoid, 396–7
non-delegable duty to avoid, 191–3
past history of party wall, 56–7

vicarious liability, 471
for spread of fire, 180, 183–4

wall
boundary, 466
external, 148–9, 186–9, 467
fence, 467
party fence, 469, *see also* party wall, type 'a'
weather protection, rights of, 5, 78–9, 269–70, 309–10, 330–31, 402

104727